wlD

D1429270

# INDUSTRIAL AND COMMERCIAL GAS INSTALLATION PRACTICE
## (GAS SERVICE TECHNOLOGY VOLUME 3)

# INDUSTRIAL AND COMMERCIAL GAS INSTALLATION PRACTICE (GAS SERVICE TECHNOLOGY VOLUME 3)

This book is the third and last of a series of manuals devoted to the theory and practice of gas service. The volumes were the result of some years of investigation and planning, first by a Working Group of the then Gas Council and subsequently by the Gas Manuals Steering Committee of the British Gas Corporation.

The Working Group established the need for the three volumes, to deal with: (1) the basic science and practice of gas service; (2) domestic installation and servicing practice; and (3) commercial and industrial servicing practice. The Steering Committee then undertook the task of producing the manuals. The technical editing was carried out by the late George Jasper, Senior Technical Adviser (Customer Services), who recast the individual contributions into a consistent style to make a coherent and comprehensive body of information.

An update of volume 3 was carried out between 1990 and 1992 by Eric Glennon, a former Customer Service Training Officer with British Gas, and Ray Proffitt who was then the Technical Consultancy Manager with British Gas plc North Western. They retained the original format for the revision along with the names of the authors of the original chapters. However, items referring to legislation, standards, etc., were amended where necessary and much additional material was added to cover recent developments within the industry.

Two new chapters, Combined Heat and Power (10) and Gas and the Environment (13), both written by Ray Proffitt, were also added.

This latest edition has been edited and updated by Frank Saxon, Assessment Manager, Blackburn College, George Sedgwick, a former Industrial and Commercial Manager (Operations) British Gas, now also at Blackburn College, and Ray Proffitt, now a consultant industrial engineer, who has once again given us the benefit of his experience and expertise.

# Industrial and Commercial Gas Installation Practice

## (Gas Service Technology Volume 3)

*Revised by*
Frank Saxon, George
Sedgwick and Ray Proffitt

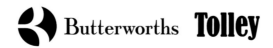

Butterworths **Tolley**

This title first published by
Benn Technical Books

First Edition 1980
Second impression 1981
Third impression 1982
Fourth impression 1988
Fifth impression 1989
Sixth impression 1990
Second Edition 1992
Third Edition 2000

Published by Butterworths Tolley
2 Addiscombe Road, Croydon CR9 5AF

© Reed Elsevier (UK) Ltd 2000

Typeset by Letterpart Ltd, Reigate, Surrey

Printed in Great Britain by
Bath Press, Bath

ISBN 0 75450 223-6

# Contents

# Preface

Industrial and Commercial Gas Installation Practice, or Gas Service Technology Volume 3 as it was originally known, was first edited by the late George Jasper. In the intervening years changes have occurred within the gas industry in Britain which would have astonished colleagues from an earlier time. Not only have British standards been constantly introduced and updated but also the introduction of European Standards has begun. Many new codes of practice have been introduced by The Institute of Gas Engineers and the Health and Safety Executive. It has now become a legal requirement that only competent people can install or work on gas-fired industrial or commercial equipment and pipework.

For most types of activities it is a requirement that the operative be registered with CORGI (Confederation of Registered Gas Installers). To gain membership of CORGI applicants must achieve success in a series of gas safety assessments (Nationally Accredited Certification Scheme for Individual Gas Operatives ACS) in the areas of work in which they operate, this certification must be renewed every five years.

The editors feel that much of the information needed by students studying for these qualifications will be found in this volume or its companion volumes. The reader should also consult current legislation, standards and, of course, manufacturers' instructions. Wherever possible reference has been made to statutory obligations, in particular the Gas Safety (Installation and Use) Regulations 1998, which should be studied in conjunction with this volume.

The installation and servicing methods and procedures described in this book are believed to be accepted good practice at the time of writing. However the editors and contributing authors cannot accept responsibility for any problems arising from the use of this information.

Each chapter in the book has been revised where necessary and any new Codes of Practice or standards have been included. Major updates carried out for this volume include:

- new information and equipment for metering including temperature and pressure correction;

- testing and purging of large installations IGE/UP/1 and IGE/UP/1A;
- the chapter on Commercial Catering has been extended to include mobile catering units and the use of Liquefied Petroleum Gas (LPG);
- the Environment chapter has been updated to reflect modern thinking in plant efficiency and environmental protection.

We would like to thank manufacturers for the use of photographs and diagrams and these are acknowledged in the text. Particular thanks for specialist information are due to the following:

Mike Godber, Technical Engineer, Advantica Technology Ltd for assistance with metering correction factors and equipment;

The Institution of Gas Engineers for information from their Utilization Procedures UP/1 and UP/1A relating to non-domestic testing and purging;

The Heating and Ventilation Contractors Association for information from their publication 'Standard for Kitchen Ventilation Systems' (DW/171);

Blackburn College for the use of facilities and resources.

Finally we acknowledge the help received from Michael Webb, senior editor at Butterworths Tolley, and Chris Leggett of typesetters Letterpart Limited, in putting together this third edition.

Frank Saxon
George Sedgwick
Ray Proffitt
*December 2000*

# Large Installations

Chapter 1 is based on an original draft by E. Glennon

## Introduction

Large diameter pipework in non-domestic premises should be installed to comply with the requirements of the Gas Safety (Installation and Use) Regulations 1998 and the Guidance Notes on the Installation of Gas Pipework, Boosters and Compressors in Customers' Premises, Institute of Gas Engineers document IGE/UP/2.

They deal with the materials and methods for installing pipework downstream from the meter control to the burners on the equipment. They apply to pipes of 28 mm or larger diameter for pressures up to 7 bar.

Work on large installations poses a number of problems. Some of these are also associated with pipework less than 28 mm in domestic and commercial premises and have been dealt with in Vol. 2 and reference to BS 6891 revised edition 1998.

Pipe sizing is generally carried out as described in Vol. 1 although gas flow calculators are available and offer a convenient method of calculating gas flows, velocities and pressure drops.

Velocities can be critical, particularly in unfiltered supplies where dust or debris at high velocities can cause erosion or damage to valves or controls. Generally on supplies filtered to 250 µm a maximum velocity of 40 m/s is permissible.

Pipes are usually steel, copper, ductile iron or polyethylene (PE). In high rise buildings steel pipe is recommended for all vertical risers above 15 m.

## Pipework Layouts

The route selected for the pipework should be as short as possible without being too obtrusive. The design should allow for possible future extensions.

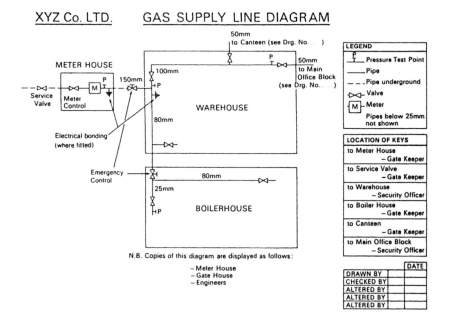

*Fig. 1   Gas supply line diagram*

The installation should include valves as necessary to provide:

● section isolation
● purging
● use in emergency.

For commercial or industrial premises having two or more floors to which gas is supplied by a service pipe larger than 50 mm diameter, valves must be fitted to enable each floor to be isolated. Where a single floor is divided into self-contained areas, the pipework to each area must also be valved.

In addition, line diagrams showing all installation pipework not less than 25 mm is a statutory requirement in certain premises where the gas service is 50 mm or more and should be provided to enable isolating valves to be quickly located in case of emergency. One copy of the diagram must be fitted as near as possible to each primary meter. Other copies may be placed at the gate-house and the services engineers' office as appropriate. The diagram must be updated whenever alterations are made to the installation.

An example of a supply diagram is shown in Fig. 1. It shows sufficient detail to identify the isolating valves but need not include every final connection. A key should always be provided. Diagrams are usually A4 size and protected by glass or plastic. The provision of the diagram is the responsibility of the installer and the occupier, in the case of a factory.

## Pipework Buried Underground

Supplies running from one building to another may be buried or carried in ducts. A typical case is where the meter house is isolated from the main building. Where installation pipework is buried it should generally conform to the recommendations for service pipes (Vol. 2, Chapter 4) and reference should be made to relevant publications and standards. These include the Institution of Gas Engineers publication IGE/TD/3, 'Distribution Mains' and IGE/TD/4, 'Gas Services'.

All pipes must be adequately protected against corrosion. Up to 50 mm, pipes are obtainable with factory applied wrapping or sheathing. Joints or exposed sections of pipe must be covered, after testing for soundness, usually with self adhesive PVC tape or a bandage impregnated with a petroleum grease. Larger, uncoated pipes must be wrapped after being laid. On some sites cathodic protection may be necessary or the excavation may be filled with a passive material, chalk or sand.

Exposed pipework may be painted rather than wrapped unless situated in a very corrosive atmosphere.

The route for underground pipework must be selected with the following points in mind:

- it must avoid close proximity to unstable structures
- it must be kept clear of walls which retain materials above the level of the ground in which the pipes will be laid
- pipes must not pass under load-bearing walls, foundations or footings
- in the proximity of any structure known to have unventilated voids
- areas where there may have been a recent infill, where this is impossible welded steel or PE pipe should be used
- pipes should be laid at minimum distances from buildings as shown in Table 1 until a direction change is required to enter that building.

**TABLE 1**

| | Minimum distance from building (m) | | |
|---|---|---|---|
| | | Pressure | |
| Material | Low | Medium | Intermediate |
| Steel | 1 | 1 | 1 |
| PE | 1 | 3 | 3 |
| Copper | 1 | n/a | n/a |

The amount of cover above the pipe must be at least that given in Table 2. In special circumstances this may be reduced if the back fill is suitably reinforced.

**TABLE 2  Depth of Cover on Buried Installation Pipes**

| Pipe Size | Minimum Depth of Cover | |
| --- | --- | --- |
| | Roadways and Grass Verge | Paved Foot Walk |
| 50 mm and below | 375 mm | 375 mm |
| Above 50 mm | 750 mm | 600 mm |

Where valves are fitted underground they should be provided with access to spindles and lubrication points. Valve pits should have covers of adequate strength or surface boxes to BS 5834. The position of the valve should be indicated by a marker plate, Fig. 2. This plate is also used to show the position of syphons or purge points.

V denotes valve
S denotes syphon
PP denotes purge point
13 denotes distance of marker plate from V, S or PP
(this is in metres)

Minimum dimensions 150 mm × 150 mm

*Fig. 2    Marker plate to show valves, syphons or purge points*

Although syphons or dip pipes are not necessary on natural gas supplies, they may be fitted to pipework which will be hydraulically tested to provide a means of removing the water.

## Pipework Above Ground

Pipes entering buildings should pass through the wall or floor in sleeves sealed at one end by a non-hardening, non-combustible material to prevent the passage of water, vermin or gas but still allow normal movement of the pipe.

The internal diameters of sleeves should allow an annular space of at least 3 mm around the pipe to enable satisfactory insertion of the pipe. Pipework below 28 mm should be installed in accordance with BS 6891.

Pipework should not be laid through electrical intake chambers, transformer rooms or lift shafts. It should be spaced at least 25 mm away from any electrical or other service. On large diameter installations spaces up to 250 mm may be required.

The pipes should be adequately protected against corrosion both internally and externally. Precautions should be taken to prevent the entry of dirt, debris and welding scale into the pipework during installation.

Gas pipes should be electrically cross bonded to other services as described in Vol. 1, Chapter 9. When any pipes are disconnected, a temporary continuity bond should be attached before the supply is broken (Vol. 2, Chapter 3).

If pipework is exposed in a high position, for example on the roof of a tall building, it should be protected by fitting lightning conductors to BS 6651.

Gas pipework should be easily identifiable in accordance with BS 1710. Where there are no other piped gas supplies it is sufficient to paint the pipes with Yellow Ochre 08 C 35 to BS 4800. Where there are other piped gases, for example on a chemical works, it is desirable to provide more precise identification by a secondary colour band over the base of Yellow Ochre. For natural gas the secondary band colour is now called Primrose 10 E 53 to BS 4800. Alternatively, the name of the gas or its chemical symbol may be used.

## Pipework in Ducts

Any ducts which contain gas pipework must be constructed to comply with the Gas Safety (Installation and Use) Regulations 1998 and Building Regulations. Special consideration shall also be given to pipework in ducts where fire separation is important. BS 8313 and BS 5588 give specific guidance.

Gas pipework in false or suspended ceilings and below suspended floors shall be treated as pipework in ducts. Voids in a cavity wall shall not be used unless specifically designed as a ventilated duct.

The installation of gas pipework in ducts containing other services such as hot and cold water, heating pipes, fuel pipes, electrical conduits and cables may be carried out. Ductwork used for air distribution around buildings shall not have gas pipework routed in or through it. Specific services which shall not be installed in ducts containing gas pipework include:

- ventilation ducts and vacuum pipes that operate at sub-atmospheric pressure which are not welded or all brazed construction
- services containing oxidising or corrosive fluids.

In addition certain restrictions apply where ducts contain combined services; BS 8313 gives further guidance.

Unventilated ducts or voids shall not have gas pipework installed within them unless:

- it is continuously sleeved throughout the unventilated duct or void and the sleeve ventilated at one or both ends to a safe place
- it is filled with a crushed inert infill material such as slate chippings or dry washed sand.

Ducts containing gas pipework should be ventilated to ensure that minor leaks do not become unsafe and are diluted to below 25% LFL. Free air openings sized in accordance with Table 2a would normally meet this requirement. Ventilation of ducts should be provided by natural ventilation and free air openings shall lead to safe places, preferably direct to outside air.

Gas pipework in ducts shall be checked with a suitable portable gas detector, prior to and on completion of any work within the duct.

**TABLE 2a**

| Cross sectional area of Duct (m²) | Minimum free area of each opening (m²) |
|---|---|
| Not exceeding 0.05 | Cross sectional area of duct |
| 0.05 and not exceeding 7.5 | 0.05 |
| Exceeding 7.5 | 1/150th of the cross sectional area of duct |

## Jointing

*Steel*

**TABLE 2b**

| | Jointing Method | | | | | | | | |
|---|---|---|---|---|---|---|---|---|---|
| | Below ground | | | High rise buildings or ducts | | | Above ground | | |
| | Pressure in bar | | | | | | | | |
| Pipe Size (mm) | 0 to 0.075 | 0.075 to 2.0 | 2.0 to 7.0 | 0 to 0.075 | 0.075 to 2.0 | 2.0 to 7.0 | 0 to 0.075 | 0.075 to 2.0 | 2.0 to 7.0 |
| Up to 25 | Screw or Weld | Screw or Weld | Weld | Screw or Weld | Screw or Weld | Weld | Screw or Weld | Screw or Weld | Screw or Weld |
| 26 – 50 | Screw or Weld | Screw or Weld | Weld | Screw or Weld | Screw or Weld | Weld | Screw or Weld | Screw or Weld | Screw or Weld |
| 51 – 80 | Weld | Weld | Weld | Weld | Weld | Weld | Screw or Weld | Screw or Weld | Weld |
| 81 – 100 | Weld | Weld | Weld | Weld | Weld | Weld | Screw or Weld | Weld | Weld |
| >100 | Weld | Weld | Weld | Weld | Weld | Weld | Weld | Weld | Weld |

Where pipe is screwed to BS 21, malleable iron fittings are used as described in Vol. 2, Chapter 1. Alternatively, flanges may be fitted either by screwing or welding on to the pipe. Where pipework is welded the number of flanged joints should be kept to a minimum and they should all be welded on to the pipe. Minimum welding standards that apply are BS 2640 (gas welding) or BS 2971 (arc welding) for pressures up to 2 bar, for higher pressures 10% of welds should be non-destructively tested.

Flanges should conform to BS 4504, Part 1. Flanges to BS 10 are obsolescent (1962) although many are still in use on the district. There are a number of different types shown in Fig. 3. These include:

(a)   integral flanges – part of a valve or fitting
(b)   welding neck – for welding to pipe end
(c)   plate – for welding to pipe wall
(d)   screwed boss – for screwing to threaded pipe
(e)   slip-on boss – for welding to pipe wall
(f)   loose – for welded-on lapped pipe ends (Form A)
      blank – for sealing off a pipe or fitting (Form B).

Gaskets for flanges may be either 'full-face' or 'inside bolt circle' (IBC).

Full-face gaskets extend to the outer edge of the flange and are used on flanges which have a perfectly flat face, Fig. 3(d).

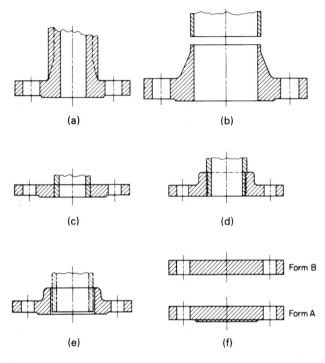

*Fig. 3    Flanges (BS 4504): (a) integral flange; (b) welding neck; (c) plate flange; (d) C.I. screwed boss; (e) slip-on boss; (f) blank flanges*

IBC gaskets have an outside diameter equal to the bolt pitch circle less one bolt hole diameter. They are used on flanges with a centre raised section, Fig. 3(a).

Semi-rigid couplings, flange adapters or compression fittings may be used; guidance on their use is contained in the section entitled 'Flexible Connections' and more specifically in the Institute of Gas Engineers IGE/UP/2 document.

### Copper

On copper tube, capillary and compression fittings are used (Vol. 2, Chapter 1) conforming to BS 2871 Part 1 and shall not be used for 1st and 2nd family gases where the size is above 67 mm and 35 mm for 3rd family gases.

Compression joints may only be used where they are accessible, they may not be enclosed in a sleeve or a sealed duct.

Capillary and compression joints may not be used in vertical ducts in high rise buildings.

*Polyethylene (PE)*

This pipe is extensively used by Transco and pipework contractors, it is also described in Vol. 2, Chapters 1 and 4. It is only suitable for use below ground, because it deteriorates on exposure to daylight.

Jointing is by electrofusion and compression couplings with reinforcing inserts. The methods are described in detail in Vol. 2, Chapter 4.

## Associated Components

*Valves*

The selection of manual valves is covered in Institute of Gas Engineers document IGE/UP/2. An emergency control valve shall be fitted in the pipework as near as reasonably practicable to the point of entry into the building. In some cases, the meter control may satisfy this requirement. The valve shall be fitted with a label 'Gas Emergency Control'.

The Gas Safety (Installation and Use) Regulations and the Institute of Gas Engineers IGE/UP/2 document provide specific requirements on the method of operation and labelling of valves. In addition to the emergency control valve, isolation valves shall be fitted:

- in any building with 2 or more floors and a supply pipe 50 mm or above, the branch supply to each floor should have a valve fitted
- premises with self-contained areas where the supply pipe is 50 mm or above
- any branch of 50 mm or above when the operating pressure is greater than 75 mbar.

Valves should also be fitted for plant isolation upstream of all other plant controls. They should be provided with a means of disconnection, for example a union or flange, immediately on their outlet.

Valves should be clearly identified, accessible and easy to operate. The installation of automatic valves operated by gas, fire or smoke detection systems must be considered carefully particularly where equipment or appliances are fitted without flame safeguards. In such cases installations shall be designed to prove system integrity prior to the restoration of the gas supply. The fitting of non-return valves is dealt with in Chapter 4.

*Filters*

Where there is a likelihood of dust causing erosion or adversely affecting the operation of the plant a filter should be fitted. This would normally remove particles larger than 250 μm although in some cases finer filters may be necessary.

Filters should be fitted immediately upstream of the plant to be protected. Where the plant is not more than 20 m downstream of a meter governor filter, additional filters may not be necessary.

*Purge points*

Plugged or capped purge points should be fitted at isolating valves and in other appropriate positions to allow the installation to be safely tested and purged in accordance with the recommended procedures. Each purge point shall incorporate a valve to control its operation.

## Pipe Supports

Pipework must be adequately supported to BS 3974. Owing to their considerable weight, large diameter pipes require much stronger supports than smaller installations. In factories pipework may be subjected to extremes of temperature and vibration, both of which may cause movement. The coefficient of linear expansion for mild steel is 0.011 mm/m°C, so the supports must allow for thermal movement without damaging any insulation or corrosion protection applied to the pipework.

The spacing of the supports is given in Table 3.

TABLE 3  Maximum Distances Between Pipe Supports

| Pipe Supports Maximum Spacing (m) | | | | |
|---|---|---|---|---|
| Nominal Bore Steel (mm) | Outside diameter Copper (mm) | Screwed Steel and Iron | Welded Steel | Copper |
| 15 | 15 | 2.0 | 2.5 | 1.2 |
| 20 | 22 | 2.5 | 2.5 | 1.8 |
| 25 | 28 | 2.5 | 3.0 | 1.8 |
| 32 | 35 | 2.7 | 3.0 | 2.5 |
| 40 | 42 | 3.0 | 3.5 | 2.5 |
| 50 | | 3.0 | 4.0 | Not recommended |
| 65 | | 3.0 | 4.5 | Not recommended |
| 80 | | 3.0 | 5.5 | Not recommended |
| 100 | | 3.0 | 6.0 | Not recommended |
| 150 | | Not recommended | 7.0 | Not recommended |
| 200 | | Not recommended | 8.5 | Not recommended |
| 250 | | Not recommended | 9.0 | Not recommended |

Generally the structure of the industrial building lends itself to the carrying of pipe supports. Brackets and hangers can be attached to rolled steel trusses, girders and columns by lugs, bolts or welding. On concrete sections, supports may be secured by masonry bolts or fitted to plates and held by bolts through the structure.

*Horizontal pipework*

Rollers and chairs mounted on brackets are a simple method of supporting uninsulated pipe while allowing lateral movement. Examples are shown in Fig. 4. Rollers are usually cast iron although small sizes are available in bronze.

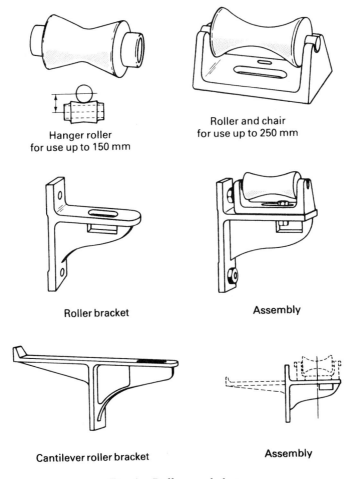

Fig. 4   *Rollers and chairs*

For insulated pipes, sliders or cradles with rollers and base plates may be used, Fig. 5. The pipe is first clamped to the cradle which rests on a flanged roller mounted on a flanged plate. When positioned the pipe is insulated.

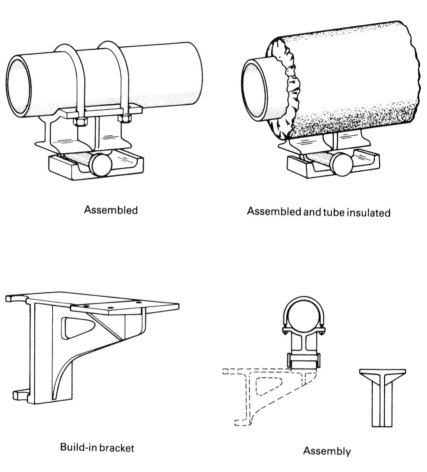

Assembled                Assembled and tube insulated

Build-in bracket                           Assembly

*Fig. 5    Cradle and roller*

Where banks of pipes are run horizontally together they may be supported by low level trestles, Fig. 6, or suspended from a roof truss or girder as in Fig. 7. In both these examples shown, rollers have been dispensed with. The pipes are held in position by flat guide plates and slide directly on the supporting steel section.

Typical pipe support brackets which may be used for single or multiple runs of pipe are shown in Fig. 8. The diagram illustrates the methods of securing the brackets to the structure.

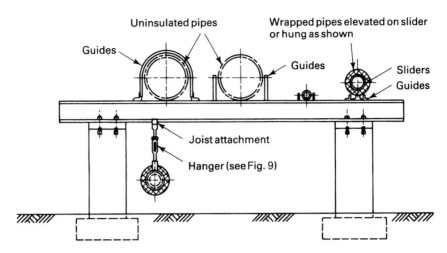

For carrying multiple runs of pipelines at low levels.

*Fig. 6   Low-level trestles*

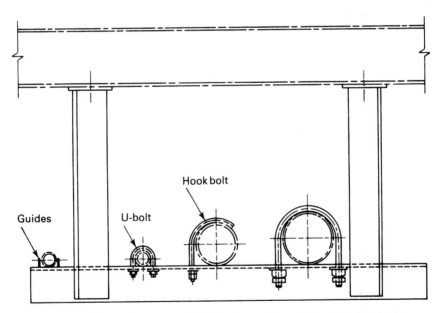

This simple form of support, attached below another structure, enables banks of pipe to be carried in the horizontal plane.

*Fig. 7   Suspended support*

A simple hanger for uninsulated pipes is shown in Fig. 9. Because it is suspended by a domed washer, the sling rod has a limited amount

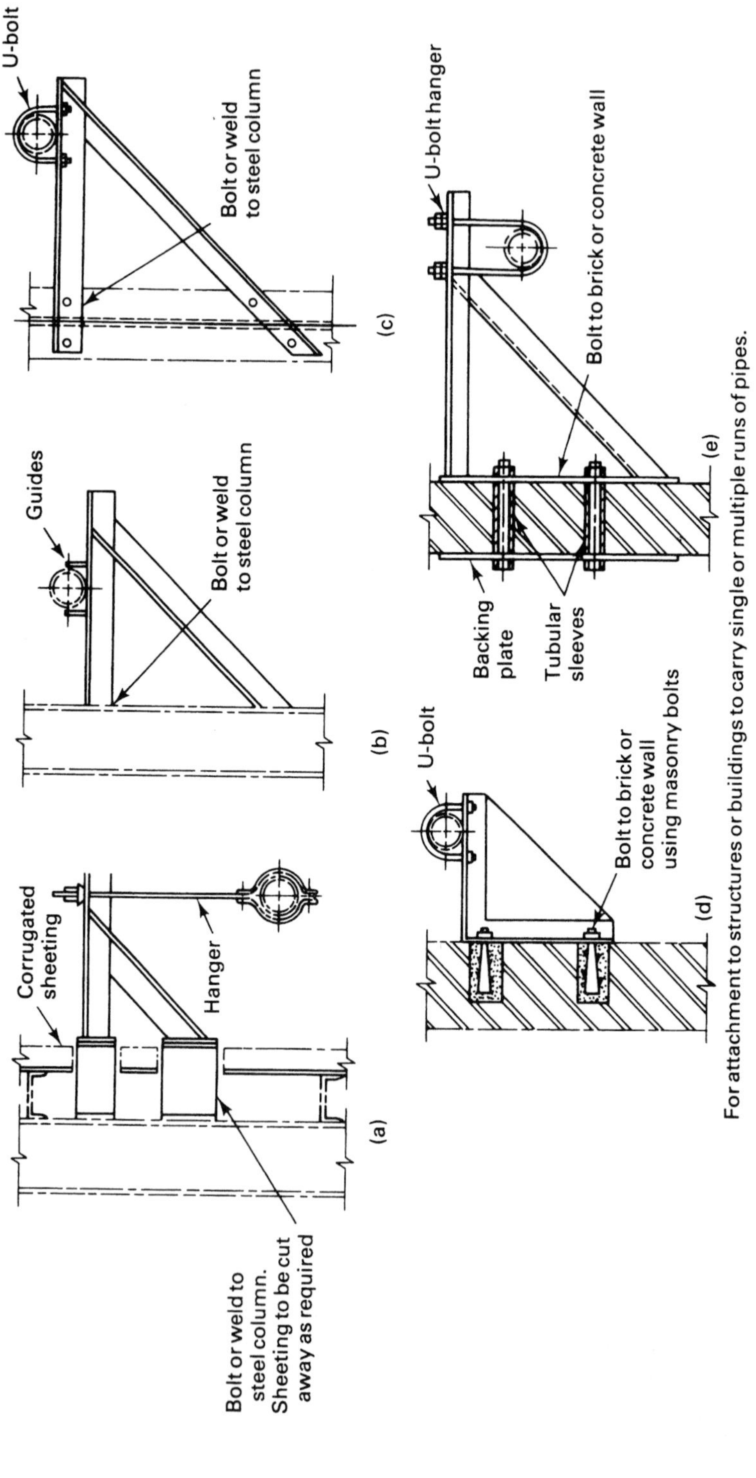

Fig 8   *Pipe support brackets: (a), (b), (c) bolted or welded to steel column; (d) masonry bolts; (e) bolted through concrete column*

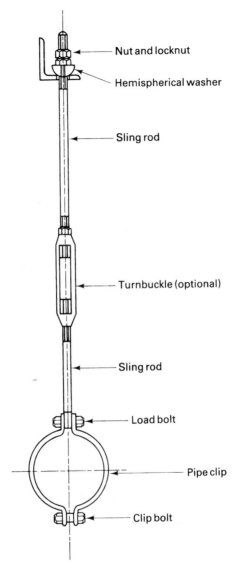

*Fig. 9   Hanger for uninsulated pipe*

of pivoting movement to allow for small movement of the pipework. A more complex, spring-loaded hanger suitable for insulated pipe is shown in Fig. 10. This can accommodate positional changes in a pipe. The spring housing carries the full load on the hanger and the hanger should only be used where load changes are relatively small.

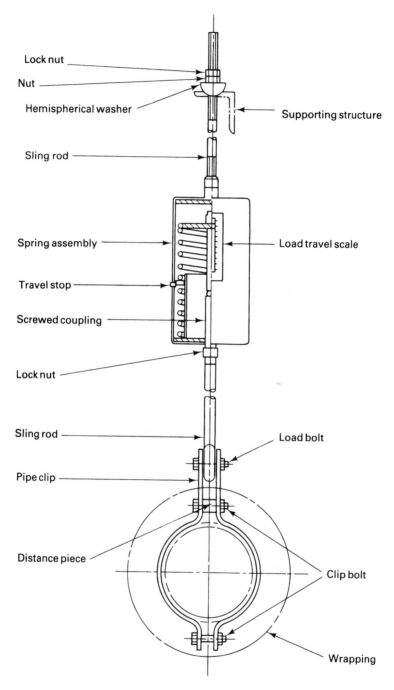

*Fig. 10   Spring-loaded hanger for insulated pipe*

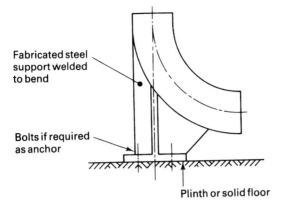

Fabricated steel
support welded
to bend

Bolts if required
as anchor

Plinth or solid floor

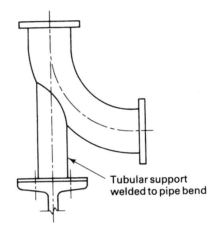

Tubular support
welded to pipe bend

*Fig. 11    Duckfoot bends*

### Vertical pipework

Support for vertical pipes may be increased by 25% subject to the
following:

- supported at its base by fabricated duckfoot bend
- various points along the riser.

Where consideration is given to suspension from top of riser alone,
all of the following criteria must be met:

- pipework must be of welded steel construction
- construction of the building is such that it can support the riser's
  total weight

- pipework and joints are strong enough to support the riser's weight
- no riser consists of no other components apart from flanges
- horizontal movement is suitably restrained.

Duckfoot bends are secured to the pipe line by welding or flanges, Fig. 11. They are fitted at the bottom of a riser and are often mounted on a concrete plinth which may be separate from the main structure.

Spring hangers, Fig. 12, support the pipework by spring-loaded

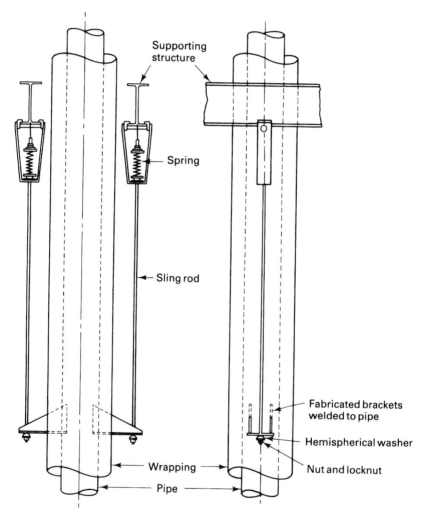

*Fig. 12   Spring hangers for vertical pipe*

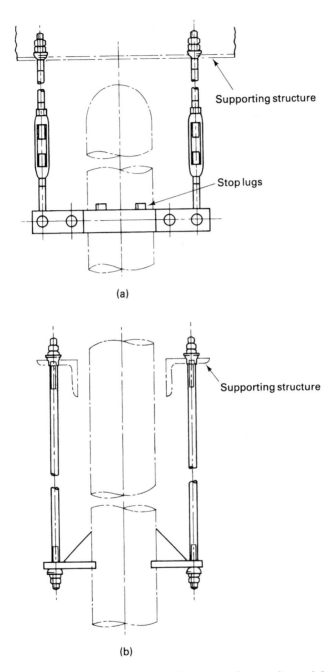

(a)

(b)

*Fig. 13    Hangers for vertical pipes: (a) hanger with some limited free movement; (b) support for heavier riser*

sling rods attached to brackets which have been welded to the pipe wall. The pipework may be insulated or wrapped after the brackets have been fixed.

Figure 13 shows two alternative methods of supporting vertical pipework. They may both be used with insulated or uninsulated pipe and offer some limited free movement. The hanger at (b) is generally more suitable for heavier applications.

## Fabricating Pipework

Fabrication of pipework is often carried out where repetitive pipework sections are needed. Manufacturers are able to provide sections or complete installations which are fabricated to drawings provided by customers. Installers can then assemble the fabricated sections on site and make any necessary make up pieces to complete the installation.

## Flexible Connections

Pipework in industrial situations may be subjected to:

- movement
- expansion
- vibration
- strain.

Because of this, some means must be found to counteract these forces and flexible or semi-rigid connections may be used.

Most large buildings are prone to some structural movement and high rise buildings, for example, are designed to sway slightly with wind pressure. So allowance must be made for the installation pipework to move relative to the incoming supply. Corrugated bellows-type steel connections are often used on lateral offtakes from vertical risers.

Thermal expansion is most likely to have its greatest effect on long lengths of exposed pipework or connections to furnace burners. Underground pipework is not normally affected. Semi-rigid couplings can accommodate lateral movement up to 10 mm. They should only be used as expansion joints if all joints are couplings. Flexible bellow-type connections are satisfactory if fitted with restraining ties.

Vibration is normally due to moving machinery, for example, gas compressors or mixers, gas engines or automatic process machinery. To prevent vibration being transmitted throughout the installation pipework, armoured flexible metallic tubing is commonly used. Couplings are unsuitable for continuous vibration.

Strain may be produced in a number of ways. Where it is likely to be caused by misalignment of final connections, couplings may be used. Two couplings joined by a short length of pipe can accommodate a lateral displacement.

Where strain is due to torsion or rotation it is best prevented by the use of swivel joints.

Flexible connections in the form of sheathed flexible tubing are generally used where burners or appliances must be moved to gain access for servicing or cleaning. They can also connect supplies to floating restaurants or bars moored in tidal waters. Where equipment is used in several different locations, flexible connections can correct any misalignment at each point of use.

When using flexible connections the following points should be noted:

- current codes of practice should be studied to determine the circumstances under which each form of connection may be used
- the length of the connection should be as short as possible
- it must not pass through any rigid structures, for example, walls or floors
- only semi-rigid couplings may be used in buried pipework
- connections should otherwise be in accessible locations
- it must be protected against heat and mechanical damage
- it must be checked for leakage at regular intervals
- where electrical continuity is required it may be necessary to fit separate bonding across semi-rigid couplings or flexible tubing
- where couplings are fitted above ground they should be provided with some form of restraint to prevent the pipes separating, tie rods or chains may be used
- individual pipes must be securely anchored to ensure that the flexible connections are not strained
- there should be an isolating valve upstream of the flexible connection
- mechanically jointed semi-rigid couplings and flange adaptors shall not be used at pressures above 2 bar when installed above ground.

Hydraulically formed couplings and adapters may be used above ground at higher operating pressures.

Figures 14 and 15 show mechanically jointed semi-rigid coupling and flange adaptor respectively. Fig. 16 shows the hydraulically formed joint which when pressurised by oil in chambers A and B forms a permanent and flexible joint. The semi-rigid coupling with tie rods shown in Fig. 17 restricts the angular deflection in the plane of the ties but prevents pipe separation.

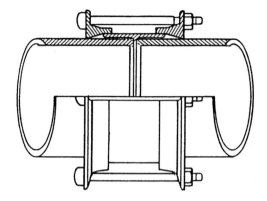

*Fig. 14    Mechanically jointed semi-rigid coupling*

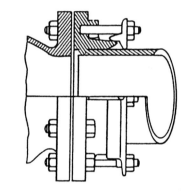

*Fig. 15    Flange adaptor*

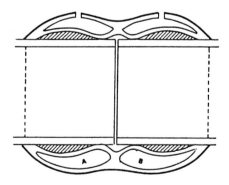

*Fig. 16    Hydraulically formed joint*

Figure 18 shows swivel joints but care must be taken with accurate alignment and must only be installed in exposed pipework.

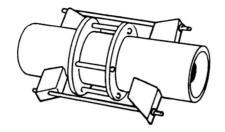

*Fig. 17   Semi-rigid coupling with tie rods*

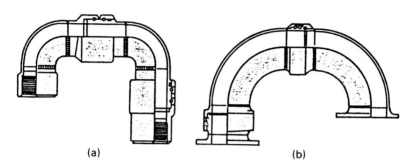

(a)                                              (b)

*Fig. 18   Swivel joint*

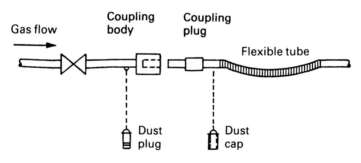

*Fig. 19   Quick release couplings*

Figure 19 shows quick release couplings and must be of the type having self sealing valves in both plug and body. Methods of installing flexible tubes are shown in Fig. 20, whilst Fig. 21 and Fig. 22 show how vertical travel and horizontal travel can be accommodated.

Table 4 indicates the suitability of the types of flexible connections for various applications.

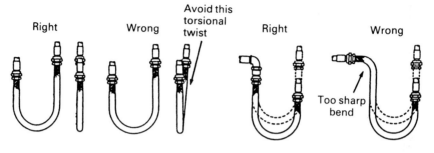

*Fig. 20    Methods of installing flexible tubes*

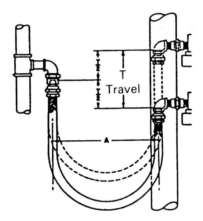

*Fig. 21    Flexible tube installation – vertical loop for maximum vertical travel*

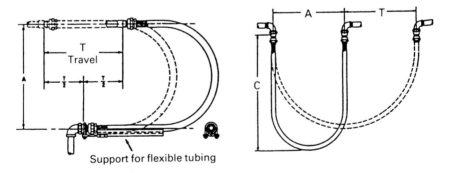

*Fig. 22    Flexible tube installation for horizontal travel*

**TABLE 4**

| Application | Type | Suitability | Remarks |
|---|---|---|---|
| Thermal Expansion | Semi-rigid Coupling and Flange Adaptor | Not Suitable | Not Suitable. |
| | Bellows | Suitable | Ideally suited for this duty but restraining ties should be fitted across the joint and the pipework shall be supported either side to prevent the expansion joint having to carry a mechanical load. |
| | Swivel Joint | Not Suitable | Not Suitable. |
| | Flexible Tube | Not Suitable | Not Suitable. |
| Misalignment | Semi-rigid Coupling and Flange Adaptor | Suitable | Couplings can provide up to 6° deflection, flange adaptors 3° of annular deflection. Lateral displacement accommodated by two couplings with short pipe between. |
| | Bellows | Limited | Only specially designed bellows are suitable for this application. |
| | Swivel Joint | Not Suitable | Not Suitable. |
| | Flexible Tube | Suitable | Suitable and not subject to any angular limitation. Can accommodate any amount of lateral displacement given appropriate length. |
| Structural Movement | Semi-rigid Coupling and Flange Adaptor | Suitable | They are suitable provided they are used in pairs with a short length of pipe between. |
| | Bellows | Suitable | Purpose designed bellows are the preferred type for use in ducts and for branch connections from risers in multi-storey buildings. |
| | Swivel Joint | Suitable | In the appropriate combinations they may be used on exposed pipework (not common). |
| | Flexible Tube | Suitable | Short length may be used for this application but purpose designed bellows preferred. |
| Vibration | Semi-rigid Coupling and Flange Adaptor | Not Suitable | Not Suitable. |
| | Bellows | Limited | Purpose designed bellows are available to allow for the effects of vibration. |
| | Swivel Joint | Not Suitable | Not Suitable. |
| | Flexible Tube | Suitable | Manufacturers should be consulted for advice on minimum length required for the amplitude and frequency of vibration expected. |
| Rotation and Torsion | Semi-rigid Coupling and Flange Adaptor | Suitable for torsional strain only | May be used for installations which would cause torsional strain or where rotary alignment is required. Not suitable for rotational joints. |
| | Bellows | Not Suitable | Not Suitable. |
| | Swivel Joint | Suitable | Specifically designed for rotation and will prevent torsional strain. |
| | Flexible Tube | Not Suitable | Not Suitable. |

| Application | Type | Suitability | Remarks |
|---|---|---|---|
| Mobility | Semi-rigid Coupling and Flange Adaptor | Not Suitable | Not Suitable. |
| | Bellows | Not Suitable | Not Suitable. |
| | Swivel Joint | Suitable | May provide the necessary mobility of pipework by using suitable combinations of joints, for example dog leg arrangements will allow horizontal or vertical movement. |
| | Flexible Tube | Suitable | Allows both vertical and horizontal movement. For vertical movement the flex should hang from fixed points looking vertically down. For horizontal movements the flex may be installed in a vertical loop for short travel or in a horizontal loop when maximum travel required. Horizontal flex must be supported at the ends to avoid sagging, strain and failure at end fittings. |
| Portability | Semi-rigid Coupling and Flange Adaptor | Not Suitable | Not Suitable. |
| | Bellows | Not Suitable | Not Suitable. |
| | Swivel Joint | Not Suitable | Not Suitable. |
| | Flexible Tube | Suitable | Sheathed flexible tubes are suitable. |

## Tools

### Pipe gripping tools

These are generally a larger version of those used on smaller pipework and described in Vol. 1, Chapter 12. Manufacturers' designs differ slightly although the general principles are the same.

Straight and stillson type pipe wrenches are available in lengths from 150 mm to 1500 mm to deal with pipes from 20 mm to 200 mm diameter, Fig. 23. The hardened alloy steel jaws are replaceable, handles may be steel, malleable iron or aluminium.

Chain wrenches, Fig. 24 give a very positive grip and can be used in close quarters. The jaws are replaceable and, in some designs, reversible. They are available in similar sizes to the stillsons.

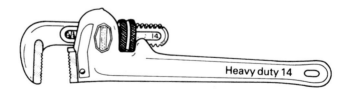

*Fig. 23   Straight pipe wrench*

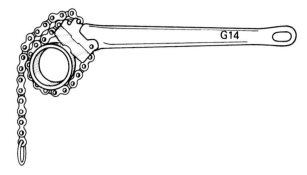

*Fig. 24    Chain wrench*

## Pipe cutting tools

Hacksaws may be used on smaller diameter pipe and power hacksaws are used in workshops. Wheel cutters are normally used on larger

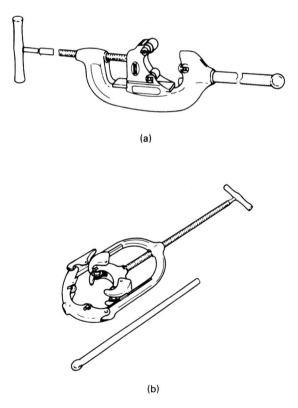

(a)

(b)

*Fig. 25    Pipe cutters: (a) single wheel; (b) four wheel*

diameter pipes. They may have from one to four cutting wheels which are designed for cast iron, ductile iron or steel pipe. Fig. 25(a) shows a single wheel cutter for pipes up to 100 mm diameter and (b) is a four wheel hinged cutter for pipes up to 300 mm diameter. When using cutters it is necessary to remove the burrs formed on the inside and outside of the pipe.

The rollers on the single wheeled type minimise the outer burr but the internal burr must be removed by a reamer. There are several designs and Fig. 26 shows a spiral ratchet pipe reamer for pipes 64 mm to 100 mm.

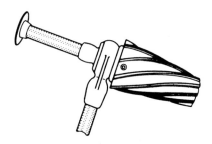

*Fig. 26    Spiral ratchet pipe reamer*

Various types of power operated cutting machines are available. The type shown in Fig. 27 clamps on to the pipe and the machine head, carrying a small circular saw blade, is rotated once around the pipe. The high speed saw gives a square cut without burrs. Bevel cutters may be used in place of the saw to prepare pipe ends for welding.

*Pipe threading tools*

Geared receder dies, Fig. 28, are used for threading pipe from 50 mm to 150 mm. They may be operated manually by a ratchet lever or be connected to a small, portable power drive, Fig. 29. Larger power units, mounted on stands may be fitted with a variety of cutters, die heads or receders.

Many types of screwing machines are available for pipes up to 100 mm. Above this size the geared receders are used.

*Pipe bending tools*

Bending is usually carried out by hydraulic machines (Fig. 30). These are designed to bend pipe up to 150 mm and are larger and more robust versions of the smaller models, Vol. 1, Chapter 12. Most

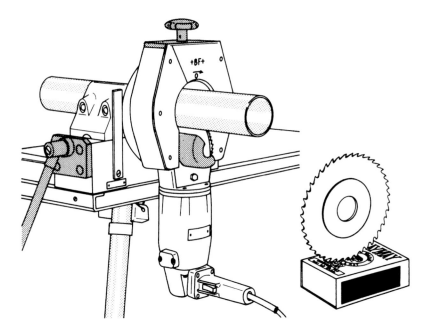

Fig. 27    Pipe cutting saw

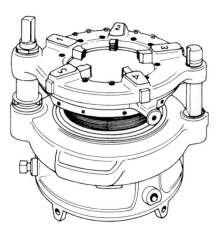

Fig. 28    Geared receder die stock

machines are manually operated. Power pumps can substantially
reduce the time taken to form a bend and are commonly used for
repetitive work.

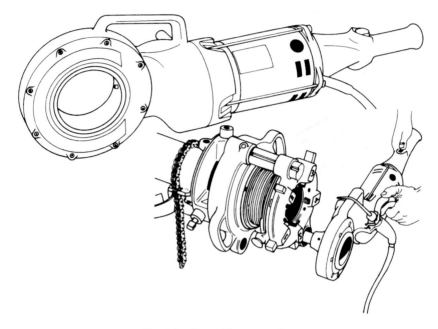

*Fig. 29    Portable power drive*

*Pipe lifting tools*

Large diameter pipes are heavy and it requires considerable effort to
manhandle them into position, particularly as they are often fitted at
high level. In many cases cranes or wire hoists are available on site
and pipes and equipment can be moved with the aid of slings. A
common type of sling is the woven nylon webbing loop, Fig. 31(a),
which is simply passed around the object and looped on to the hook.
Special slings are designed to lift large pipes, Fig. 31(b).

When cranes are not available chain hoists may be used either
hooked on to roof trusses or supported by shearlegs. The chain hoist
has a reduction gear driven by a continuous loop of chain. The
gearing drives on to another length of chain with a hook attached for
carrying the load, Fig. 32.

It is important that the safe working load (SWL) of the block and
slings is not exceeded. If the hoist is attached to any structural
members of the building, ensure that these can adequately support
the load. All chains, ropes or lifting tackle should be checked before
use and examined regularly by a competent person to meet safety and
insurance requirements.

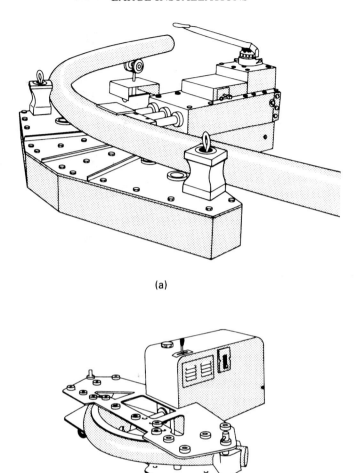

(a)

(b)

*Fig. 30   Hydraulic bending machines: (a) manual; (b) power operated*

**Safety Precautions**

The safety precautions to be observed when using tools and installing pipework have been described in the previous volumes. They apply equally to large installation work and the following points should particularly be borne in mind:

- do not use a power tool unless you are completely familiar with its operation
- always read the instruction booklet

(a)

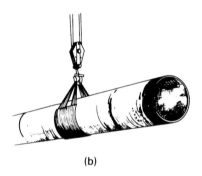

(b)

*Fig. 31    Slings: (a) nylon webbing; (b) pipe sling*

- make certain that the tool is the right one for the job
  – adequately powered
  – in good condition
  – correctly assembled
- never use a tool with the guard removed
- if a tool does not work, do not tamper with it, repairs should be carried out by an expert
- always unplug the electrical supply before carrying out any adjustments to the tool
- where pipes are stacked on site, secure them against movement with chocks and timbers
- because pipe wrenches are often used at high level and heavy pressures are exerted, the teeth must be clean and sharp, a slip could cause a serious accident

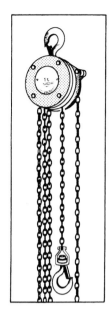

*Fig. 32    Chain hoist*

- always wear protective clothing when required; safety helmets, goggles, ear protectors and dust masks are often necessary
- always report any accident whether or not it results in injury
- get immediate first aid treatment for any injury, even a minor wound.

**Commissioning**

When the installation is completed the following operations should be carried out:

- check that all pipework is adequately and securely supported
- ensure that test points have been fitted at appropriate places and particularly at the end of pipe runs to enable testing and purging to be carried out satisfactorily
- check that valves have been fitted with handles or hand wheels giving a clear indication whether the valve is open or closed
- before testing for soundness isolate the pipework from any existing installation or meter by plugging or capping off. Valves should not be used as the sole means of isolation

- the joints on underground pipework should not be reinstated until after testing but pipes must be securely anchored and precautions taken to safeguard personnel while excavations are open
- pipework must be tested for soundness in accordance with 'Soundness testing procedures for industrial and commercial gas installations, IGE/UP/1 or IGE/UP/1A'
- after testing connect to the existing gas supply and test the final connections with leak detection fluid
- ensure that pipework is correctly wrapped or painted and colour coded
- purge the installation in accordance with 'Purging procedures for non-domestic gas installations, IGE/UP/1 or IGE/UP/1A'
- any outlets not in immediate use should be capped or plugged
- back fill and reinstate any excavations and wrap or paint any exposed points or connections
- check that drawings of the installation have been fixed by the primary meter and at other appropriate locations
- remove surplus material and clear up the site.

**Soundness Testing and Purging**

The procedure for the Soundness Testing and Purging of Commercial and Industrial installations is currently covered by two documents: IGE/UP/1A and IGE/UP/1. Personnel carrying out work defined within these two documents must be competent; in general terms this means personnel should make reference to current relevant documents, be adequately trained and have practical experience which would enable them to carry out the work in a safe manner.

Throughout this section references are made to both IGE/UP/1 and IGE/UP/1A tables and page numbers. These documents are the only means of approved reference and it is essential that current copies are obtained.

In simple terms soundness testing is carried out to prove the system has no leaks and that a smell of gas will not occur when the system is in use. However this is almost impossible to achieve and therefore some leakage tolerances are allowed, such tolerances utilising IGE/UP/1 are based on specific principles. Instruments used for soundness testing must be capable of being read accurately to identify the leakage from each metre of pipework. This is known as the minimum detectable leak rate and providing we also know the system volume and an acceptable leakage rate per/hour we can determine the required test period in minutes.

Although based on the same principles IGE/UP/1A provides a more simple method of determining the test period and also provides maximum allowable pressure drops for existing systems. The simplicity of using the document and the effect of carrying out a soundness test for an incorrect period can be seen in the following Example.

*Example 1*

An existing pipework system is installed externally except for a small occupied area of 15 m³, the system volume is 0.88 m³ and the normal operating pressure is 21 mbar. Utilising IGE/UP/1A the soundness test period is determined as 6 minutes and the maximum allowable pressure loss is 0.8 mbar.

Contractor A has been employed to repair a known gas escape on the system described above. After repairing the escape and without making reference to either IGE/UP/1 or IGE/UP/1A the system is tested. The engineer tests the system by his usual method for 10 minutes and will accept no more than 0.5 mbar pressure loss, after the 10 minute test he registers a leak of 0.8 mbar and fails the system.

In this case the engineer has applied a more stringent standard than is required. Although this is not totally incorrect it may lead to unnecessary losses in production and costs to his customer.

On a different occasion Contractor B has been employed to repair a known gas escape on the system described above. After repairing the escape and without making reference to either IGE/UP/1 or IGE/UP/1A the system is tested. The engineer tests the system by his usual method for 2 minutes and it will accept no more than 0.5 mbar pressure loss. After the 2 minute test he registers a leak of 0.5 mbar and passes the system.

In this case the engineer has applied a less stringent standard than is required, the test has been carried out for only a third of the required time and therefore the system may have a leak in excess of the maximum allowable.

Throughout this revised section on soundness testing and purging, worked examples are shown. They are based on the meters and pipework systems shown in Fig. 33 and utilise information obtained from a page number in either IGE/UP/1 or IGE/UP/1A.

**IGE/UP/1A** – Institute of Gas Engineers document 'Soundness Testing and Direct Purging of Small Low Pressure Industrial and Commercial Natural Gas Installations'. This document as it implies was introduced in 1998 to provide procedures for smaller installations containing natural gas. It outlines procedures in a similar manner to those previously described in British Gas publications ISM 7 & 8 with enhancements reflecting current good practice. It is impractical within this publication to fully explain all aspects of the document and therefore information is concentrated on the general requirements of typical systems which fall within the scope of IGE/UP/1A; for more specific guidance reference should be made to the current edition of IGE/UP/1A.

The procedures apply to:

- soundness testing and purging of new and existing pipework, extensions and replacement pipework and where pipework is to be taken out of service either temporarily or permanently
- systems where a complete loss of pressure has occurred, air being present or a leak is known or suspected
- installation pipework downstream of a primary meter control valve, where the inlet meter pressure does not exceed 75 mbar and the normal system operating pressure is 21 mbar
- installations with pipework not exceeding 150 mm in diameter
- system volume including any meter, pipework and associated fittings not exceeding 1 m$^3$.

**Soundness Testing** – the procedures can be applied to the soundness testing of pipework on new and existing systems that fall within the scope of the document. Existing systems should be tested at their normal operating pressure, new systems and extensions at 1.5 times this pressure.

**Prior to undertaking any soundness testing:**

- examine and estimate system volume
- establish the test pressure
- select suitable gauge
- determine permitted pressure drop
- determine test period.

**Examine and estimate system volume** – it is essential that the system to be tested is examined to identify any potential problems and to ensure that an accurate estimate of the system volume can be established – a figure of 10% should be added to the system volume to account for associated fittings. Where primary or sub meters form part of the system their volume must also be included in the total volume for the system. System volumes should be calculated to the nearest two significant figures.

Fig. 33 shows simple outline drawings of typical systems. For simplicity only the meter control, meter outlet and appliance isolation valves are shown. Meter volumes are shown in IGE/UP/1A Page 7 and volumes for diaphragm meters can be read directly from the table. However the following calculation is required for Rotary and Turbine meters.

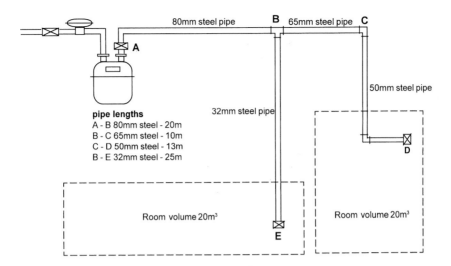

**Installation with diaphragm meter**

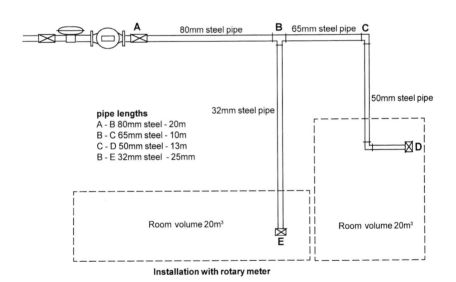

**Installation with rotary meter**

*Fig. 33    Systems used in worked examples*

*Example 2*

Meter Rotary (RD) dia. 80mm – flange to flange 0.3m
IGE/UP/1A Table 1 Page 7 ($0.79d^2 \times L$)

$0.79 \times$ dia. in m $\times$ dia. in m $\times$ Flange to flange in m = Meter volume $m^3$

$0.79 \times 80/1000$ (0.08m) $\times 80/1000$ (0.08m) $\times 0.3$m = Meter volume $m^3$

$0.79 \times 0.08 \times 0.08 \times 0.3$ = Meter volume of $0.0015m^3$

System pipework volumes can be obtained from IGE/UP/1A Table 2 Page 8 – on examining the system it is found that the meter is housed externally, the pipework system is installed on external walls with the exception of pipework and equipment installed in two occupied areas both 20 $m^3$ in volume.

*Example 3*

System contains Meter U65 – Meter vol. $0.1m^3$ from IGE/UP/1A Table 1 Page 7

Associated meter installation pipework 0.5m of 80mm $0.5 \times 0.0054 + 10\%$ = $0.0029m^3$

Pipework 20m of 80mm + 25m of 32mm + 10m of 65mm + 13m of 50mm + 10% = Pipework vol. $m^3$

From IGE/UP/1A Table 2 Page 8

$(20 \times 0.0054) + (25 \times 0.0011) + (10 \times 0.0038) + (13 \times 0.0024)$

$0.108m^3 + 0.027m^3 + 0.038m^3 + 0.031m^3 + 10\% = 0.224m^3$

Meter volume + Meter pipework + System pipework = Total system volume

$0.1m^3$    +    $0.0029m^3$    +    $0.224m^3$   = $0.33m^3$ Total system volume

**Establish the test pressure –**
Ensure within the scope of the document
New systems = 1.5 × System operating pressure
Existing systems = Normal system operating pressure

**Select suitable pressure gauge** – water or electronic.

**Determine permitted pressure drop** – on new systems there should be no perceptible movement when using a water gauge. With an electronic gauge this should not exceed 0.25 mbar. On existing systems the maximum permitted pressure drop is obtained from IGE/UP/1A Table 6 Page 4. In addition the smallest occupied space through which any of the system pipework passes must be identified as it is likely any leaks would be smelt in this area first. The figures given in IGE/UP/1A Table 6 Page 14 are the maximum allowable pressure drops in mbars during the test period providing there is no smell of gas.

The figures given in IGE/UP/1A Table 6 Page 14 can be calculated if required using Example 4. Calculated maximum allowable pressure drop figures must be for the maximum volume figure within the ranges shown in IGE/UP/1A Table 6 Page 14. In Example 3 the total system volume is 0.33 m$^3$ and as this is between 0.3 m$^3$ to 0.35 m$^3$ we must use 0.35 m$^3$.

*Example 4*

$$\frac{\text{Min readable on gauge 0.5mbar} \times \text{Smallest occupied space m}^3 \times \text{test period in min}}{\text{System volume in m}^3 \times 60 \text{ minutes}}$$

= Maximum allowable pressure drop in mbars

$$\frac{0.5 \times 20 \times 3}{0.35 \times 60} = \frac{30}{21} = 1.4\text{mbar}$$

**Determine test period** – the period for new installation is shown in IGE/UP/1A Table 3 Page 10 and the period for existing installation in Table 4 Page 11. Examples 5 and 6 show how the test period is established based on the calculated system volume. The test period for any system must be at least 2 minutes.

*Example 5*

**New System Rotary meter (80mm) and system pipework**

Rotary meter volume from Example 2 = 0.0015m$^3$

Associated meter installation pipework 0.5m of 80mm:
0.5 × 0.0054 + 10% = 0.0029m$^3$

Pipework system volume from Example 3 = 0.224m$^3$

Total system volume: 0.0015m$^3$ + 0.0029 + 0.224m$^3$ = 0.23m$^3$

Test pressure: Normal operating pressure 21mbar × 1.5 = 31.5mbar

Pressure Gauge – Water or Electronic

Pressure drop allowable – Water gauge = No perceptible movement
Electronic = 0.25mbar

Test period from IGE/UP/1A Table 3 Page 10: Total system volume 0.23m$^3$ = 3 min

*Example 6*

**Existing System U65 Diaphragm meter and system pipework**

U65 meter volume from IGE/UP/1A Table 1 Page 7 = 0.1m$^3$

Associated meter installation pipework 0.5m of 80mm:
0.5 × 0.0054 + 10% = 0.0029m$^3$

Pipework system volume from Example 3 = 0.224m$^3$

Total system volume: 0.1m$^3$ + 0.0029 + 0.224m$^3$ = 0.33m$^3$

Test pressure: Normal operating pressure of system 21mbar

Pressure gauge – Water or Electronic

Pressure drop allowable – Determine and calculate volume of smallest occupied space = 20m$^3$

Obtain pressure drop allowable from IGE/UP/1A Table 6 Page 14, read off from vertical column. Total system volume of 0.33m$^3$ and horizontal column volume of smallest occupied space 20m$^3$, obtain pressure drop allowable of 1.4mbar

Test period from IGE/UP/1A Table 4 Page 11: Total system vol 0.33m$^3$ = 3 min

**Electrical continuity** – ensure suitable electrical temporary continuity bonds are available and that continuity is maintained at all times when joints are to be broken. Pipework should be cleaned off to provide a good contact.

## Carrying Out a Soundness Test

Soundness testing should only take place when weather conditions and temperatures within the system are stable.

### New Systems

- valves to and from the system shall be spaded off, plugged or capped and left in the open position. Ensure any valves within the system are in the open position. Bypass any system valves or controls which may prevent equalisation throughout the system
- connect pressure gauge, raise pressure in system with air to test pressure
- allow system to stabilise for test period or 6 minutes, whichever is the longer, then isolate the air supply
- monitor the gauge for the test period (the system can be regarded as sound providing it meets the criteria previously stated for new systems). If a leak is detected it must be located, repaired and the test repeated
- remove any temporary bypasses, re-pressurise the system to the test pressure and re-test any affected joints with leak detection fluid
- complete test record forms (Fig. 34) or certificate.

### Existing Systems

- ensure all system valves are in open position, close appliance isolation valves, connect pressure gauge, carry out let-by test. Raise pressure in system with gas to normal operating pressure, turn off valve and reduce system pressure by 50%, monitor pressure for the let-by test period. If the system pressure does not increase by 0.5 mbar during this period the valve has proved satisfactory. An increase of greater than 0.5 mbar would require the valve to be repaired or replaced. Let-by test periods are shown in IGE/UP/1A Table 5 Page 12
- raise the system pressure with gas to normal operating pressure and allow stabilisation period of 6 minutes then isolate valve
- monitor the gauge for the test period. The system can be regarded as sound providing it meets the criteria previously

**SOUNDNESS TESTING PROCEDURES FOR NEW SMALL COMMERCIAL
AND INDUSTRIAL GAS INSTALLATIONS UP TO 1 m³ CAPACITY WITH
A MAXIMUM PIPE DIAMETER OF 150 mm**

Customer's Name ........................................................................ Address ..............................................................................................

................................................................................................................................................................................................................

Job Ref: ............................................................ Tester ...................................................... Supervisor ......................................................

Working Pressure ......................................................................mbar   (if this exceeds 21 mbar contact Supervisor)

Pipe Size.................................................................................................   (if this exceeds 150 mm diameter contact Supervisor)

Conditions for the test must be steady, therefore the answer to the next two questions must be 'Yes' or 'Not Applicable'.

If the answer is 'No' contact the supervisor.

If part of the pipework is in the open, are the weather conditions stable (note steady sun
could be unacceptable if the pipe is in the hot sun)                                          Yes/No/Not Applicable*

Is the building temperature steady?                                                             Yes/No/Not Applicable*

**PROCEDURE**

NOTE: IT IS ESSENTIAL THAT A WATER GAUGE BE USED FOR THIS PROCEDURE.

1. Estimate total volume of the installation to be tested from Table 1 and 2.
                                                             Total Volume ..........................................................................................dm³

2. Determine the test period from Table 3.                 Test Period ......................................................................................min.

3. Spade, plug or cap off all ends of relevant pipe sections to be tested.

4. Raise the installation to the test pressure of .............mbar.

5. Allow the temperature to stabilise for 6 minutes maintaining the pressure.

6. Isolate the source of pressure and observe the water gauge for the test period.

7. If there is any perceptible gauge movement there is an unacceptable leak and it must be dealt with.

                                                             Gauge movement (if any) ..........................................................................mbar

8. When a test has been carried out with no perceptible gauge movement, the installation can be regarded as safe, provided
   that all pipe joints in ill-ventilated areas have been tested with leak detection fluid and any leaks dealt with.

                                                             Checked                              Yes/No/Not Applicable*

9. Put the installation into service (See Purging Procedures for Non-Domestic Gas Installations IGE/UP/1A).

10. Are there any joints in ill-ventilated or unoccupied areas and ducts.
    Yes/No*.                                               If yes, advise Supervisor.
    Check these areas with a suitable gas detector. (The use of leak detection fluid is not acceptable). No perceptible movement
    from 0% LEL on the LEL scale.
                                                             Checked                              Yes/No/Not Applicable*

11. Provided this test is successful and that there is no smell of gas anywhere, the installation can be regarded as sound
    (Subject to 12).
                                                             Is there a smell of gas?                          Yes/No*

    NOTE: FOR A SUCCESSFUL TEST THE ANSWERS TO 8 AND 10 MUST BE 'YES' OR 'NOT APPLICABLE',
    THE ANSWER TO 11 MUST BE 'NO'. IF THIS IS NOT THE CASE – CONTACT THE SUPERVISOR.

    **NOTES:**

(a) In item 4 governors etc., are temporarily by-passed, for example, using pressure points to avoid 'trapping' pressure.

(b) When the gas is turned on the joints prior to the meter governor should be tested using soap solution.

Tester's Signature  ......................................................................  Date............................................

12. Three to four days after commissioning repeat checks (as in 10) on any joints in ill-ventilated or unoccupied areas and ducts.

    Satisfactory?    Yes/No*                                                     If No, advise Supervisor

Tester's Signature  ..............................................................................  Date............................................

*Delete as applicable.

*Fig. 34   Soundness test record form*

gauge is chosen. The suitability of a gauge is determined by its pressure range and the maximum time it can be used. Gauges must also be capable of reading accurately the minimum pressure drop for a particular gauge type. This is termed the minimum detectable/readable leak rate and is shown in $m^3$ per metre of pipework. Detectable/readable figures are shown in IGE/UP/1 Appendix 2 Page 38. Suitable gauges can be selected from IGE/UP/1 Table 2 Page 8, the table shows the minimum readable figure in mbars for specific gauge types. Manufacturers of gauges provide instruction on their use and recommendations on periodic calibration checks. Figures shown in IGE/UP/1 Table 3 Page 9 should be considered as maximum time intervals between calibration checks of gauge accuracy.

**Determine maximum permitted release/leak rates** – IGE/UP/1 Appendix 2 Page 37 provides the maximum volumetric release rate in $m^3$/hr for different gases, installation types and locations. This figure is also known as the permitted leakage rate and forms part of the test period calculation.

The maximum permitted release rate figure shown for new systems or extensions must be used when calculating the test period for new or extended systems.

Existing systems can be installed in a variety of locations and have different permitted release rates. This must be taken into account when considering the amount of gas we may be releasing or permitting to leak into these areas. These areas are generally defined as:

**Potentially hazardous areas** – existing systems have often been installed for a number of years and therefore may be in areas which would be unacceptable for new or extended systems.

These areas may have little or no ventilation and it is possible for small leaks to develop, resulting in smells of gas being identified or for potential concentrations to build up which if ignited could result in fires or explosions. In such areas we cannot rely solely on the result of a soundness test and all joints within the area shall be checked with either a leak detection fluid or a suitable gas detector. Consideration should be given to testing as new, any section within the system where adequate checks cannot be made. Any leaks identified in potentially hazardous areas must be repaired.

**Occupied areas** – in these areas it is assumed that with normal standard ventilation a small leak would not be detected by smell. As the release rate figure is the maximum that can be released into 1 $m^3$ of space in 1 hour we must identify the room in which a smell would occur first because of its size. This is known as the smallest occupied space through which the system pipework passes. Example 7 shows how the maximum leak rate figure is calculated from the volume of the smallest occupied space.

*Example 7*

In Fig. 33 the pipework system is mainly run externally, except for two occupied areas both 20m³ in volume. As they are both the same volume 20m³ is identified as the smallest occupied space.

Determine maximum release rate figure for an occupied space from IGE/UP/1 Appendix 2 Page 37

Max release rate occupied areas × Smallest occupied space = Permitted leak m³/hr

$$0.0005\text{m}^3/\text{hr} \qquad \times \qquad 20\text{m}^3 \qquad = \qquad 0.01\text{m}^3/\text{hr}$$

\* The calculated permitted release rate figure for occupied areas must not exceed 0.03m³/hr

**Large open work and exposed areas** – as the wording implies these are areas in which a small leak would not be detected by smell as the pipework would be outside or well ventilated. In many instances part of these systems may be installed in potentially hazardous areas. Where this occurs the potentially hazardous areas shall be checked as previously described. Any leaks that are found must be repaired before applying the higher permitted leak rate for large open areas.

**Underground pipework (buried)** – pipework underground is not normally subject to the same influences as pipework above ground and shall be tested separately. Any leakage in excess of 0.5 mbar shall be measured and compared to the maximum permitted release rate for underground pipework. It is essential not to confuse buried pipework with pipework installed in ducts. Special care is required when underground pipework is installed near to other services as leaks have a tendency to track along them. Care must also be taken when pipework is in the proximity of ducts and cellars.

**Determining the test period** – this must be calculated from a number of known factors and must be not less than 2 minutes. In Examples 8 and 9 the meter and pipework system volumes previously obtained for Fig. 33 are used.

**Electrical continuity** – ensure suitable electrical temporary continuity bonds are available and that continuity is maintained at all times when joints are to be broken. Pipework should be cleaned off to provide a good contact.

## Carrying Out a Soundness Test

Soundness testing should only take place when weather conditions and temperatures within the system are stable.

## Example 8

New system operating at 21mbar contains U65 meter and pipework

Total system volume: Meter volume + Meter pipework + System pipework

Total system volume: $0.1m^3 + 0.0029m^3 + 0.224m^3 = 0.33m^3$

Test pressure new system = 75mbar – Gauge High SG IGE/UP/1 Table 2 Page 8

Minimum detectable leak rate = 0.0005 IGE/UP/1 Appendix 2 Page 38, this figure has been chosen because the system is new and we must use a gauge that has a suitable pressure range

Max permitted release rate new system = 0.0014 IGE/UP/1 Appendix 2 Page 37

$$\frac{\text{Min detectable leak rate} \times \text{Total system vol m}^3 \times 60}{\text{Permitted release rate}} = \text{Test period in minutes}$$

$$\frac{0.0005 \times 0.33 \times 60}{0.0014} = 7.07 \text{ minutes}$$

Test period = 8 minutes (calculated time always rounded up to next highest minute)

## Example 9

Existing system operating at 21mbar contains U65 meter and pipework. The pipework system is mainly run externally, except for two occupied areas both 20m³ in volume. As they are both the same volume 20m³ is identified as the smallest occupied space.

Total system volume: Meter volume + Meter pipework + System pipework

Total system volume: $0.1m^3 + 0.0029m^3 + 0.224m^3 = 0.33m^3$

Test pressure new system = 21mbar (see Gauge type Water IGE/UP/1 Table 2 Page 8)

Minimum detectable leak rate = 0.0005 (IGE/UP/1 Appendix 2 Page 38)

Maximum permitted release rate existing system
$0.0005 \times 20m^3 = 0.01$ (IGE/UP/1 Appendix 2 Page 9)

$$\frac{\text{Min detectable leak rate} \times \text{Total system vol m}^3 \times 60}{\text{Permitted release rate}} = \text{Test period in minutes}$$

$$\frac{0.0005 \times 0.33 \times 60}{0.01} = 0.99 \text{ minutes}$$

Test period = 2 minutes (calculated time always rounded up to at least 2 minutes)

## New Systems

- valves to and from the system shall be spaded off, plugged or capped and left in the open position. Ensure any valves within the system are in the open position. Bypass any system valves or controls which may prevent equalisation throughout the system
- connect pressure gauge and raise pressure in system with air or inert gas to test pressure. If the test pressure is to exceed 2 bar

the system must be raised to this pressure gradually at 10% intervals, allowing time for stabilisation at each stage. Personnel shall not be within the test area when the test pressure is in excess of 7 bar

- allow system to stabilise for test period or 15 minutes whichever is the longer and then isolate pressure source. Should test periods exceed 60 minutes, a maximum stabilisation period of 60 minutes would normally be acceptable
- monitor the gauge for the test period. The system can be regarded as sound providing there is no perceptible movement on the gauge. If a leak is detected it must be located, repaired and the test repeated
- remove any temporary bypasses, re-pressurise the system to the test pressure and re-test any affected joints with leak detection fluid
- complete test record forms (Fig. 34) or certificate.

## Existing Systems

- ensure upstream valve closed, check valves within the system are in the open position, close appliance isolation valves
- connect pressure gauge, carry out let-by test. Raise pressure in system to 50%, monitor gauge and check for let-by for the same time as the system test period. If the system pressure does not increase by more than the minimum reading of the gauge used, the valve has proved satisfactory. An increase of greater than the minimum reading would require the valve to be repaired or replaced
- raise the system pressure to the normal operating pressure, allow system to stabilise for test period or 15 minutes whichever is the longer and then isolate pressure source. Should test periods exceed 60 minutes a maximum stabilisation period of 60 minutes would normally be acceptable. Precautions for systems above 2 bar need to be taken as previously outlined
- monitor the gauge for the test period. The system can be regarded as sound providing movement on the gauge used does not exceed the minimum readable figure in IGE/UP/1 Table 2 Page 8. Any leak detected which is greater than this figure should be timed and the actual leakage calculated. If the calculated leakage is greater than the determined permitted leak rate then any leaks must be located, repaired and the test repeated. Example 10 shows how the actual leakage rate is calculated from the graph in IGE/UP/1 Appendix 2 Page 44

- system pipework within potentially hazardous areas, ventilated ducts, confined space or unoccupied areas shall be checked with leak detection fluid or suitable gas detector and there shall be no evidence of a leak
- complete test record forms (Fig. 34) or certificate.

*Example 10*

Existing system operating at 21mbar contains U65 meter and pipework. The pipework system is mainly run externally, except for two occupied areas both 20m$^3$ in volume. As they are both the same volume 20m$^3$ is identified as the smallest occupied space.

Total system volume = 0.3m$^3$

Minimum detectable leak rate = 0.0005 (IGE/UP/1 Appendix 2 Page 38)

Maximum permitted release rate existing system
= 0.0005 × 20m$^3$ = 0.01 (IGE/UP/1 Appendix 2 Page 37)

Test period = 2 minutes

Actual pressure drop during test period = 3mbars

To obtain the equivalent leak rate figure from the graph, read off the system volume on the base horizontal line. As the base line is 0–10m$^3$ we must change our volume to make it readable: 0.3m$^3$ × 10 = 3m$^3$ . Now trace a line vertically upwards to meet the diagonal line on the graph; from this point trace a horizontal line towards the leakage equivalent column and read off the figure in m$^3$.

The figure obtained is 0.0015m$^3$. As our volume was multiplied by 10 we must also change this figure to 0.0015 ÷ 10 = 0.00015m$^3$/hr

$$\frac{\text{Equivalent leak rate} \times \text{Actual pressure drop in mbars during test period} \times 60}{\text{Minimum readable on gauge} \times \text{Actual test period}}$$

= Actual leakage in m$^3$/hr

$$\frac{0.00015 \times 3 \times 60}{0.5 \times 2} = 0.027\text{m}^3/\text{hr}$$

The calculated actual leakage is 0.027m$^3$/hr. This is greater than 0.01m$^3$/hr; therefore the test on the system has failed.

An alternative method which will give an approximate answer is:

$$\frac{\text{Volume in m}^3 \times \text{Pressure drop in mbars} \times \text{mins in/hr}}{\text{Approx. Atmospheric pressure mbars} \times \text{Test period}} = \text{Leakage rate in m}^3/\text{hr}$$

Using the figures in Example 10

$$\frac{0.3 \times 3 \times 60}{1000 \times 2} = 0.027\text{m}^3/\text{hr}$$

## Testing of Systems Where the System Volume is Not Known

Testing by this method must be considered with caution. Consideration needs to be given to the reasons why the volume cannot be calculated. If the pipework cannot be seen, little would be known with regard to its condition and routing if an accurate survey of the system has not been carried out. Testing by this method incorporates the use of a suitable test meter or calibrated pump. The procedures previously stated for existing systems must be followed up to the point where stabilisation has taken place. The system shall now be connected to a suitable test meter and provided with an independent gas supply. This supply shall be capable of maintaining the system at the test pressure without any perceptible movement on the gauge and throughout the period the test meter is recording leakage from the system. Recorded average leakage should now be compared to the permitted leakage rate for the system as shown in the following Example 11.

*Example 11*

System with unknown volume

The pipework system is mainly run externally, except for two occupied areas both $20m^3$ in volume. As they are both the same volume, $20m^3$ is identified as the smallest occupied space.

Recorded leakage through a test meter of $30dm^3$ in 15 minutes

Convert to an hourly rate in $4 \times 30dm^3 = \dfrac{120dm^3/hr}{1000dm^3} = 0.12m^3/hr$

Permitted leakage rate occupied areas: $0.0005 \times 20 = 0.01m^3/hr$

The recorded leakage is $0.12m^3/hr$. As this is greater than the permitted leakage of $0.01m^3/hr$, leaks must be located and repaired.

**Purging** – the purging procedures outlined using either IGE/UP/1 or IGE/UP/1A are very similar. Examples describing the method of calculating purge volumes and times are the same for either document. Methods of venting and determining successful purges are slightly different in both documents and some additional considerations need to be made when purging systems outside the scope of IGE/UP/1A. Comments previously made regarding the practicalities of providing a totally comprehensive guide also apply to purging and more specific guidance on purging should be obtained from the current edition of IGE/UP/1 or IGE/UP/1A.

Purging of new systems and meters is mainly undertaken to ensure the system contains only fuel gas and that all air has been displaced. Existing meters and systems require purging for a number of reasons. These include the re-commissioning of supplies after repairs or alterations and also when fuel gas is to be removed from the system. Purging to remove fuel gas from the system is carried out to make it safe to work on or to leave it in a safe condition when not in use. Inert gas purging should be carried out whenever it is considered unsafe to undertake a direct or air purge. Due to the possibility of asphyxiation care must be taken when inert purging into confined spaces.

Purging is generally carried out to:

- displace air or an air/gas mixture with a fuel gas, normally referred to as a Direct Purge
- displace air with an inert gas which is then displaced by a fuel gas or to displace a fuel gas by an inert gas, normally referred to as an Inert Purge
- displace fuel gas or a gas/air mixture with air
- displace one fuel gas with another fuel gas.

*Considerations prior to purging*

- consideration shall be given to minimising the release of methane into the atmosphere
- purging should be planned and procedures written where considered appropriate. Due account should be taken of the system's condition ensuring that no ring mains exist. A number of simultaneous purges can be undertaken at one time providing sufficient personnel are on-site to undertake the purge safely. Where the main and branches are to be purged separately the main and the largest diameter branch shall be purged first. Remaining branches from the main shall be purged in order of descending pipe diameter. It is good practice to ensure staff and personnel on-site are aware the purge is taking place. Vent areas cordoned off and 'no smoking' and 'no naked lights' notices prominently displayed
- purging of meters should only be carried out by a competent person who has the approval of the owner
- direct purging shall be continuous ensuring the minimum purge flowrates shown in IGE/UP/1 Appendix 5 Page 52 are achieved. Purges that are incomplete shall be abandoned immediately and a new complete purge undertaken in accordance with IGE/UP/1
- vent outlets shall be located in open air at least 2.5 m above ground and a minimum of 5 m downwind of any ignition source. Electrical switchgear and isolators close to purge or vent points shall not be operated, potential ignition sources shall not be near to purge or vent points and extreme care shall be taken to avoid the drift of gases into buildings or air ventilators
- in circumstances where a fuel gas to air purge is required, air should be admitted as close as possible to the upstream isolation point and at a pressure similar to that at which the system normally operates

- vent points on a system should be at least 25% of the diameter of the largest pipework. Any number of vent points can be operated together providing they are supervised and adequate communications are available to ensure a safe purge. Purge hoses should be of a diameter which would not create a pressure loss in the vent pipe. In general terms a purge hose equal in diameter to the vent point diameter would be satisfactory. Purge hoses shall be suitable for fuel gas, sound and anti-static. Vent pipes should be 20 mm for up to 40 mm diameter pipework systems and 40 mm for pipework systems above 40 mm. A typical vent pipe is shown in Fig. 35.
- ensure suitable electrical temporary continuity bonds are available and that continuity is maintained at all times when joints are to be broken. Pipework should be cleaned off to provide a good contact.

**Additional considerations IGE/UP/1** – the wider scope of IGE/UP/1 means that some further considerations need to be made in addition to those described for IGE/UP/1A. Although described as additional to the requirements of IGE/UP/1A, they provide useful guidance on some specific safety matters relating generally to purging:

- consideration shall be given to the impact of releasing potentially large amounts of methane into the atmosphere. As the document covers larger systems greater amounts of gases may be released and therefore where practical purge gases should be flared off. In circumstances where this may not be possible, precautions need to be taken due to the potential hazards caused by gases drifting into adjacent areas. Special consideration needs to be given to basement areas and confined areas where fuel or inert gases heavier than air are used
- it is essential not to confuse oxygen cylinders with those containing inert gases. Contents of cylinders shall be confirmed either by the use of oxygen detectors or written confirmation from the cylinder supplier. Any electrical equipment must be confirmed intrinsically safe. Gas detectors shall be checked for calibration and tested annually. Fire extinguishers of a suitable type shall be available for use if required
- except for the most simple purges, written procedures should be drawn up and followed. Consultation with responsible persons on-site would be required to ensure compliance with any specific site rules, legislation, risk assessments and work permits. Activities taking place during purging may be at various or remote locations, therefore a suitable communication system shall be

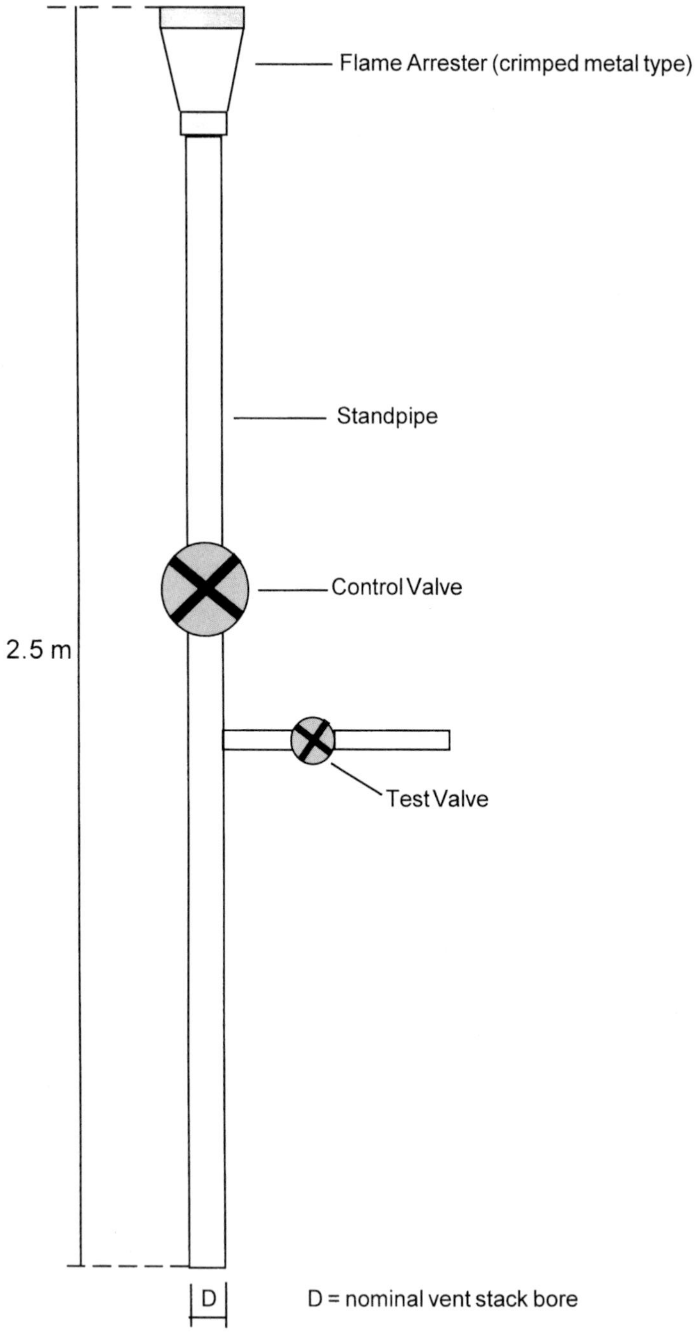

*Fig. 35   Typical vent pipe arrangment including flame arrester to BS 7244*

available. Other work in vent areas shall be prohibited and adequate warning notices displayed stating 'No Smoking' or 'Naked Lights'

● precautions need to be taken when carrying out air or inert gas purging. It is essential that the supplier's distribution main is blanked or spaded to prevent entry of these purge gases

● heavy ends (this term refers to the possible vaporisation of liquid LPG). It is possible after an apparently complete purge for flammable vapours to be produced. Where there is any suspicion of this happening, additional readings shall be taken. Readings shall continue at 15 minute intervals until a satisfactory level is obtained. Specialist advice shall be obtained if any difficulties are experienced

● calculated purge volumes shall be measured using a meter or orifice plate (on some small systems this may not be required). Measurement of a purge volume shall not be relied upon to confirm a complete purge – this can only be achieved by satisfactory testing of the vent gases. IGE/UP/1 Table 6a Page 22 provides vent gas test criteria for Natural Gas. Table 6b Page 22 shows safe purge end limits and flammability limits for typical fuel gases.


## Calculation of Purge Volumes

Where primary or sub meters form part of the system their purge volume must also be included in the total purge volume for the system.

Typical purge volumes for rotary, turbine, ultrasonic and diaphragm meters can be obtained from Table 5:

**TABLE 5**

| Meter type $m^3/h$ | $Ft^3/h$ | Volume $(m^3)$ | Badged Capacity/rev or Cyclic Volume |
|---|---|---|---|
| U6 | 212 | 0.008 | 0.071 ft$^3$ (2 dm$^3$) |
| U16 | 565 | 0.025 | 0.2 ft$^3$ (5.66 dm$^3$) |
| U25 | 883 | 0.037 | 0.35 ft$^3$ (10 dm$^3$) |
| U40 | 1412 | 0.067 | 0.71 ft$^3$ (20 dm$^3$) |
| U65 | 2295 | 0.100 | 1.11 ft$^3$ (31.4 dm$^3$) |
| U100 | 3530 | 0.182 | 2.0 ft$^3$ (56.6 dm$^3$) |
| U160 | 5650 | 0.304 | 2.5 ft$^3$ (70.8 dm$^3$) |
| RD or Turbine | | 0.79 d$^2$L where: d = diameter of meter connection (m) L = flange to flange dimension (m) | |
| Domestic Ultrasonic | | 0.0024 | |

In order to calculate purge volumes each part of the system to be purged must be included:

- Diaphragm meters – Capacity per revolution of the meter Table 5 × 5
- Rotary, turbine or ultrasonic meters – approximate volume of meter × 1.5
- System pipework – System pipework volume × 1.5
- Purge equipment – Purge hose volume × 1.5

Examples 12–14 show how to calculate purge volumes for systems, meters and pipework systems shown in Fig. 33 are again used. Although it is possible to purge the system by carrying out one purge, two purges are shown as this would minimise the manpower required.

*Example 12*

Total purge volume where a U65 Diaphragm meter and pipework are to be purged

Purge 1 includes U65 meter, associated meter pipework and pipework from A–D excluding B–E

Meter U65 Volume per rev of 31.4dm$^3$ × 5 = 157dm$^3$ divide by 1000 = 0.157m$^3$

Meter installation pipework 0.5m of 80mm:
0.5 × 0.0054 + 10% (0.0029m$^3$) × 1.5 = 0.0043 m$^3$

Purge 1
System pipework 20m of 80mm steel + 10m of 65mm steel + 13m of 50mm steel = Pipework volume m$^3$

        (20 × 0.0054)       +    (10 × 0.0038)   +   (13 × 0.0024)

0.108m$^3$ + 0.038m$^3$ + 0.031m$^3$ + 10% = 0.194m$^3$

Pipework system vol of 0.194m$^3$ – Pipework purge vol 0.194m$^3$ × 1.5 = 0.291m$^3$

System pipework purge vol + Associated meter pipework vol = Total pipework purge vol m$^3$

0.291m$^3$ + 0.0043m$^3$ = 0.295m$^3$

Purge hose vol (6m of 25mm hose) Hose purge vol 0.0002m$^3$ × 1.5 = 0.0003m$^3$

Meter purge vol + Total pipework purge vol + Hose purge vol = Total purge vol m$^3$

   0.157m$^3$   +        0.295m$^3$    +   0.0003   =     0.45m$^3$

Purge 2 includes only pipework from B–E

Purge 2
System pipework 25m of 32mm steel × 1.5 = Pipework purge vol m$^3$

25 × 0.0011 × 1.5 = 0.0412m$^3$

Purge hose vol (6m of 25mm hose) Hose purge vol 0.0002m$^3$ × 1.5 = 0.0003m$^3$

Pipework purge vol + Hose purge vol = Total purge vol m$^3$ Purge 2

   0.0412m$^3$   +   0.0003m$^3$   =     0.04m$^3$

## Example 13

Total purge volume where rotary, turbine or ultrasonic meter and pipework are to be purged

Purge 1 includes 80mm Rotary meter, associated meter pipework and pipework from A–D excluding B–E

Meter Rotary (RD) from Example 1
Meter volume of $0.0015m^3$ – Meter purge volume $0.0015m^3 \times 1.5 = 0.002m^3$

Meter purge vol + Total pipework purge vol + Hose purge vol = Total purge vol $m^3$

$0.002m^3$     +          $0.295m^3$          +      $0.0003$      =      $0.29m^3$

Purge 2 includes only pipework from B–E

Pipework purge vol + Hose purge vol = Total purge vol $m^3$ Purge 2

$0.0412m^3 + 0.0003m^3 = 0.04m^3$

## Example 14

Total purge volume pipework only

Purge 1 includes only pipework from A–D

Total pipework purge vol + Hose purge vol = Total purge vol $m^3$ Purge 1

$0.295m^3$          +      $0.0003$      =          $0.29m^3$

Purge 2 includes only pipework from B–E

Pipework purge vol + Hose purge vol = Total purge vol $m^3$ Purge 2

$0.0412m^3 + 0.0003m^3 = 0.04m^3$

**Note – the purge hose volumes have been taken as PE Coiled pipework in IGE/UP/1 Appendix 1 Page 31**

**Calculation of Purge time**

In order to determine the maximum purge time we must calculate the total system volume, determine the largest nominal bore pipework that will be included in any purge and the minimum purge gas flowrate figure for that diameter of pipe. Example 15 shows how to calculate the purge time utilising the minimum purge flowrate figures shown in IGE/UP/1 Appendix 5 Page 52.

**Carrying out the Purge** – the methods described are for typical systems in which natural gas is to be admitted or removed. There are some differences between the two documents when carrying air to fuel gas or fuel gas to air purges – where the requirements of IGE/UP/1 differ they are shown in *italics*. Inert and fuel to fuel purging is outside the scope of IGE/UP/1A and these methods are described separately.

*Example 15*

Using figures obtained in Example 14 the largest nominal pipe diameter is 80mm; IGE/UP/1 Appendix 5 Page 52 shows the minimum purge flowrate for 80mm pipework is $11m^3$/hr

$$\frac{\text{Total Purge 1 volume } 0.29m^3 \times 60 \text{ min/hr}}{\text{Minimum purge flowrate } 11m^3/hr} = \text{Maximum purge time 1.58 minutes}$$

$$\frac{\text{Total Purge 2 volume } 0.04m^3 \times 60 \text{ min/hr}}{\text{Minimum purge flowrate } 1.7m^3/hr} = \text{Maximum purge time 1.41 minutes}$$

For convenience the time could be calculated in seconds:

$$\frac{\text{Total Purge 1 volume } 0.29m^3 \times 3600 \text{ sec/hr}}{\text{Minimum purge flowrate } 11m^3/hr} = \text{Maximum purge time 94 seconds}$$

$$\frac{\text{Total Purge 2 volume } 0.04m^3 \times 3600 \text{ sec/hr}}{\text{Minimum purge flowrate } 1.7m^3/hr} = \text{Maximum purge time 84 seconds}$$

*Air to fuel gas*

1.  Ensure soundness test has been carried out immediately prior to purging, section isolation and vent point valves must be closed, any valves within the system should be open. Valves must be clearly marked 'Do not operate, purging in progress'.
2.  Open vent points and fuel gas isolation valve, start to test vent gas when 50% of the purge time has elapsed.
    *Testing of vent gases to be commenced once the calculated purge volume has been passed.*
3.  A satisfactory purge has been achieved when using a suitable gas detector or oxygen analyser readings of 90% fuel gas or less than 4% oxygen are obtained in the vent gas. Close all vent points, seal off vent points with plugs or caps. Any joints that have been disturbed and joints within ducts, unoccupied areas or confined spaces shall be checked with leak detection fluid or a suitable gas detector. Any leaks found shall be repaired.
    *If these levels are not achieved within the purge time the purge is incomplete, the system must be immediately isolated from the gas supply and purged with nitrogen in accordance with IGE/UP/1.*
4.  Appliances that are connected to the system must be relit, commissioned or sealed off from the system.

*Fuel gas to air*

1.  Ensure an adequate air supply is available to complete a successful purge, the air connection should be as close as

possible to the section isolation valve. Air should be admitted into the system and its pressure monitored throughout the purge. The air pressure should not exceed the system normal operating pressure and shall not fall below 50% of this pressure throughout the purge.

2.  Ensure section isolation and vent point valves are closed, any valves within the system should be open. Appliances shall be isolated. Valves must be clearly marked 'Do not operate, purging in progress'.

3.  Carry out let-by test on section isolation valve as described in 'Soundness Testing Existing Systems'.

4.  Isolate and disconnect system from the fuel gas supply, cap, plug or spade open ends, open vent points and air supply and start to test vent gas when 50% of the purge time has elapsed. *Testing of vent gases to be commenced once the calculated purge volume has been passed.*

5.  A satisfactory purge has been achieved when using a suitable gas detector or oxygen analyser readings of 40% LFL of fuel gas or a level greater than 20.5% oxygen are obtained in the vent gas. Close all vent points and disconnect air supply, seal off vent points and air inlet valve with plugs or caps and any joints that have been disturbed shall be checked with leak detection fluid or a suitable gas detector.
    *If these levels are not achieved within the purge time the purge is incomplete. The system must be immediately isolated from the gas supply and purged with nitrogen in accordance with IGE/ UP/1.*

### Direct purging into a well ventilated area

In some instances it is permitted to release purge gas direct into a well ventilated area without the use of a purge hose, vent pipe or flame arrestor. This method should only be considered on small pipework systems and appliance control trains.

The following restrictions also apply:

- total volume of the system shall not exceed 0.02 m$^3$, the internal room volume must be at least 30 m$^3$ and be well ventilated. Mechanical ventilation should be in operation and windows and doors open. Diaphragm meters must not form part of the system
- the purge point and the purge control valve shall be within the ventilated area and there shall be no source of potential ignition within 3 m. The area shall be monitored to ensure fuel gas concentrations do not exceed 10% LFL. Purging shall cease

immediately if readings of this level are achieved and an appropriate purge to atmosphere carried out utilising a purge hose and vent point.

## Inert purging

*Air to inert gas to fuel gas*

1.  Ensure soundness test has been carried out immediately prior to purging, section isolation and vent point valves must be closed, any valves within the system should be open. Valves must be clearly marked 'Do not operate, purging in progress'.
2.  Ensure an adequate nitrogen supply is available to complete a successful purge. The nitrogen connection should be as close as possible to the section isolation valve. Nitrogen should be admitted into the system and its pressure monitored throughout the purge. The nitrogen pressure should not exceed the system normal operating pressure and shall not fall below 50% of this pressure throughout the purge.
3.  Isolate and disconnect system from the fuel gas supply with cap, plug or spade. Open vent points and nitrogen supply. Testing of vent gases to be commenced once the calculated purge volume has been passed. A satisfactory air to inert purge has been achieved when, using a suitable oxygen analyser, readings of less than 8% oxygen are achieved. Close all vent points and disconnect the nitrogen supply. Seal off with plugs or caps.
4.  Remove blanks or spades from fuel gas isolation valve, open vent points and fuel gas isolation valve and start to test vent gas when the calculated purge volume has been passed.
5.  A satisfactory purge has been achieved when, using a suitable gas detector or oxygen analyser, readings of 90% fuel gas or less than 4% oxygen are obtained in the vent gas. Close all vent points and seal off vent points with plugs or caps. Any joints that have been disturbed and joints within ducts, unoccupied areas or confined spaces shall be checked with leak detection fluid or a suitable gas detector. Any leaks found shall be repaired.
6.  Appliances that are connected to the system must be relit, commissioned or sealed off from the system.

*Fuel gas to inert to air*

1.  Ensure an adequate nitrogen supply is available to complete a successful purge – the nitrogen connection should be as close

as possible to the section isolation valve. Nitrogen should be admitted into the system and its pressure monitored throughout the purge. The nitrogen pressure should not exceed the system normal operating pressure and shall not fall below 50% of this pressure throughout the purge.

2.    Isolate and disconnect system from the fuel gas supply with cap, plug or spade. Open vent points and nitrogen supply. Testing of vent gases to be commenced once the calculated purge volume has been passed. A satisfactory gas to inert purge has been achieved when, using a suitable gas detector, readings of less than 7.5% fuel gas are achieved. Close all vent points and disconnect the nitrogen supply. Seal off with plugs or caps.

*Fuel gas to fuel gas*

This may be carried out in circumstances where it is considered safe by burning off the existing fuel gas until the replacement fuel gas is identified. It is essential when undertaking this type of purge to consider the burning characteristics of the two fuel gases. It is unlikely the two fuels will burn correctly on a test burner or appliance and therefore suitable burners or appliances must be available for both fuels. If this is not practical a direct purge should be carried out.

# Industrial and Commercial Gas Meters

---

Chapter 2 is based on an original draft by R. H. Wharton

---

## Introduction

Significant changes have taken place in recent years with regard to gas supplies and British Gas is no longer the only supplier. Gas supplies are now available from numerous suppliers in domestic, industrial or commercial premises. In general the meter is still installed and maintained by British Gas. However information on gas used is passed onto the customers chosen supplier.

Contracts signed by customers with these suppliers often have specific supply conditions. It would be good practice to ensure customers have agreed with the supplier any proposed changes which affect the meter or the amount of gas consumed.

The basic principles of the common types of gas meter were outlined in Vol. 1, Chapter 8. This showed meters to be measuring devices which:

- recorded the total quantity of gas which has passed through them
- indicated the rate of flow at any moment of time.

So the information provided by a meter could be used as a:

- basis for charging gas used
- means of measuring the gas rate of a burner or the rate at which any other plant, such as gas engines, burns gas.

Under the Gas Act 1995, meters used as the basis for a charge must be stamped or 'Badged', except where a customer is supplied under the terms of a special contract. However, it is the policy of British Gas plc to use badged meters for the supply of gas to special contract customers. It is a statutory requirement for meters to be calibrated, this currently discharged by the Technical Directorate of OFGAS.

The accuracy of the most commonly used meters is as follows:

Diaphragm meters   ± 3% from Q min to 2 Q min
                   ± 2% from 2 Q min to Q max

RD and Turbine meters   ± 1% from 0.2 Q max to Q max
                        ± 2% from Q min to 0.2 Q max.

## Types of Meters

Meters may be classified as:

- inferential
- positive displacement.

In the inferential types, the quantity of gas flowing may be inferred from any of the following measurements:

- speed rotation of a turbine
- pressure loss across an orifice
- difference between static and kinetic pressures
- change of temperature in a wire
- height of a rotating float in a tapered tube.

In positive displacement meters, a definite volume of gas is measured by displacement through one of the following:

- bellows or diaphragms
- compartments submerged in a liquid
- spaces between impellers or vanes.

## BADGED METERS

### Diaphragm Meters

*Tin-case meters*

These meters are no longer made. All tin-case meters should now have been replaced by British Gas under a programme of work to replace them all with more suitable meters.

*Steel-case Meters*

The 'U' series of steel-case meters is shown in Table 1.

**TABLE 1  Range of Unit Construction Meters**

| Designation | Badged rating | | Connections | |
|---|---|---|---|---|
| | $m^3/h$ | $ft^3/h$ | Type | Nominal Bore |
| U6 | 6 | 212 | | 25 mm (1 in) |
| U16 | 16 | 565 | Threaded | 32 mm (1¼ in) |
| U25 | 25 | 883 | | 50 mm (2 in) |
| U40 | 40 | 1412 | | 50 mm (2 in) |
| U65 | 65 | 2295 | | 65 mm (2½ in) |
| U100 | 100 | 3530 | Flanged | 80 mm (3 in) |
| U160 | 160 | 5650 | | 100 mm (4 in) |

*Fig. 1  'U' series meter*

The U6 meter, shown in Fig. 1, is described and illustrated in Vol. 1, Chapter 7. It was developed from the D07 meter for domestic use and its success prompted the redesign of the larger meters.

Because they were designed to operate at lower pressure drops, the tin-case meters may be allowed to operate at an overload rate.

'U' series meters, however, must not be subjected to loads in excess of their badged ratings. Overloading the meter would cause it to operate at too high a speed, resulting in excessive pressure drop. It would also invalidate the maker's guarantee.

## Rotary Displacement Meters

### Roots-type Meter

This was described in Vol. 1, Chapter 8. It measures gas by trapping fixed volumes between two impellers rotating in opposite directions and the casing, as shown in Fig. 2. The impellers are geared together so that they rotate in opposite directions. A sectional view of a typical meter is shown in Fig. 3, Fig. 4 shows an actual meter whilst Fig. 5 and Fig. 6 indicate the mechanisms of positive displacement meters made by different manufacturers.

Roots-type meters are available in a range of sizes from 800 ft$^3$/h (23 m$^3$/h) to 102,000 ft$^3$/h (2,893 m$^3$/h).

### Rotary Vane Meters

These are also known as 'vane and gate' or 'rotary vane' meters.

Rotary meters are positive displacement meters and operate on the principle of a series of chambers of known volume being filled and rotated to discharge downstream. The motive force to rotate the chambers is provided by a differential between inlet and outlet pressures. The meter has three key elements: (1) an annular measuring chamber formed by two concentric chambers; (2) a four vane assembly which is free to rotate and (3) a sealing gate which is connected through gears to the vane assembly and rotates in the same direction. The gate prevents gas from bypassing the measuring chamber.

A demand for gas creates a lower pressure at the outlet of the meter causing a differential across one of the vanes resulting in rotation of the assembly to discharge the next chamber. Constant volumes are thus transported from inlet to outlet and the volume measurement recorded on the index. The operation is shown in Fig. 7(a) and (b).

Rotary vane meters are available in a range of sizes from 2,000 ft$^3$/h (57 m$^3$/h) to 14,100 ft$^3$/h (400 m$^3$/h).

Figure 7 shows a sectional view of the meter and Fig. 8 shows the actual meter.

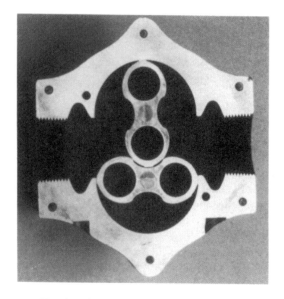

*Fig. 2    Operation of Roots-type meter*

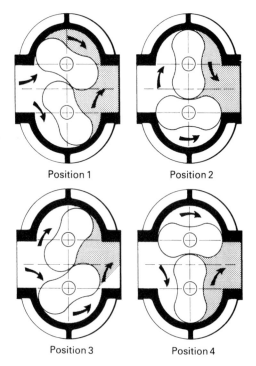

Position 1    Position 2

Position 3    Position 4

*Fig. 3    Sectional view of Roots meter (Dresser)*

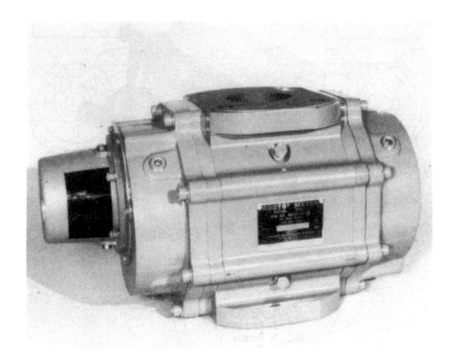

*Fig. 4   Rotary positive displacement meter (Dresser)*

*Fig. 5   Mechanism of Roots meter*

*Fig. 8    Rotary vane meter*

special protection from the weather. In sizes up to 16,000 ft$^3$/h (454 m$^3$/h), they may be supported by the pipework or 'line-mounted'. Above this size they must rest on a plinth or base.

Where large step load changes are anticipated with boosters or compressors, rotary positive displacement meters should be used with caution. This is because in extreme cases with sudden load increase, a temporary low pressure at the meter outlet may result in pilot burner outage. In addition, a sudden load decrease may cause the meter to overrun because of a temporary over-pressure condition.

In summary, the advantages of the rotary positive displacement meter are:

- large rangeability
- good linearity
- suitable for low flow rate
- performance hardly influenced by upstream disturbance or by distorted velocity profile
- it can be installed close to a bend without loss of accuracy.

The disadvantages are:

- detrimental effect of severe pulsating flows
- can generate cyclic flow pulsations
- filtration is essential; pipeline debris can cause the meter to jam
- relative cost.

## Turbine Meters

Turbine meters measure the speed at which gas flows through an annular gas-way of known cross-sectional area.

The gas-way is formed by a diffuser cone situated centrally in the body and a lightweight air foil turbine is placed so that the blades rotate around the annulus, Fig. 9. The speed of the rotation of the turbine is proportional to the velocity of the gas through the gas-way and also proportional to the volume of gas flowing.

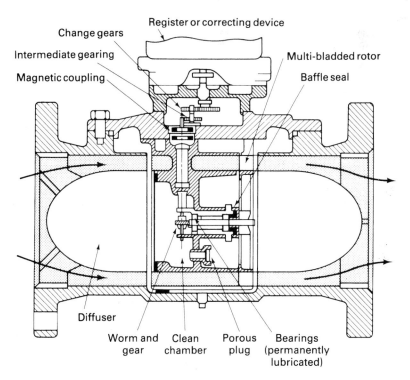

*Fig. 9   Principle of turbine meter (IGA)*

The drive from the turbine spindle is transferred to the index mechanism by a magnetic coupling, so eliminating the need for a stuffing box.

Because the gas is made to flow through a narrow annulus, its velocity is increased. This provides the necessary energy to drive the mechanism and gives more accurate registration.

As the turbine rotates freely, it has a tendency to overrun when flow stops. So the meter is best suited to situations where the gas flow is generally constant. Some models have an aerodynamic brake to minimise overrun.

Meters are available in a range of sizes from 8,800 ft³/h (249 m³/h) to 230,000 ft³/h (6,514 m³/h). Fig. 10 shows a sectional view of a turbine meter.

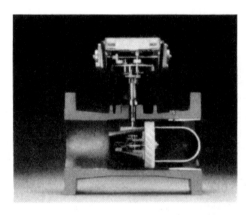

*Fig. 10    Sectional view of turbine meter*

Figure 11(a), (b) and (c) shows three different makes of turbine meters, (a) IGA; (b) NFC; (c) Valor.

In common with the rotary displacement meters there is no statutory maximum permissible pressure drop through turbine meters. Pressure drop increases with the rate of flow up to about 5 mbar at the badged rating.

If this is too high for low pressure installations then meters should be sized on the basis of an acceptable pressure drop. For example, a meter might have a pressure drop of 4 mbar at 140,000 ft³/h (3,965 m³/h) but a pressure drop of only 2 mbar at 100,000 ft³/h (2,832 m³/h). Manufacturers will give details of the pressure drop through their meters at various rates of flow. A graph showing the relationship for turbine and rotary displacement meters is given in Fig. 12.

*Fig. 11(a)    Turbine meter (IGA)*

*Fig. 11(b)    Turbine meter (NFC)*

*Fig. 11(c)    Turbine meter (Valour)*

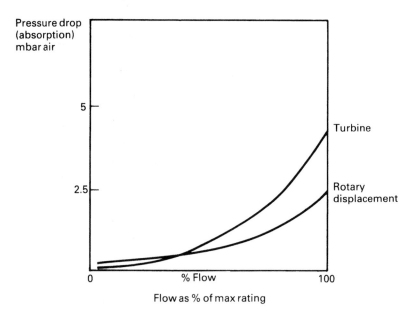

*Fig. 12    Graph of pressure drop with flow for turbine and RD meters*

Because of the tendency to overrun, the turbine is not suitable for loads where gas only flows for short periods during the day, or where the flow cycles rapidly on and off.

At low pressure, accuracy is maintained over a turndown of 15:1. This range increases as the pressure is increased and meters must be selected to have an acceptable accuracy at the minimum and maximum flow rates required.

All turbine meters can operate at pressures up to 7 bar and some are suitable for up to 100 bar. The small meters rated up to 14,100 ft³/h (400 m³/h) are made of aluminium whilst the larger sizes are made of steel. Some meters have integral straightening vanes while others have them supplied separately by the manufacturers. These take the form of a pipe section of 4 diameters in length containing a tube bundle at one end. The straightening device is fitted to the inlet of the meter with the tube bundle farthest from the meter. The device is supplied with each badged meter.

If the meter stops rotating for any reason, gas can continue to flow. So a bypass is not strictly necessary for continuity of supply. It is, however, necessary to allow the meter to be shut down for maintenance or exchange.

These meters are capable of driving ancillary equipment through either a mechanical linkage or an electrical pulse. There are, however, limitations to the type of instrument which may be driven mechanically without affecting the registration and both manufacturers and the Department of Energy instructions must be followed.

Turbine meters may be fitted outside without special protection and may generally be lined-mounted. Where meters are fitted with ancillary equipment it is advisable to mount them on a support.

In summary the advantages of turbine meters are:

- accuracy
- large rangeability with pressure
- versatility
- moderate pressure loss
- maintain flow when faulty
- compact (usually same size as pipework).

The disadvantages are:

- relative cost
- sensitive to flow disturbances and velocity profiles
- periodic calibration recommended to offset bearing wear
- regular bearing lubrication required.

## NON-BADGED INFERENTIAL METERS

### Turbine Meter Anemometer

Essentially similar in design to the rotary meter described in Vol. 1, Chapter 8, this meter is used as a non-domestic secondary meter, Fig. 13. These have not been made for some time but are still found in customers' premises.

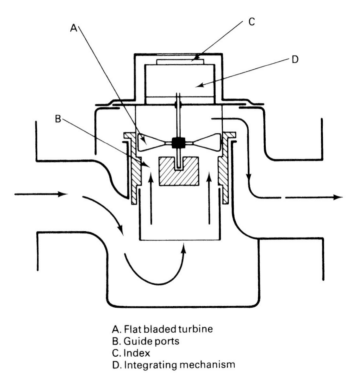

A. Flat bladed turbine
B. Guide ports
C. Index
D. Integrating mechanism

*Fig. 13    Vane or turbine anemometer*

The rotor has flat, aluminium blades and is usually mounted vertically in a central measuring unit which may easily be withdrawn for servicing or repair. Gas is directed on to the rotor blades through guide ports in the base of the unit. A drive from the rotor spindle operates an integrating mechanism and the gas consumption is shown on the index on the top of the meter.

These meters were available in a range of sizes up to 7,000 ft³/h (200 m³/h). Although not designed to be badged, their accuracy may be within ± 2% over a 10:1 turndown. They can operate at pressures up to 1.7 bar and must be fitted level in horizontal pipework.

## Orifice Meters

Meters which indicate gas flow by measuring differential pressures were introduced in Vol. 1, Chapter 6 and Chapter 8. Of these, the orifice meter is the most commonly used, mainly because of its simplicity and low overall cost. It is capable of operating at high pressures and is frequently used for measuring the flow of gas through transmission networks but results in a significant pressure drop across the meter.

Orifice plates are made as thin as possible and bored out to produce a sharp, square-edged, circular orifice. They are usually fitted between flanges with the orifice concentric with the pipe bore, Fig. 14. When a thicker plate is used to maintain rigidity, the orifice is chamfered on the outlet edge. The material used for the plate is a non-corrodible metal such as stainless steel or monel metal.

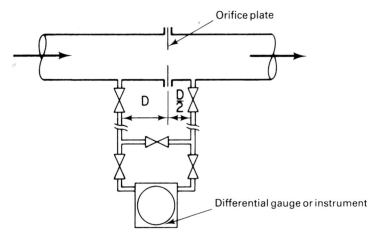

*Fig. 14   Orifice meter*

Accuracy of measurement depends on the condition of the plate and regular maintenance is necessary to ensure that it does not become dirty, distorted or eroded. It is also necessary to have straight, unobstructed pipework for about 20 diameters upstream and 15 diameters downstream of the meter. Orifice meters are unsuitable for situations where the flow rate varies considerably or pulsates. However, different flow rates can be metered by simply changing the orifice and/or the differential instrument. Maximum flow can be increased by up to 16 times in the same diameter pipe. Accuracy of $\pm$ 2% can be maintained over a turndown of about 4:1.

Meter failure has no effect on continuity of supply and a bypass is not essential. It is, however, useful for pressurising downstream pipework and for maintenance.

The design of orifice meters is specified in BS 1042, Part 1.

### Other Instruments

These brief notes are included for information if required.

*Venturi Meter*

This has the advantage of creating a lower pressure loss than the orifice plate and is used on low pressure systems, see Fig. 15. The flow rate through the meter can be determined by a differential pressure gauge or by a positive displacement meter, used as a 'shunt' meter and measuring about 1% of the total gas flow.

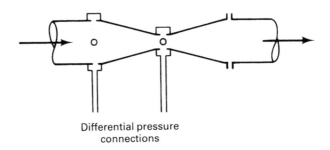

Differential pressure
connections

*Fig. 15    Venturi meter*

*Pitot Tube*

This is seldom used as a permanent meter installation, see Fig. 16. It has the advantage of portability and is generally used for checking flows on mains networks and in some pipework in industrial premises. It requires very careful use to obtain accurate readings. The point of the tube must be in line with the flow of gas and positioned at the point of mean velocity or the cross-section of the pipe traversed in accordance with BS 1042, Sections 2.1 and 2.2.

*Insertion Meters*

As an alternative to the pitot tube, a small turbine wheel on the end of a rod may be inserted into the pipe. The rotation of the turbine is conveyed to an external indicating device and the meter can measure

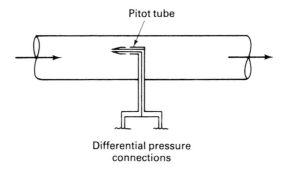

Fig. 16    Pitot meter

the average velocity of the gas flow. The general velocity range is between 0.31 m/s and 31 m/s. It is not highly accurate.

### Hot Wire Anemometer

In one form, the gas flow passes over an electric heater, A, situated between two resistance thermometers, B and C. The heater is controlled by the thermometers and raises the temperature of the gas by about 1°C. The amount of energy required to do this is measured by a watt hour meter which is calibrated to give a direct reading of gas volume, see Fig. 17.

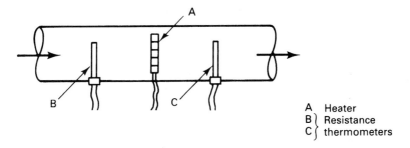

A   Heater
B ⎫ Resistance
C ⎭ thermometers

Fig. 17    Hot wire anemometer

### Rotameter

Rotameter is the generic term used for variable area flowmeters, but more correctly Rotameter is the name of one company that makes such meters.

This instrument is commonly used in laboratories and on appliance testing rigs in workshops, see Fig. 18. It consists of a float in a transparent, tapered, vertical tube. Gas flows through the annular gap around the float and so creates a pressure difference. This supports the float which has small slots in its outer edge so that it rotates as it is suspended on the upward flow of gas. The height of the float in the tube is proportional to the rate of flow of the gas, which may be read on a scale at the side of the tube.

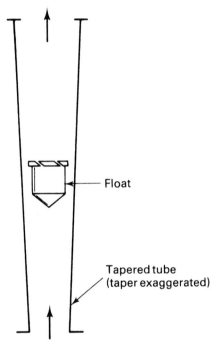

*Fig. 18    Rotameter*

*Vortex Shedding Meters*

The principle of this meter is shown in Fig. 19. When a fluid passes an obstruction turbulence is created which is independent of temperature and pressure. The rate of flow of gas is proportional to the frequency that the vortices are shed from the obstruction. Accuracy of measurement may be better than ± 1% and the rangeability is about 50:1, at pressures greater than 7 bar. They must not be installed near compressors or regulators and should be fitted with about 15 diameters of straight pipe including flow straighteners upstream and about 5 pipe diameters downstream.

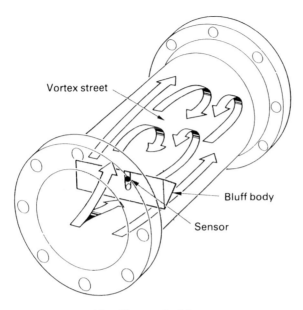

*Fig. 19    Vortex shedding meter*

*Ultrasonic Meters*

When used on high pressure gas flows (greater than 20 bar) this meter is accurate to within ± 1% and has a measurement range of more than 100:1. Ultrasonic meters work on the fact that the time taken for a pulse travelling with the gas flow is less than the time for a pulse to travel against the flow. By having two ultrasonic pulse probes mounted at 45° to the pipe, the difference in transit time of the pulses is proportional to the gas velocity, see Fig. 20.

**Installation of Meters**

The installation of gas meters used as a basis for charging is covered by:

- BS 6400 Specification for Installation of Domestic Gas Meters (2nd family gases).
- IGE/GM/1 (1998) Gas Meter Installations for Pressures not Exceeding 100 bar.
- OFGAS Codes of Practice COP/1a – COP/1b – COP/1c.

Legal requirements of the following documents must be taken into account and should be studied in association with the recommendations:

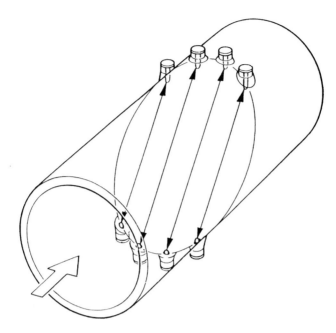

*Fig. 20    Ultrasonic meter*

The Gas Act 1986 as amended by the Gas Act 1995
The Gas Safety (Installation and Use) Regulations 2000
Gas (Meter) Regulations 1983
The Measuring Instruments (EEC Requirements) (Gas Volume Meters) Regulations 1988.

The installation of non-domestic low pressure diaphragm meters is, in many respects, similar to the installation of domestic meters. The requirements given in Vol. 2, Chapter 4 are applicable in many cases.

Rotary displacement and turbine meters must be installed in accordance with manufacturer's instructions.

Points of particular importance are as follows:

- filtering
- levelling
- oiling
- pressurising.

*Filtering*

Gas-borne solids, in the form of dust or rust particles can affect the operation of meters therefore it is important to prevent their ingress into the meter.

Diaphragm meters are not generally affected but a coarse filter is recommended, particularly if the gas pressures are greater than 75 mbar.

Rotary displacement meters require filters to stop particles larger than 250 micron. With some meters filtration down to 50 micron is necessary. Where filters are fitted upstream of the governors, it is advisable to fit a 'top-hat' filter on the upstream of the meter. This should face the gas flow, as shown in Fig. 21.

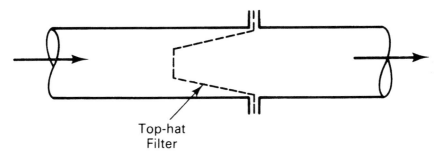

Top-hat
Filter

*Fig. 21    Top-hat filter*

Turbine meters should be protected with filters capable of removing particles larger than 250 micron to prevent damage to the blades and erosion.

Orifice plates may be eroded by dust and deposits can affect registration. A fine filter is required capable of stopping particles larger than 50 micron.

Filters should have ease of access for cleaning or exchanging. They should also be clearly identifiable from the outside of the pipework. Trapped dust should not be allowed to fall back into the system.

Meters can also be damaged by slugs of liquid and suitable receivers and drain points may be necessary.

*Oiling*

Oiling or lubrication of rotary displacement meters and turbine meters shall be carried out in accordance with the manufacturer's recommendations. Periodic maintenance checks recommended by manufacturers should be carried out to ensure correct operation.

*Pressurising*

Most types of meters may be damaged if there is an excessive pressure difference across the meter. On diaphragm meters this can puncture diaphragms and strain linkages. Where pressures are high (greater than 75 mbar) pipework downstream of the meter must be pressurised through a bypass or temporary bypass before the meter outlet and inlet valves are very slowly opened, in that order. The bypass or temporary bypass should then be closed slowly and removed or sealed in the closed position.

As a general guide when commissioning meters with bypasses fitted the inlet and outlet valves should be closed and the bypass opened, the outlet valve should then be slowly opened allowing equalisation of pressure. The inlet valve can then be opened and the bypass closed. In circumstances where the fitted bypass is a means of maintaining supplies, flow rates should be reduced to a minimum during commissioning. Commissioning of meters without bypasses should commence with the outlet or downstream valves closed, the inlet valve should then be slowly opened allowing the system pressure to equalise steadily.

## Location of Meters

*Low Pressure Meters*

Inside a building a primary meter should be located:

- as near as possible to the point of entry of the service pipe
- as near as practicable to the site boundary
- with adequate access for repair, maintenance or exchange
- to permit the installation or servicing of control devices or ancillary equipment
- in a well-ventilated position
- protected from any corrosive material or atmosphere
- remote from any heat source
- to avoid causing an obstruction
- where there is no risk of damage.

Outdoor locations should be chosen to give free access for repair, maintenance or exchange. The site must be well drained and not subject to flooding. The installation may need to be fenced and protected from vehicles and vandalism. Meters should not normally be exposed to temperatures below − 5°C or above 35°C.

Where the meter is located in a purpose-built house the following points should be noted:

- planning permission may be required and local by-laws must be observed

- the materials must be – weather and waterproof
  – fireproof to BS 476
  – strong and durable
- there should be a substantial foundation
- the housing must allow access for repair, maintenance or exchange of the meter or any item of ancillary equipment
- access must be provided to permit the use of lifting gear, where required
- the housing should be fitted with outward opening doors and be adequately ventilated
- any electrical equipment installed must be to the required safety standards
- the housing should not be occupied by personnel or used to store materials.

*High Pressure Meters*

On non-domestic premises these should be located as close to the boundary fence as possible, to reduce the length of high pressure pipework to a minimum, particularly indoors. The installation should preferably be outside and away from buildings or in a separate meter housing.

If this is not possible, a special room should be constructed within the building to house the installation. It should have external, outward opening doors, gas-tight internal walls and adequate ventilation to outside air.

If the meter is separate from the pressure governors and in the same building as the industrial process, the foregoing precautions may not be necessary.

In addition to the requirements for low pressure meters the following points should be observed.

- external sites should be clear of high-tension electric cables and large trees
- care must be taken to prevent noise from the installation causing a nuisance
- any outdoor installation should be fenced with access for repair, maintenance or fire-fighting
- any housing should have two sets of outward opening double doors, preferably on opposite sides of the building
- an explosion relief may be desirable, this could be incorporated in the roof design
- relief valves should have vent pipes terminating outside the building

- the total ventilation area should be not less than 2% of the floor area and equally dispersed at the top and bottom of the walls.

## Bypassing Meters

A full-flow bypass may be required where complete continuity of supply is essential and failure of the meter would affect the flow.

Similarly, where the full rate of flow must be maintained during maintenance operations, a bypass or temporary rider is required.

Where full flow is not required, a low-flow bypass may be fitted to assist commissioning. It is essential under current supply regulations that the fitting of bypasses to meters is only undertaken with the approval of the gas supplier and transporter.

## Warning Notices

Permanent notices which call attention to the special features of the installation should be mounted in a prominent position near to the meter, it is essential that notices of a statutory nature are displayed in accordance with the Gas Safety (Installation and Use) Regulations 2000. Their purposes are to:

- indicate that a service syphon is fitted
- indicate the action to be taken by the customer in the event of a gas escape
- prohibit smoking, naked lights and any other potential sources of ignition
- indicate where a common service pipe supplies two or more primary meters (this must include the number of meters and preferably their location)
- indicate the number and preferably the location of any secondary meters and give an instruction for the gas to be turned off at all appliances or meters before turning off the primary meter inlet valve
- give instructions for restoring the supply by opening the bypass valve when all appliances and secondary meters have been turned off
- prohibit the use of boosters, compressors or gas engines without permission of the gas transporter
- indicate that a non-return valve must be fitted in the gas supply where the gas is used in conjunction with compressed air or other gases

- warn that the meter may be damaged if the inlet valve is not fully open before starting any booster, compressor or gas engine, or if the valve is closed while the plant is running (devices to protect the meter are described in Chapter 4)
- provide suitable line diagrams indicating pipework and isolation valve positions.

In the case of high pressure installations, to:

- indicate the presence of gas at high pressure and any pressure limits agreed with the customer
- prohibit interference with the installation except to deal with a gas escape or open a meter bypass.

## Gas Volume Conversion

conversion device | Device using electronic circuitry that computes and indicates the volume, at standard conditions, of gas that has passed through a gas meter installation, and using as inputs the volume at measurement conditions as measured by a gas meter and other parameters such as temperature and pressure. The deviation from the ideal gas law is compensated by the compressibility factor (see also 'conversion system').

*Note: It is also possible for the conversion device to include the calibration curve of the meter.*

conversion system | System comprising a conversion device or a flow computer, a pressure transducer and a temperature sensor.

flow computer | Device that utilizes a high frequency signal from a gas meter and calculates a standard cumulative flow of gas based on this signal, pre-programmed parameters and other gas properties determined by suitable transducers.

LF transmitter/HF transmitter | A generic term meaning the device connected to the gas meter that provides electrical pulse outputs directly proportional to the actual volume passed by a gas meter, for example 1 pulse represents 1 unit of volume.

| standard reference conditions | Standard conditions of temperature, pressure and compressibility, to which gas is converted to account for measurement of those values. There are a number of different standard reference conditions in use at present (see IGE/GM/1 Edition 2). In the UK, metric standard conditions apply which is 15°C, 1,013.25 mbar. |
|---|---|
| CV | Calorific value $MJm^{-3}$. |
| E | Total energy. |

*Introduction*

The vast majority of non-domestic meters used within the gas industry measure volume flow. They make no allowance for temperature and pressure effects that can greatly alter the true amount of gas passed for a given volume flow.

To correct the meter reading for these effects, an on-line volume conversion device can be used, or fixed factor correction can be applied to the reading at a later date.

Primarily, an electronic conversion system offers greater measurement accuracy. When compared with the fixed factors, however, the cost of the conversion device may be greater than the benefit obtained from the improvement in measurement accuracy.

Historically, conversion devices have been known as meter correctors and were required to meet BS 4161: Part 8: 1987; this in turn has led to the widespread use of terminology such as corrected volume.

However the draft European standard prEN 12405: 1996 uses different terminology – electronic volume conversion devices for use with second family gases up to a pressure of 75 bar gauge over a working temperature range of – 25°C to + 60°C. It is possible for the meter conversion device to include provision for the correction of the error curve for the meter.

*Principle of Measurement*

Gas is sold in units of energy, typically megajoules (MJ). A typical gas meter will measure and totalise the volume of gas passed at the conditions existing at the meter. The relationship between this volume and energy is given by:

$$E = V \times CF \times CV$$

E = total energy
V = unconverted volume of gas recorded by the gas meter at the temperature and pressure conditions existing at the gas meter
CF = conversion factor used to convert the volume measured by the gas meter to equivalent volume at Standard Conditions (15°C and 1.01325 bar)
CV = Calorific value $(MJm^{-3})$

CF may be a fixed factor from the Gas (Calculation of Thermal Energy) Regulations, or a value derived from site gas conditions using a conversion system.

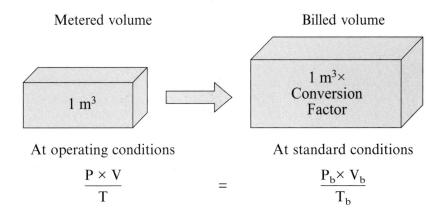

Metered volume                    Billed volume

$1 \, m^3$

$1 \, m^3 \times$ Conversion Factor

At operating conditions            At standard conditions

$$\frac{P \times V}{T} \quad = \quad \frac{P_b \times V_b}{T_b}$$

Thus, to correctly bill a consumer, all the above factors need to be known. However, the first selection procedure is to determine which of the above need to be measured and which can be estimated (by the use of some form of fixed factors). Conversion can be achieved by a conversion device or fixed factor. It is assumed that a volumetric gas meter is used to determine the quantity of gas consumed. It is assumed that the value of CV is measured and/or calculated at some other location, and the value made available during the accounting process.
Fixed factors are:

| | |
|---|---|
| T | = 1.0098 |
| P | = $\{1013.25 + [Mg - A] [1013.25]^{-1}\}$ |
| Z (P ≤ Z bar) | = 1 |
| Z (P > 2 bar) | = $0.9978 \{1 - [2.26 \times 10^{-6}] [1013.25 + (Mg - A)]\}^{-1}$ |
| Mg | = meter pressure (mbar) |
| A | = attitude correction (mbar) given in Tables in Regs |
| T | = assume gas temperature is 12.2°C and |

Fixed factor 1.02264 assumes:

- gas meter pressure = 21 mbar meter attitude = 66 m

so,

| | | |
|---|---|---|
| P | = | $(1013.25 + 21 - 8.114)\ (1013.25)^{-1}$ |
| | = | 1.01272 |
| T | = | 1.0098 |
| Z | = | 1 |

Thus,

| | | |
|---|---|---|
| PTZ | = | $(1.01272)\ (1.0098)\ (1)$ |
| | = | 1.02264 |

*UK Situation*

The following provides relevant information necessary to determine the appropriate method of volume conversion for a particular site and also covers the principles used to determine the method of conversion. The information given is based on industry practice within the UK. Other countries not only may have differing legislation, but significantly different climatic conditions that may influence the choice of, or whether or not to use, a volume conversion system.

The use, or otherwise, of electronic conversion devices is regulated under the Gas Act via the Gas (Calculation of Thermal Energy) Regulations which state that there is no legal reason either to fit, or not to fit, a full PTZ volume conversion system. However, there are, currently, some restrictions on the use of 'temperature only' conversion systems.

The choice of whether to use fixed factors or to fit a volume conversion system and the type (i.e. only, PTZ etc.) may be made by the following:

*(Transco's Network Code demands that, for meters reasonably expected to pass more than 2.93 GWh/annum (100,000 therms/annum) full PTZ conversion is required.)*

- it may not be possible to use fixed factors on some sites for local reasons. This would lead to the probability of large metering pressure variations. As 'P only' conversion is not permitted by the Gas (Calculation of Thermal Energy) Regulations, PTZ conversion is to be applied under such circumstances
- where it may be considered that the fitting of an electronic conversion system would lead to a cost benefit when compared

to the use of fixed factors. This may be for many reasons, including unusual metering temperatures and/or pressures and/or gas compositions

- the Gas (Calculation of Thermal Energy) Regulations demand that 'T only' electronic conversion be used only for meters reasonably expected to pass more than 73,200 kWh/annum (2,500 therms/annum)
- the choice between a converter and a flow computer will depend on local equipment and circumstances, for example:
  — load size and the resultant commercial exposure from system uncertainties
  — high pressure applications
  — requirement to continually update flow values for telemetry purposes
  — on-line updating of CV.

In the latter two cases, a flow computer will be needed.

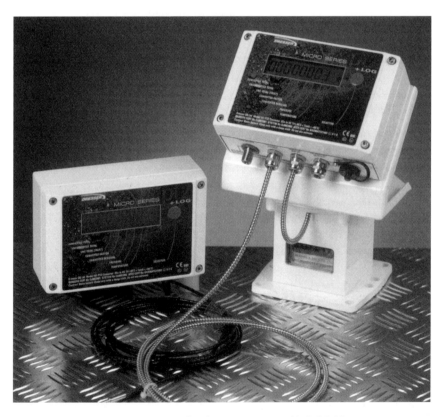

*Fig. 22   Typical volume corrector (ROOTS)*

In addition, it is likely that a flow computer system will require the use of an HF pulse output from the meter.

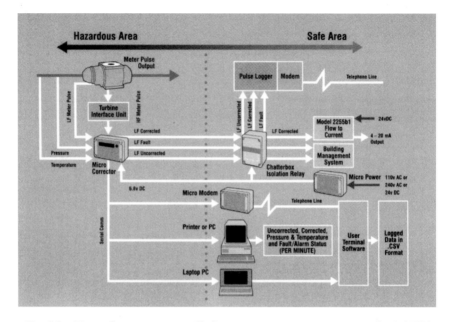

*Fig. 23   Typical corrector installed into communications system (ROOTS)*

### 'Fixed Factors'

*General*

Volume conversion may be carried out immediately on-site by the use of an electronic conversion device, or remotely by the use of 'fixed factors'. These fixed factors can be determined by reference to the Gas (Calculation of Thermal Energy) Regulations 1996.

There are two forms of fixed factor:

- an overall UK factor and
- a site-specific factor.

Both are calculated by using the same equations in the Gas (Calculation of Thermal Energy) Regulations, but different assumptions are made.

*The Overall UK Factor (1.02264)*

This is used on meters reasonably expected to pass no greater than 732 MWh/annum (25,000 therms/annum) and is 1.02264.

This is calculated using the following assumptions:

- meter pressure         = 21 mbar
- meter altitude          = 66 m
- barometric pressure    = 1013.25 mbar abs
- gas temperature        = 12.2°C
- compressibility         = 1

As the altitude increases, the atmospheric pressure decreases. The Gas (Calculation of Thermal Energy) Regulations provide a lookup table that enables the conversion of altitude to a pressure that is then deducted from the nominal atmospheric pressure, the altitude used from the lookup table being rounded up to the nearest 2.5 m. Alternatively, for the purposes of calculating the cost benefits later described in this appendix, the following may be used:

(a) pressure to be deducted (mbar) = altitude (m) (to nearest 2.5 m) × 0.1202

Using the lookup table gives a pressure reduction of 8.114 mbar for an altitude of 66 m (67.5 m being the nearest 2.5 m step from the table).

(b) Alternatively, using the formula:
67.5 × 0.1202 = 8.114 mbar

Thus, using the conversion equation:

Total conversion factor = $P_{factor} \times T_{factor} \times Z_{factor}$

$= (1013.25 - 8.114 + 21) \times (1013.25)^{-1} \times (15 + 273.15) (12.2 + 273.15)^{-1} \times 1$

$= 1.02264$

*Note 1: This equation uses absolute units. To convert from °C to K, add 273.15*

### A Site-specific factor

In the case of a site-specific factor, the altitude and meter pressure may be different from the conditions that are used for the overall UK factor. Thus, the site-specific factor is calculated using the line conditions that apply at the meter.

This is used on meters reasonably expected to pass more than 732 MWh/annum (25,000 therms/annum) and is calculated in a similar way to the UK factor but uses the following assumptions:

- meter pressure         = Mg = site-specific (mbar)
- meter altitude          = A = site-specific (m)

- barometric pressure      = 1013.25 mbar abs
- gas temperature      = 12.2°C
- compressibility factor
  (Pressure > 2 bar gauge) = see equation below
- compressibility factor
  (Pressure ≤ 2 bar gauge) = 1

Compressibility (Z) equation as in the Gas (Calculation of Thermal Energy) Regulations 1996

$$Z = 0.9978 \, \{1 - [2.26 \times 10^{-6}] [1013.25 + (Mg - A)]\}^{-1}$$

In the following example, the site-specific factor for a meter operating under the following conditions would be:

- meter pressure      = 100 mbar
- meter altitude      = 100 m
- atmospheric pressure reduction from the
  lookup table in the Gas (Calculation of
  Thermal Energy) Regulations 1996      = 12.021
- barometric pressure      = 1013.25 mbar abs
- gas temperature      = 12.2°C
- compressibility      = 1

Thus, using the conversion equation:

Total conversion factor = $P_{factor} \times T_{factor} \times Z_{factor}$

$$= (1013.25 - 12.021 + 100) \, (1013.25)^{-1} \times (15 + 273.15) \, (12.2 + 273.15)^{-1} \times 1$$

$$= 1.09749$$

*Note: With these calculations, there can be a small difference between the answer shown and that given by the Gas (Calculation of Thermal Energy) Regulations. This is due to rounding errors.*

## Why Fit a Gas Volume Conversion System?

When considering the installation of a gas volume conversion system, various factors need to be taken into account.

Listed below are some of the more important items that may be considered:

- legal requirements to meet statutory legislation or contractual obligations
- commercial requirements to comply with a Network Code (Transco's)

- the meter is operating with variable pressure conditions that make the use of fixed or calculated conversion factors impractical
- there needs to be a benefit from the installation that can be measured against the installation costs, such as:
  — reduced exposure to commercial risk
  — reduced uncertainty against fixed conversion factors
  — reduced billing uncertainty by more accurate measurement of gas consumed
- significant deviations at the meter site from average billing conditions.

As part of the commercial assessment for the installation of a gas volume conversion system, it is necessary to understand the effects that gas temperature, pressure and compressibility can have on the overall metering system accuracy.

It will also be necessary to consider the level of performance required from the conversion system, both at initial commissioning and during service.

All conversion equipment needs to meet the requirements of prEN 12405. This Standard demands an overall accuracy, under operating conditions, of ± 1% and, at constant 'laboratory' conditions, an accuracy of ± 0.5%.

While there are legal requirements in the UK for the uncertainty performance of primary meter elements, there are no similar requirements for the conversion system, the combined meter and conversion system, or for in-service tolerance.

## Temperature

Of the three parameters, temperature is the easiest to consider. The only temperature fixed factor allowed currently by the Gas (Calculation of Thermal Energy) Regulations is based on the 'average' UK gas temperature of 12.2°C. In fact, gas temperatures may vary dramatically due to individual circumstances. For example, meters close to a regulator having a large pressure drop can easily be subject to gas temperatures close to 0°C throughout the year. Conversely, an indoor installation in close proximity to a heat source may be subject to temperatures in excess of 25°C.

The gas temperature can also vary throughout the year with significant winter and summer variations from the 'average' UK gas temperature of 12.2°C.

- for every 3°C above 12.2°C, the result of not having a conversion device will be a 1% over-estimate of the true gas consumption

- typically a 3°C change results in a 1% error of true gas consumption.

For the calculation of temperature factors, convert the measured temperature °C into the absolute units (K) by the addition of the absolute zero offset of 273.15. Thus:

12.2°C = 285.35 K.

Therefore, a 3 degree shift is approximately 1% of 285.35.

*Pressure*

Pressure is more complex than temperature due to the number of different reasons for the pressure to vary and the number of different choices allowed for 'fixed factors' by the Gas (Calculation of Thermal Energy) Regulations.

The three reasons that meter pressure can change are:

- errors in the meter pressure regulator set point (for low pressure and domestic applications. This is, usually, 21 mbar with a tolerance of + 1 mbar)
- droop in the regulator pressure as the flow increases
- barometric variations.

(a) **Regulator Performance**

The deviations in the meter regulator pressure from the set point is determined by regulator performance with respect to its 'droop' characteristic i.e. how far its pressure will fall as the flow passed through it increases, and the accuracy of the initial setting. See IGE/GM/6 for further reference.

As the altitude increases, the atmospheric pressure decreases. The Gas (Calculation of Thermal Energy) Regulations provide a lookup table that enables the conversion of altitude to a pressure that is then deducted from the nominal atmospheric pressure, the altitude used from the lookup table being rounded up to the nearest 2.5 m.

(b) **UK Weather Conditions**

These can give barometric pressure variations over the range 950–1050 mbar (although mostly they are around 970 to 1030 mbar or ± 3.0%). However, standard conditions assume a pressure of 1013.25 mbar. In addition, because of the complex weather patterns in the UK, there is no clear correlation between barometric pressure and the seasons (this is not true within large land masses, where winter pressures will always tend to be higher than summer pressures). However, there can be significant variations over shorter periods of time (usually one to two days) with the corresponding effect on metering pressures. Thus, where large loads are involved, it is necessary to measure the gas pressure in absolute units to take account of the barometric pressure variations.

(c) **Compressibility**

This is the most complex of the three parameters to assess due to being dependant on not only the gas pressure and temperature, but also on the gas composition. If sophisticated gas properties' software is available and the site's gas composition is known, the relevant information may be readily calculated.

Due to the complex nature of compressibility calculations, the time and cost for an experienced operator to calculate $Z$ independently may be greater than the cost difference between a PT and PTZ conversion device. Thus, having followed the guidelines in this section, it may be that the appropriate solution is to automatically purchase a conversion device that includes the $Z$ function.

- If the meter gas pressure is not greater than 500 mbar then a fixed value of 1 is to be entered into the conversion device as the value of $Z$
- If the gas composition is unknown, reference is made to the PGT for the relevant information

  *Note: In the UK, Transco will usually be the provider of such gas composition data.*

- If the gas composition can be established and is relatively stable i.e. the site is fed exclusively from a single terminal, this composition is used to set the parameters within the conversion device
- If the site composition changes due to it being fed from a number of terminals, then a weighted average composition may be obtained. The time period over which the average is obtained should be not less than 1 year.

Figure 24 shows typical arrangements for the installation of pressure tappings and transducers, additional points to bear in mind are:

- meter manufacturer's pressure points shall be used, where they are not installed on the meter they should be positioned immediately upstream of the meter but away from pressure pulsations
- 2 thermowells should be fitted at an angle of 60° of centre and should preferably be within 0.6 m of the meter outlet and no more than 3 diameters

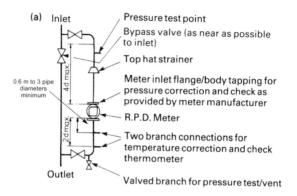

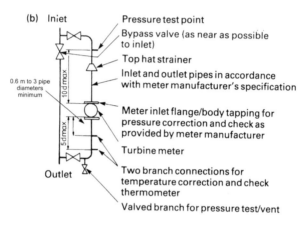

*Fig. 24  (a) & (b) Installation of transducers for electronic correctors*

- pipework design upstream and downstream of any meter installation should be in accordance with the manufacturer's recommendations.

## References

Lists of all legislation, Standards, Codes of Practice and related Procedures in Gas Volume Conversion follow:

## Legislation

> Gas Act 1986 as amended by the Gas Act 1995
> Gas (Calculation of Thermal Energy) Regulations 1996
> Gas Safety (Installation and Use) Regulations 1998
> Gas Safety (Management) Regulations 1996
> Pressure Systems Safety Regulations 2000.

The Institution of Gas Engineers

| | |
|---|---|
| IGE/GM/1 Edition 2 | Gas meter installations for pressures not exceeding 100 bar |
| IGE/GM/4 | Flowmetering practices for pressures between 38 and 250 bar |
| IGE/GM/5 | Selection, installation and use of electronic gas volume conversion systems |
| IGE/GM/6 | Specification for low pressure diaphragm and rotary displacement meter installations with badged meter capacities exceeding 6 m³/h (212 ft³/h) but not exceeding 1076 m³/h (38,000 ft³/h) |
| IGE/GM/7 | Electrical connections to gas meters |
| IGE/SR/25 | Hazardous area classification for Natural Gas Installations. |

BSI Standards and Publications (abbreviated titles)

| | |
|---|---|
| BS 4161 | Gas meters Part 8 Specification for electronic volume corrector |
| BS 4683 | Electrical apparatus for explosive atmospheres |
| BS 5345 | Code of practice for selection, installation and maintenance of electrical apparatus for use in potentially explosive atmospheres (other than mining applications or explosive processing and manufacture) |
| BS 5490 | Specification for classification of degrees of protection provided by enclosures |
| BS 5501 | Electrical apparatus for potentially explosive atmospheres |
| BS 6651 | Protection of structures against lightning |
| BS 6941 | Electrical apparatus for explosive atmospheres with type of protection N |
| BS 7430 | Earthing |
| BS 7671 | IEE Wiring Regulations |
| BS EN 50081–1 | Electromagnetic compatibility. Generic emission standards: residential, commercial and light industry |
| BS EN 50081–2 | Electromagnetic compatibility. Generic emission standards: industrial environment |
| BS EN 60079 | Electrical apparatus for explosive gas atmospheres. |

*Other References*

| | |
|---|---|
| AGA 8 | Compressibility and supercompressibility for natural gas and other hydrocarbon gases. |

## Isolating Valves

It should be possible to isolate any metering installation, and particularly a high pressure installation, from a safe distance.

Where valves are fitted close to meters, they should be of a type which will not cause undue flow disturbances.

Valves must have suitable characteristics for their particular purpose, bearing in mind that:

- meter valves remain open for long periods and must then shut off completely on the first operation
- bypass valves remain closed for long periods and must then open quickly in an emergency.

The specific requirements for valves is described in IGE/GM/1.

## Commissioning Meters

*Pre-installation Check*

Before fitting any meter the following points should be checked:

- pipework flanges are in line, square to the pipe and the correct distance apart
- filter bodies have been cleaned out and the correct elements fitted
- upstream pipework between the meter and the filter is dry, clean and free from debris
- rotary meter rotors or impellers are free running
- meter flange faces are undamaged and meters positioned in the correct direction of flow.

After fitting, check the meter and pipework supports.

*Handling*

All meters are precision instruments and must be treated with great care, they must never be dropped or mishandled. Always lift in accordance with manufacturer's instructions, particularly when using slings and mechanical lifting aids.

Avoid strain on the index and damage to the index glass. Replacing a glass may require a visit from the official meter examiner.

During transport or storage the meter connections should be sealed off.

*Testing*

For installations operating at pressures below 75 mbar the meter installation should be tested for soundness in accordance with the current edition of IGE/UP/1 or IGE/UP/1a.

*Purging*

The procedure is given in IGE/UP/1 or IGE/UP/1A.

Provision for purging should be provided in the pipework. Vent points should be fitted between isolating valves so that meters, filters and governors may be purged individually.

Information on purging of systems is described in Chapter 1.

*Commissioning*

All meters should be pressurised slowly. The procedures for commissioning high pressure meters are given in IGE/GM/1, 1998 and vary with the type of installation. Manufacturer's instructions should be carefully followed.

The index should be checked to ensure that the meter is operating. Rotary displacement meters should also be checked for noisy operation or high differential pressures which indicate that the meter is not operating freely. The oil sump levels should also be checked.

*Exchanging*

Before commencing any work ensure that the replacement meter is correct with respect to:

- badged rating
- operating pressure
- case dimensions
- matching flanges or other connections.

Ensure maximum ventilation and no naked lights in the vicinity. Check that breathing apparatus and other safety equipment is available, as required.

With rotary displacement meters check the new meter for clean rotors, and freedom of internal parts. Check free rotation of turbine meters. To exchange:

- open the bypass valve or fit temporary rider
- close the meter isolating valves and check that the outlet pressure is maintained
- check that inlet and outlet valves are holding

- on large or high pressure installations, vent and purge the meter; small meters may be purged by venting in a safe area after removal
- remove the meter and seal the inlet and outlet connections
- install the replacement meter
- purge the meter with the inert gas if necessary
- check the meter and connections for soundness
- open the inlet valve and pressurise the meter slowly
- open the outlet vent point and purge the meter of air or inert gas
- close the vent and open the meter outlet valve
- close the bypass valve and check that outlet pressure is maintained
- check the meter for noise and smooth operation of index
- seal the bypass valve.

A temporary continuity bond should always be attached before a component or any pipework is disconnected. The bond should remain in position until all the connections have been remade.

CHAPTER 3

# Industrial Processes and Plant

Chapter 3 is based on an original draft by A. J. Spackman

## PROCESSES

Gas is used in many industries for a wide variety of processes whose temperatures range from about 70°C up to 1650°C. At the lower end of the scale are the drying operations carried out on paints, inks and a number of granular materials including foods and pharmaceuticals. At the upper end, high temperatures are required for melting metals and glass and for heating heavy clay, refractories, limestone and metals prior to hot working.

### Drying

Drying is the removal of water, or some other solvent, from a product. It may be carried out by heating the product to about 70 to 90°C and allowing the solvent to evaporate freely. This takes a long time, so commonly radiant heating or forced convection are applied. In forced convection, hot air and sometimes products of combustion from an air heater are circulated through and over the product. Water vapour or solvent is taken up in the air, so drying the product. The rate of drying depends on the:

- temperature of the hot air
- input and output humidities
- rate of hot air circulation.

In some cases moisture is removed and the air recycled, as in an air conditioning plant. Where toxic or flammable solvent vapours are involved, special precautions, specified by the Health and Safety Executive, must be taken.

Some materials, including paints, enamels, oils and resins may undergo chemical changes when heated to temperatures of about 120 to 240°C. The processes are known as 'curing', 'stoving' or 'baking' and they produce harder surface coatings than are obtained with

101

normal air dried paints. Temperatures of this level can also be used to cause physical changes like heat setting in the textile industry.

Cores made of sand for foundry work are dried at 120 to 240°C and the moulds require temperatures of up to 200 to 400°C.

## Cooking

This is usually considered to be a 'commercial' operation rather than an 'industrial' one. However, the bulk baking of bread, biscuits, pies and pastry may be carried out in large factories and often in travelling ovens 60 to 120 m long. Typical temperatures are:

- cakes 150 to 180°C
- biscuits 190 to 240°C
- bread 150 to 260°C
- pies about 260°C.

The confectionery trade produces sweets at temperatures of 110 to 150°C, while potato crisps and chips are fried at about 180 to 210°C.

Although not a food, varnish is 'cooked' or boiled in a special pan with a fume offtake at temperatures of 270 to 320°C.

## Vitreous Enamelling

This process was described briefly in Vol. 1, Chapter 12. It consists of first preparing the surface of the metal by degreasing and by dipping in an acid solution to etch it so that the frit is keyed on.

The frit, which is composed of the glazing materials mixed with water and some clay, is applied to the metal by dipping or spraying. The components are then dried in an oven.

When dried, the frit is fired in a furnace at 650 to 700°C for cast iron components and 760 to 930°C for steel plates.

## Ceramics

Most pottery is fired twice. The first or 'biscuit' firing is to vitrify the clay and the second or 'glost' firing is to fire on the glaze. Glazed sanitary ware may be produced by a single firing.

The products are first dried to drive off a large proportion of the moisture. This may be in a separate oven or in the preheating section of a continuous kiln.

The temperatures for firing are:

- biscuit – earthenware 1150 to 1250°C
  – porcelain 1250 to 1400°C
- glaze 1050 to 1150°C.

Muffle furnaces were commonly used for pottery to protect the ware, but with natural gas, open furnaces may be employed.

Bricks are fired once at about 990 to 1430°C, in intermittent or continuous kilns, which may be of the car bogie type or Hoffman type.

## Glass

There are many different kinds of glass, each made for a particular purpose. The main constituent is silica, or sand, but various other ingredients are added to obtain the different properties. The three most common forms of glass are:

- crown glass – silica 72%, soda 15%, lime 9%, oxides 4%
- flint glass – silica 55%, lead oxide 32%, potassium and sodium oxides 12%, other oxides 1%
- heat resisting glass – silica 74%, boric oxide 16%, potassium sodium and aluminium oxides 10%.

Crown glass is an easily worked glass with a low melting point and is used for windows, bottles, electric lamp bulbs and tubes.

Flint glass, also called 'lead crystal', is a fine, clear glass used for optical lenses, table glasses and decorative ware. It is also used for cut glass work and engraving.

Heat-resisting glass, or boro-silicate glass, has a high melting point and a low coefficient of expansion. It may be made thicker than ordinary glassware to give strength to ovenware dishes.

## Glass Melting

Glass is melted in tanks or in pot furnaces at temperatures of 1300 to 1600°C. Tank furnaces may contain up to several thousand tonnes of soda glass supplying a float glass plant or several hundred tonnes of glass for a container (bottle) plant. Small kilns may heat one or more pots containing about 50 kg of glass for scientific or lead crystal glassware. The basic operations in manufacturing glass are:

- measuring and preparing the ingredients, these include some broken glass or 'cullet'
- melting the batch, known as 'founding'
- heating to above melting point to remove the bubbles of trapped gases and refine the glass, known as 'fining'
- conditioning the glass by reducing the temperature to that required for working.

In a tank furnace, the operation is continuous with the frit and cullet fed in at one end and the conditioned glass withdrawn at the other. The burners are arranged to give the temperatures required at each stage of manufacture.

Pot furnaces are intermittent and may be of capacity from a few kilograms up to 100–200 kilograms.

## Glass Fabricating

The methods used for producing glass articles are:

- blowing
- pressing
- rolling, drawing or floating.

### Blowing

Glass blowing was invented by the Romans and hand blowing is still used for high quality articles in small numbers. The required quantity of glass is withdrawn from the pot on the end of a 1.5 m tube. By controlled blowing, rotating and swinging in an arc, the glass blower produces the desired shape. This may then be placed in a hinged mould and blown to fit the mould. Other processes may be necessary to modify the shape and add or remove parts. Glass stems or handles may be attached and unwanted glass cracked off by scratch marking and applying heat. Gas flames, sometimes with oxygen, are used for these operations. After cracking off, the surface may be ground or fire-polished to give a smooth, even finish.

Bottles and electric light bulbs are blown on automatic machines which simulate the actions of the glass blower. The glass is fed from the tank down to the machine through a forehearth or channel (feeder) which is heated to ensure that the glass reaches the machine at the correct working temperature. Air blast burners are used for this purpose. The 'gob' of glass from the forehearth is first hollowed by a plunger in the 'parison mould' and then blown to shape in the final mould. The machines are equipped with groups of gas burners to reheat, cut and polish the products as required.

### Pressing

Pressing may be carried out by hand but is now commonly used for the manufacture of domestic hollow-ware, car headlamp glasses and similar products by automatic machines. The gob of glass from the forehearth is dropped into a mould and squeezed into shape by a plunger. The moulds are positioned around the circumference of a rotating table so that each mould, in turn, receives its glass, is pressed and is then finished and emptied.

### Rolling, Drawing and Floating

These are all methods of producing sheet glass, the bulk of which is used for window glazing. Figured glass is produced from patterned rollers and, in reinforced glass, wire netting is introduced before rolling. Special glass is now made to filter sunlight so that shades and blinds are unnecessary. However, most sheet glass is made by the float process.

## Glass Annealing

When the fabrication is finished, glass must be annealed to remove the stresses set up by uneven cooling otherwise the glass item would crack. This is carried out in an oven, called a 'lehr' at temperatures of about 450 to 650°C. The lehr may be a simple box oven or a long, travelling oven, depending on the quantity and the time required for treatment. Some optical lenses may need several months to cool down to prevent distortion and stress.

Glass which has not been properly blown or annealed may have built-in stresses which will cause it to shatter, often for no apparent reason. This property is used in the manufacture of 'toughened glass' for motor car windscreens and windows. The outer surface of the glass is heated to red heat and then cooled rapidly by air jets or by dipping in oil. This sets up high comparative stresses, and, when broken, the glass fractures into small chunks, instead of splinters.

## Heat Treatment of Metals

A simple explanation of heat treatment is given in Vol. 1, Chapter 12.

Metals are given heat treatment in order to bring about changes in the structure of the metal. These changes may be limited to the surface structure or may go right through to the core of the object. The changes are brought about by heating and cooling to a specified programme appropriate to the properties of the metal and the purpose for which the article was designed.

Heat treatment is carried out in a variety of furnaces and baths at temperatures from about 200°C to 1000°C. The furnaces may be either directly or indirectly fired and the baths may contain oil, molten salts or lead. Fluidised beds may also be used for heating or cooling the products.

Working flames are used for treating some articles made from steel. By applying a flame directly to the surface, or a specific area of the product, that part may be hardened, to resist wear, whilst the core remains tough, to withstand loads and shocks. For example, a gear wheel must be tough, but its teeth must be hard.

Gear wheels are flame hardened by mounting them on a rotating table over a quenching bath of oil. A number of burners are positioned so that, as the wheel rotates, the faces of each tooth are evenly heated. When the required temperature is reached, down to an appropriate depth, the gas is reduced to a low rate and the wheel and table automatically lowered into the quenching bath. The main core of the wheel does not get hot enough to change its characteristics.

There are four methods of flame hardening semi-finished steel items:

- spot method, in which a specific area is heated and then quenched; used for rocker arms and chain links
- spinning method, where the component is rotated between a number of burners and then lowered into the quench bath; used for gear wheels
- progressive method, in which the flame head moves over the article, followed by the quenching medium; used on slideways of machine tools, wheel tyres and rims and large gear teeth
- a combination of rotation and progression; used on spindles and rolls.

**Ferrous Metals**

The iron-carbon diagram, Fig. 1, shows the temperatures at which changes occur in the structure of the various forms of iron or steel. These temperatures are known as the 'critical temperatures' and the diagram represents the relationship between the composition of the metal and its critical temperatures under very slow heating and cooling conditions. In addition to the phases shown on the diagram, steel can also exist in other forms, one of which is 'martensite'. This is very hard and brittle and is formed when austentite is cooled quickly from above the critical temperatures. The process may be reversed by heating the martensite and cooling it slowly. So the formation of martensite makes an important contribution to the hardening and tempering of steel.

The principal heat treatment processes are:

- annealing
- normalising

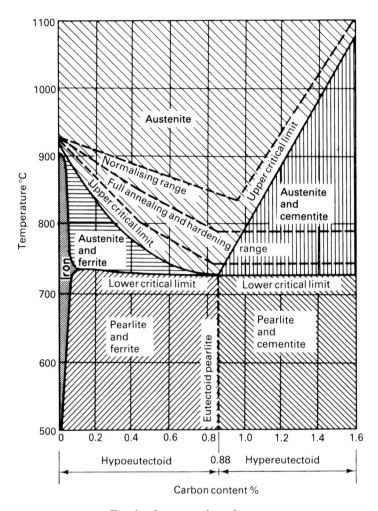

*Fig. 1   Iron – carbon diagram*

- hardening
- tempering
- case hardening.

## Annealing

Annealing is a softening process which is carried out at temperatures of between 700 and 900°C. It consists of heating to the required temperature followed by slow cooling, usually in the furnace. Annealing may be either 'full annealing' which entails heating the metal to above its lower critical temperature, or 'sub-critical annealing' when it is heated to just below the critical temperature.

Full annealing improves some of the qualities of steel by refining the grain structure.

Subcritical annealing is used for stress relieving lower carbon steels which have been welded or cold worked.

'Spheroidising' is a form of prolonged annealing at, or slightly below, the critical temperature, which improves machinability but is applied to irons with appreciable carbon levels rather than low carbon steels.

### Normalising

This process is similar to annealing but is carried out at higher temperatures, 900 to 1100°C. The steel is heated to above its upper critical temperature and then cooled in air. The slightly faster cooling rate is necessary to maintain a satisfactory grain structure.

'Patenting' is a method of softening steel wire which has been work hardened by drawing. It entails heating to about 1000°C and quenching in air or in a lead bath at 500 to 600°C.

### Hardening

Hardening consists of heating the steel to above its critical temperature, followed by quenching quickly. Water is used for fast quenching and oils are used for slower rates of quenching. Generally, faster quenching produces harder steel. Hardening produces martensite in the steel which renders it hard and brittle. For most practical purposes the steel then needs to be tempered.

Typical temperatures for hardening are 750 to 850°C.

### Tempering

Tempering consists of heating the hardened steel to temperatures of 150 to 640°C, below the critical temperature. This converts the martensite into less hard but tougher substances. The rate of cooling varies with the results desired and the composition of the steel. Tempering may be carried out in furnaces, oil, lead and salt baths.

Austempering, martempering and isothermal quenching are processes of interrupted quenching used to develop toughness and ductility and prevent internal stresses in the steel.

### Case Hardening

Low carbon steels which cannot be hardened by normal heat treatment may be given a 'case' of surface skin which has been suitably altered in composition to respond to hardening. This is done usually by:

- carburising
- nitriding
- carbo-nitriding
- cyaniding.

Carburising is a process in which carbon is introduced into the surface of the metal to a depth of 0.25 to 1.5 mm. This is by heating the metal to above the critical temperature while it is in contact with a material containing carbon. The material may be solid, liquid or gas.

The old method was to pack the work in solid charcoal in steel boxes and then heat it to about 900°C in a furnace for several hours.

A quicker method is to use molten salts containing sodium cyanide, known as 'cyaniding'. This gives better uniformity of heating and needs shorter periods of immersion.

Gas carburising is commonly used with hydrocarbons in the form of endothermic gases made from propane or natural gas being introduced into the furnace atmosphere. Temperatures used are in the region of 950 to 1050°C.

Nitriding is the formation of a hardened case by the addition of nitrogen to the surface, without quenching. Carbo-nitriding uses a controlled gas atmosphere with the addition of ammonia gas so that both carbon and nitrogen are absorbed. Temperatures are typically 850 to 950°C.

After carburising, the work must be heat treated to harden the outer case and to refine the core. The whole process of forming and hardening the case is termed case hardening.

There is a general move to use furnaces heated by electricity and containing a high vacuum level for heat treating and bright annealing of special steels.


## Non-Ferrous Metals

The main heat treatments of non-ferrous metals are annealing and ageing. Typical temperatures are given below.


### Annealing

Copper is annealed at 430 to 650°C; the major problem in the process is the prevention of oxidation and copper wire may be bright-annealed in a vacuum. Another simple method is to pass the wire in a continuous strand through an externally heated tube containing exothermic gas. No excess air is permitted to enter and the wire is cooled by a stream of water as it leaves the tube.

| Brass is annealed | at 320 to 480°C |
| Nickel and Monel | at 600 to 800°C |
| Aluminium | at 220 to 400°C. |

### Ageing

Both aluminium and magnesium alloys may be hardened by solution treatment and ageing. This is carried out at low temperatures. Generally increased hardness is obtained by ageing for longer periods at lower temperatures. Typical temperatures are:

- aluminium          120 to 230°C
- magnesium          180 to 200°C.

## Hot Working Metals

Hot working of metals is defined as working above the recrystallisation temperature and the working processes include:

- forging
- drawing
- extruding
- rolling.

For most of these processes the metal is heated in a furnace and it may be necessary to reheat the work between operations. Many different types of continuous and intermittent furnaces are used depending on the size, shape and weight of the work.

Temperatures required for hot working are typically:

- aluminium          430 to   500°C
- brass              680 to   780°C
- steel             1100 to 1300°C.

## Melting Metals

The metals most commonly melted by natural gas are:

- tin                260 to   340°C
- typemetal          270 to   340°C
- lead               330 to   390°C
- zinc               430 to   480°C
- aluminium          650 to   760°C
- magnesium          680 to   700°C
- brass              930 to   980°C
- copper            1150 to 1250°C.

All these metals oxidise readily at their melting temperatures and this may need to be avoided to reduce the loss of metal and prevent any included oxide spoiling the work.

In addition the molten metals absorb some of the gases in the furnace atmosphere. This can result in blow-holes in the casting or changes in the structure of the metal.

Some metals have particular hazards, for example, lead vapour is toxic and magnesium and aluminium can be burnt.

For melting metals at temperatures below 500°C the pot furnace is generally used and immersion tubes are also employed.

At temperatures above 500°C, the furnaces used are the crucible and the reverberatory types. Copper may be melted at very high rates (in tonnes per hour) in tower furnaces using both preheated air and preheated natural gas.

**Flame Processing**

In addition to their use for flame hardening and glass working, gas flames may also be employed for:

- flame cutting and profiling
- soldering and brazing
- metal spraying.

*Flame Cutting*

This is similar to oxy-acetylene cutting but with natural gas replacing the acetylene. When the oxy-gas flame has started initial melting within the cut, sometimes compressed air can be used to propagate it. On profiling machines, a sensing head follows a pattern or a computer programme and directs the cutting head on the work.

*Soldering and Brazing*

Soldering may be carried out by manual or automatic soldering irons, by dipping in molten solder or with a variety of blow pipes and torches. Tin cans and boxes are produced on continuous conveyors by automatic methods. The gas burners may be low or high pressure and may use compressed air.

Brazing is similar to soldering except that it requires a higher temperature, about 700°C. Torches usually employ high pressure air. Items such as heat exchangers and car radiators are brazed continuously by capillary action in a conveyor furnace, the brazing metal being included at the assembly of the component.

*Metal Spraying*

Many non-ferrous metals may be spray coated on to other metals, including tin, lead, zinc, aluminium and copper. The process involves passing the metal, in a finely divided form, through an oxy-gas flame at high velocity. The heated particles are deposited on the surface to be treated by a separate stream of high velocity air. The process may be used to build up worn areas of high value articles like steam valves.

## Submerged Combustion

This is a variation of the working flame and is used for tank heating. The flame burns below the surface of the liquid and the heat is transferred by the bubbles of the products of combustion in direct contact with the liquid.

Under atmospheric pressure the maximum temperature reached is about 82°C. The large surface area of the bubbles gives a high rate of evaporation which helps to produce a more concentrated liquid.

This method is particularly suitable for heating corrosive liquids and those liable to deposit suspended solids or scale.

## PLANT

### Ovens and Driers

*Box Ovens*

A box oven is the simplest form of industrial oven. It consists of an insulated, sheet metal box with a door, heated by a burner in the base.

It may be either:

- double cased, that is, directly fired; or
- treble cased, that is, indirectly fired.

In the double cased oven, Fig. 2, the products of combustion circulate within the oven and are removed by the flue, together with moisture or any volatile materials from the product. The flue is often taken from the top of the oven but it can also be situated at low level in the rear wall to give improved circulation and a more even oven temperature, Fig. 2(b).

The rate of ventilation is usually controlled by a flue damper. This is designed so that, when it is closed, at least one third of the flue area still remains unobstructed.

In some industrial processes, it is not possible to use the products of combustion directly to the work, for reasons of safety where

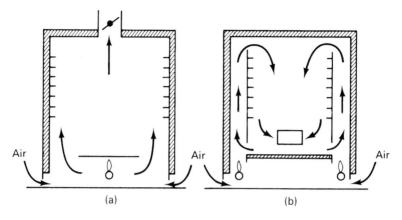

*Fig. 2   Double cased oven: (a) top flue; (b) bottom flue*

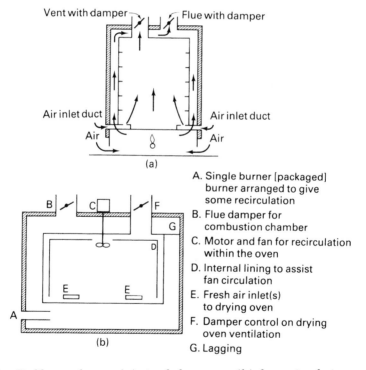

A. Single burner [packaged] burner arranged to give some recirculation

B. Flue damper for combustion chamber

C. Motor and fan for recirculation within the oven

D. Internal lining to assist fan circulation

E. Fresh air inlet(s) to drying oven

F. Damper control on drying oven ventilation

G. Lagging

*Fig. 3   Treble cased oven: (a) simple box oven; (b) fan assisted air recirculation*

volatile evaporating compounds may be liable to uncontrolled ignition. For these processes the treble cased oven is used, Fig. 3. This has an inner box with its own separate air inlet, flue outlet and appropriate ventilation control. The walls of the box are heated by the products circulating around them and so form the heat transfer surfaces. In a number of these ovens, fans and baffles may be used to circulate the air over the walls of the oven and so increase the rate of heat transfer, Fig. 3(b).

Box ovens may be heated by bar burners and fitted with thermoelectric flame monitoring systems. Electronic flame safeguard systems are now becoming more common, due to their more rapid response times, and the introduction of recognised safety standards such as IM30 or EN762.

The flue system is designed to give maximum efficiency. The flue damper may be automatically controlled by the gas control system. A motorised damper can be opened before the gas rate is increased and closed after the gas rate is reduced. This adjusts the flow through the flue to that required for the particular quantity of gas being burnt at that time, so limiting the volume of excess air entrained.

Care must be taken to interlock the burner and flue systems to avoid any hazardous situation arising.

*Forced Convection Ovens*

Independent, direct fired air heaters are now commonly used to provide forced convection for drying and heating. They can be sited in

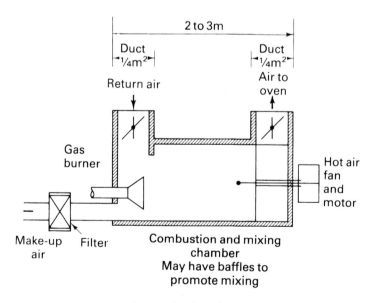

*Fig. 4   Air heater for forced convection oven*

an otherwise unused space and connected to the oven by ducting. Alternatively, they may be mounted above or below the oven itself. Existing ovens may be converted to forced convection. Efficiency can be increased by recirculating a proportion of the products of combustion. The general arrangement of an air heater is shown in Fig. 4.

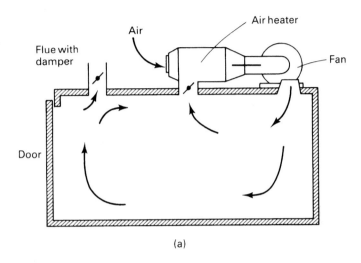

(a)

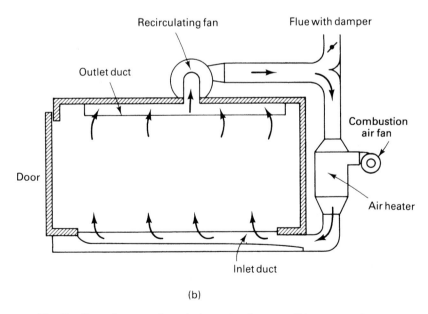

(b)

*Fig. 5   Forced convection: (a) suction burner; (b) pressure burner*

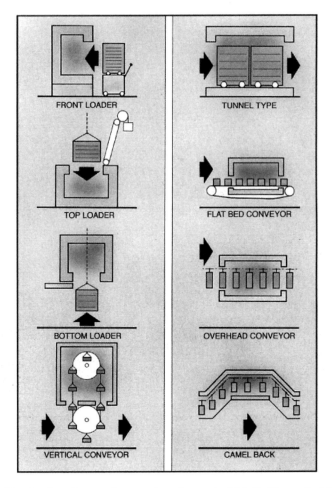

*Fig. 6    Loading and conveying the work  (J.L.S. Ovens Ltd.)*

Recirculation ovens have integral air heaters which may be either:

- pressure type, with the fan on the inlet; or
- suction type, with the fan on the outlet.

Examples are shown in Fig. 5.

Fig. 6 shows the main methods used for loading work into the ovens and for transporting the work.

## Radiant Panels

Drying processes may be carried out using infra-red radiation from gas fired radiant panels. These can be obtained with emissivities to match the particular materials to be dried.

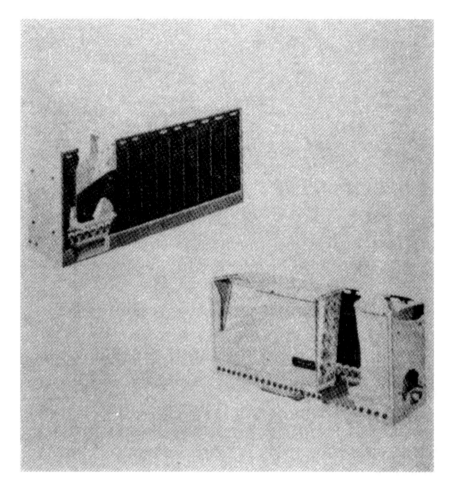

*Fig. 7   Low temperature infra-red panel*

Panels may be either:

● high temperature, giving temperatures up to 900°C
● low temperature, giving temperatures up to 350°C.

High temperature panels use the radiant burners with porous or perforated ceramic tiles described in Vol. 1, Chapter 4.

Low temperature panels usually have a black cast iron, or vitreous enamelled steel panel heated from behind by a bar burner, Fig. 7.

Catalytic conversion panels are a special form of low temperature radiant heater, which are described on page 164. In the textile industry, the introduction of these heaters has almost become stand-

*Fig. 8    Arrangement of panels in drying oven*

ard on the inlet preheat area to the stenter. The panels are fitted close to the textile feed prior to it entering the stenter proper.

Because radiation travels in straight lines, these panels are best suited to drying coatings on flat products. Banks of panels may be built up into tunnels through which the products travel on a conveyor. Typical arrangements are shown in Fig. 8.

*Conveyor Ovens*

These are generally either:

- horizontal; or
- camel-back.

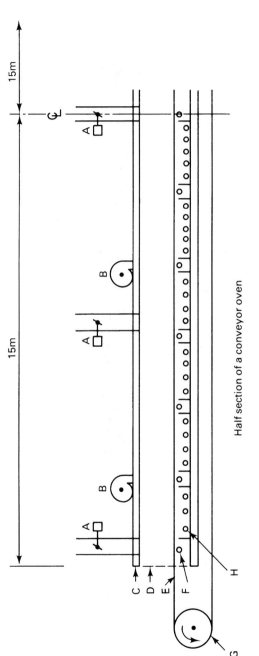

Half section of a conveyor oven

A. Motorised damper in flues.

B. Air fan.

C. Insulated case.

D. Optional curtain.

E. Conveyor, steelband, or mesh.

F. Idler rollers for conveyor.

G. Drive roller for conveyor.

H. Low pressure gas burners.

Burners fitted with individual ON/OFF taps.

Burners grouped for gas and air supplies and flue.

Thermostat control operates flue damper to suit gas flow.

*Fig 9 Horizontal conveyor oven*

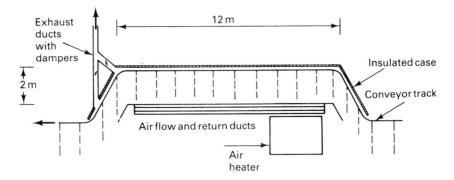

*Fig. 10    Camel-back oven*

The horizontal oven, Fig. 9, has a flat metal conveyor, often of wire mesh, passing above gas burners. The oven illustrated is a type used for baking bread, pastries or biscuits. It is fired by low pressure air blast premix gas burners in the form of pipe ribbon burners each as wide as the conveyor. These ovens, which may be up to 120 m in length, are divided into zones, each with its own gas and air controls, burners and flue. The zones may be operated at different temperatures by individual thermostats. The camel-back oven, Fig. 10, is widely used for drying and stoving in light industries. The entry and exit are sloped downwards at 45° forming seals to retain the hot air. The products are carried through the oven suspended from an overhead conveyor. The air heater which provides forced convection with recirculation is located below the centre section.

*Driers for Granular Materials*

A number of granular materials such as sand, chemicals and food-stuffs, are processed in various driers using hot air from a direct fired air heater. The common forms of drier include:

- rotary driers
- spray driers
- flash driers.

Rotary driers are similar to tumble driers. The rotating shell may be inclined so that the product moves from the feed hopper to the other end by gravity. As the shell rotates, the inner paddles or 'flights' continually turn the particles over in the stream of warm air which often travels counter to the feed.

Spray driers have the product fed in as a liquid or a slurry. It is then atomised or sprayed and dried in a stream of hot air.

Flash drying consists of dropping the wet particles into the stream of hot air and removing them when dry, usually by a cyclone through which the air passes on its way to the flue.

## Tanks and Baths

The terms 'tanks' and 'baths' are used, almost interchangeably, to describe a variety of containers, operating over a wide range of temperatures. What is called a 'tank' in one industry is called a 'bath' in another. They vary from the rectangular, galvanised iron tank holding a solution of caustic soda in water, to the small steel pot in a refractory setting for melting printer's typemetal on a linotype machine.

### Underfired Tanks

Older, less efficient tanks are heated by burners below the tank, Fig. 11. The products of combustion pass around the sides and discharge at the top, usually into a flue. The top of the tank should have a cover to retain the heat. Low temperature tanks may be insulated by floating plastic spheres on the liquid surface, which do not interfere with items being put into and taken out of the tank.

Salt and metal baths must be evenly heated to avoid solid material existing on top of molten material which is expanding and which could cause an explosion.

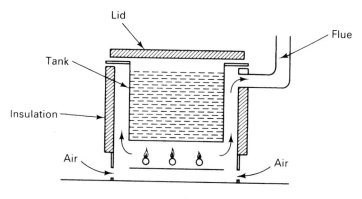

*Fig. 11    Underfired tank*

## *Immersion Tubes*

The tubes may be from 50 to 150 mm diameter. They may be heated by natural draught burners but increased efficiencies and higher gas rates can be obtained with forced draught systems.

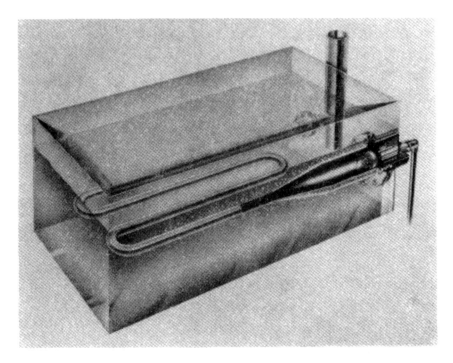

*Fig. 12   Immersion tube heated tank, for aqueous liquids*

The length and layout of the tubes depends on the size of the tank and the heat input required. A typical layout of a single tube is shown in Fig. 12. More commonly now forced draught natural gas burner systems based on a Gas Research and Technology Centre design are usual. These burner systems allow thermal inputs similar to those achieved by steam tubes. The gross thermal input for a 40 mm tube would be 47 kW and for a 150 mm tube would be 586 kW. For maximum efficiency of about 80%, the length-to-diameter ratio of the immersion tube should be 120:1. Fig. 12 shows the arrangement of the burner within the tube which is situated in the tank.

The 'Temgas' immersion heater may be fitted to existing tanks without cutting into the sides, Fig. 13. It has a combustion chamber with internal baffles, in place of the usual tube or tubes.

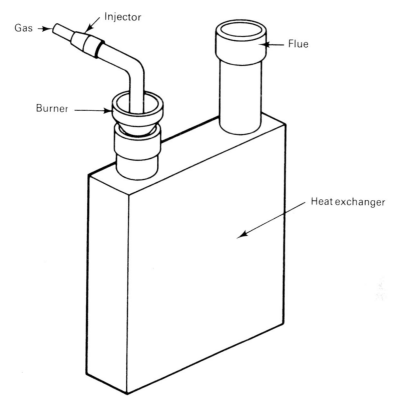

*Fig. 13    'Temgas' immersion heater*

## Direct Contact Water Heater

For applications where large volumes of hot water are required, for example in tanneries, wool scouring, food factories using hot water for hygiene purposes, the direct contact water heater may be used. This is a special application of an immersion tube where the products of combustion pass vertically up a short tower whilst the water to be heated falls counterflow over a series of perforated trays shown in Fig. 14.

If a temperature rise of 90°C was required in the water, it could be heated at the rate of between 9 and 373 litres per minute depending upon the size of heater chosen. With a temperature rise of only 20°C the range is 42 to 2,100 litres per minute. With the low temperature rise, thermal efficiencies of 98 per cent can be achieved.

Figure 15 shows a direct contact water heater with a thermal input of 1,465 kW.

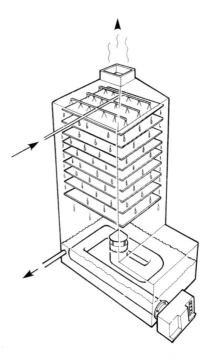

*Fig. 14   Direct contact water heater*

### Submerged Combustion

A form of submerged combustion burner is shown in Fig. 16. Because the combustion must be completed and the products expelled against the pressure of the liquid, this burner is a nozzle mix burner. It fires vertically downwards into a tube causing many small bubbles to be produced to transfer heat to the liquid. In practice in the field, most submerged combustion burners are supplied with premix from a fan system. This type of system is often used in liquids that are of a corrosive nature.

## FURNACES

### Direct-Fired

#### Reverberatory Furnaces

A reverberatory furnace is one in which the heat is supplied in a space between the charge and the roof. Consequently heat is reflected down onto the surface of the molten metal or glass from the crown of the

*Fig. 15   Direct contact water heater (Nordsea Gas Technology Limited)*

furnace. The name is however applied to any open hearth furnace in which the burners fire onto the metal, Fig. 17.

These furnaces are commonly used for remelting large amounts of scrap metal and may hold up to 50 tonnes (or up to 5000 tonnes of glass). Nozzle mixing burners are generally used.

*Oven and Box Furnaces*

These are simply refractory boxes in which the work is placed on the hearth, which forms the floor of the furnace, Fig. 18. 'Oven' furnaces are generally smaller than box furnaces and have the hearth at waist height for easy loading and unloading. Access to the furnace is usually by a counterbalanced, lift-up door. The furnaces are com-

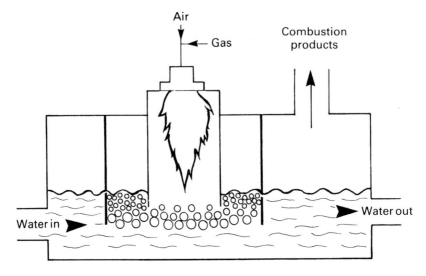

*Fig. 16    Submerged combustion burner*

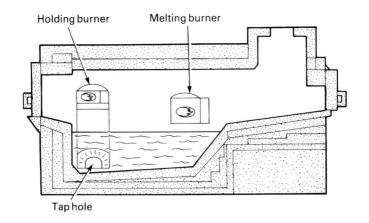

*Fig. 17    Reverberatory furnace*

monly fired by nozzle mixing burners. There are many variations of these furnaces used as kilns and for heat treatments of metals and glass.

### Bogie Hearth Furnaces

In order to make loading and unloading easier and to save time waiting for the furnace to cool down so that the load may be handled, these furnaces have a hearth which may be withdrawn on rails, Fig.

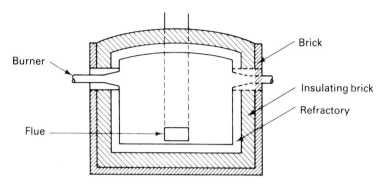

*Fig. 18    Box furnace*

19. If two hearths are used, the load on one may be removed and replaced while the second is being processed. The work is usually of substantial weight or, in the case of sanitary ware, needs all pieces specially placed on the bogie or car. The furnaces may be continuous 'once through' or 'batch type'.

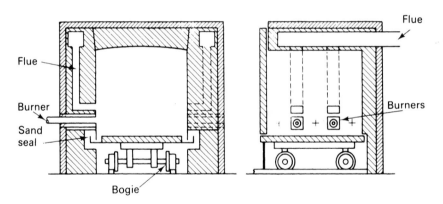

*Fig. 19    Bogie hearth furnace*

*Portable Cover Furnaces (Top Hat Furnaces)*

These use another method of making loading and unloading easier. The whole top of the furnace may be lifted off the hearth by an overhead crane. The base of the furnace may contain the burners and the flue ducts. This design is shown in Fig. 20 but some designs do have burners located in the top hat or portable cover. Two hearths may be used with one cover, which is fitted over the hearths alternatively as they are reloaded.

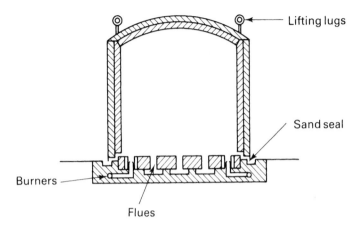

*Fig. 20    Portable cover furnace*

## Indirect Fired

### Crucible Furnaces

Crucibles are metal melting furnaces used where direct firing methods would affect the metal or when only small quantities of metal are to be melted. The crucible may be of either refractory material or metal and is emptied by:

- lifting out and pouring manually
- bailing out by hand ladle or internal pump
- mechanically tilting.

A typical lift-out furnace is shown in Fig. 21. These furnaces are commonly fired by air blast tunnel burners firing at a tangent into the annulus between the crucible and the lining. Combustion products are vented through the lid. Lift-out furnaces may be sunk into pits with their lids at floor level.

### Muffle Furnaces

In these furnaces the work is protected from the effects of the products of combustion by being contained in an inner case or 'muffle'. In high temperature furnaces the muffle is of refractory material but below 730°C either cast iron or alloy steel may be used. Fig. 22 shows a small muffle furnace, fired by air blast burners. Natural draught burners may also be used. The muffles may be fed with separately produced atmospheres to prevent oxidation or to change the surface structure of the work.

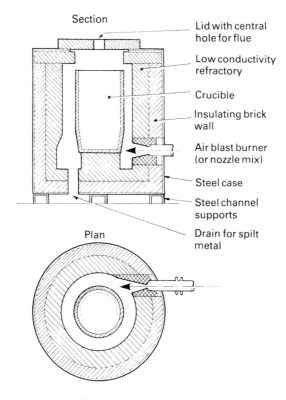

*Fig. 21    Crucible furnace*

A semi-muffle furnace is of similar construction but has no crown on the muffle.

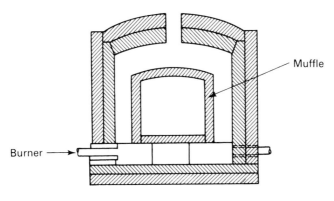

*Fig. 22    Muffle furnace*

*Radiant Tube Furnaces*

A radiant tube is an internally fired tubular heat exchanger similar to the immersion tube used in tank heating, but operating at much higher temperatures in the incandescent range. It passes through the furnace chamber and may produce furnace temperatures of 300°C to 1050°C. The radiant tubes may be of alloy steel or refractory materials and are located to achieve the temperature distribution required within the furnace. These furnaces are often used in conjunction with protective gas atmospheres, which are discussed later in this chapter.

A typical radiant tube furnace is shown in Fig. 23. Recent developments have produced ceramic radiant tubes capable of operating at much higher temperatures of about 1200°C and they are single-ended tubes employing internal recirculation of products and recuperators.

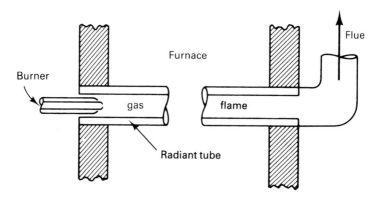

*Fig. 23    Radiant tube*

## Regeneration and Recuperation

Some of the heat in the combustion products leaving a furnace may be recovered by using it to heat water or produce steam in waste-heat boilers or to dry or preheat the work. It may, however, be returned directly to the furnace by using either:

- regenerators; or
- recuperators.

*Regenerators*

A regenerator is a series of passages through a mass of refractory material, usually chequer brickwork. Hot combustion products are passed through the regenerator, heating up the brickwork. When an adequate temperature is reached, the products are switched to a second regenerator and the incoming combustion air is heated by passing it through the hot brickwork. Fig. 24 shows an example of the layout of a regenerative furnace. There are duplicate fuel burners on both sides of the tank and these are lit alternately as the air and flue gas circulation is changed over. Regenerators are cyclic in operation although some continuous regenerators have been produced. These employ rotating drums or moving columns of ceramic pebbles. More recently compact regenerative burners have been designed and these are discussed in Chapter 4. This system of burners is commonly applied to the aluminium reverberatory melting furnaces and have had the effect of dramatically increasing the production rates. A system is shown on page 167.

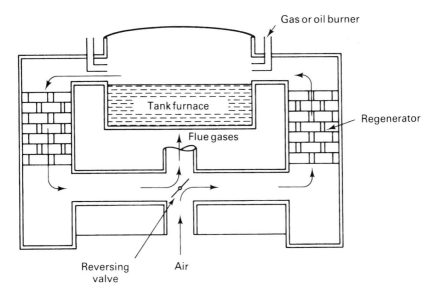

*Fig. 24   Regenerator*

*Recuperators*

A recuperator is a form of gas-to-air heat exchanger which transfers heat continuously from the combustion products to the incoming combustion air. It may comprise systems of concentric alloy steel tubes using parallel or counterflow streams of gases and air, Fig. 25.

For temperatures above 1000°C refractory sections are required.

Both regenerators and recuperators are confined to use on the larger furnaces. To recover heat on smaller and often less efficient furnaces, the recuperative burner was designed. This is a nozzle mixing burner fitted with a counterflow heat exchanger, Fig. 26. The

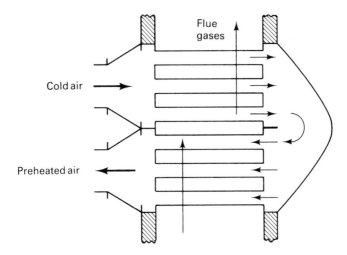

*Fig. 25    Gas flow recuperator*

combustion products are drawn in from the furnace and pass down an annulus and over the surface of an annular tube carrying air to the burner nozzle. The products are normally extracted by means of an eductor on the burner flue which also allows the furnace pressure to be controlled. It is necessary to have a relatively clean furnace atmosphere to avoid contaminating the heat transfer areas within the burner.

### Furnace Atmospheres

Oxidising, neutral and reducing atmospheres and their effects on iron were briefly described in Vol. 1, Chapter 2. However, the terms are misleading since, as was pointed out, atmospheres which do not contain $O_2$ may still be oxidising if they contain $CO_2$ or $H_2O$.

In consequence, furnace atmospheres are usually described as 'endothermic' or 'exothermic' which relate to the methods by which they are produced.

Chemical reactions which take in heat are endothermic.

Reactions which give out heat are exothermic.

*Endothermic* atmospheres are produced by heating the fuel gas with a small amount of air in a retort containing a catalyst. This is known as 'catalytic reforming'. It produces a gas which contains principally $CO$, $H_2$ and $N_2$. With the addition of some propane or methane it may be used for gas carburising.

*Exothermic* atmospheres are produced by the combustion of the fuel gas with a controlled amount of air usually in a premix

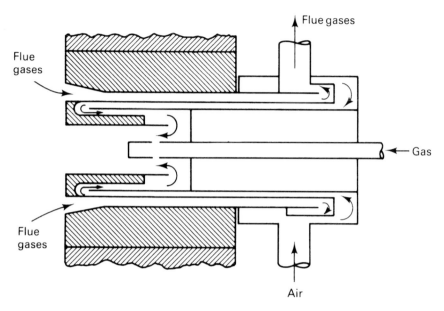

*Fig. 26   Recuperative burner*

system. 'Lean' atmospheres are produced by burning the gas with its theoretical or 'stoichiometric' air requirement, or very slightly below. This produces a gas consisting principally of $CO_2$, $H_2O$ and $N_2$. 'Rich' atmospheres are produced by partial combustion in lower air – gas ratio mixtures.

Both lean and rich exothermic atmospheres may be 'stripped', that is have the $CO_2$ and $H_2O$ removed, by several different processes.

The composition and application of endothermic and exothermic atmospheres is given in Table 1. As can be seen, rich stripped exothermic gas can be used as a substitute for endothermic gas for many purposes. This is an advantage, since endothermic gas is more difficult and more expensive to produce. As mentioned earlier there has been a general move to use furnaces that are indirectly heated and contain a high vacuum to produce brightly annealed steels and products requiring very clean finishes.

## Safety Devices

The combustion of any fuel gas in an enclosed space always presents a potential explosion hazard. Precautions and devices are necessary to ensure safety. The principal methods of providing protection are:

**TABLE 1 Controlled Furnace Atmospheres Produced from Natural Gas**

| | Atmosphere | Production | Composition % Volume | | | | | | Typical Applications |
|---|---|---|---|---|---|---|---|---|---|
| | | | CO | $H_2$ | $CO_2$ | $H_2O$ | $CH_4$ | $N_2$ | |
| Exothermic | Lean | Complete combustion of fuel gases | 0–3 | 0–4 | 12–10 | 2–3 | Nil | Balance | Bright and clean annealing copper, nickel, brasses, aluminium |
| | Lean stripped | $CO_2$ and $H_2O$ removed | 0–3 | 0–4 | Nil | Nil | Nil | Balance | Annealing. Carrier gas for carbon restoration. Ferrous metals |
| | Rich | Partial combustion of fuel gas | 9–12 | 11–15 | 5–7 | 2–3 | 1–2 | Balance | Normalising ferrous metals, brazing and sintering copper |
| | Rich stripped | $CO_2$ and $H_2O$ removed | 10–13 | 12–15 | Nil | Nil | 1–2 | Balance | Substitute for endothermic atmosphere for most purposes |
| | Modified | Partial combustion $CO_2 + H_2O$ removed. CO shift reaction $CO_2$ and $H_2O$ removed | Nil | 3–12 | Nil | Nil | Nil | Balance | Long cycle annealing, low carbon and mild steels |
| Endothermic | Endothermic | Catalytic reforming fuel gas and air | 20–25 | 30–45 | Nil | Dewpoint +15 to −15 | 0.5 to 1.0 | Balance | Hardening, brazing + sintering carrier gas for carburising and carbonitriding ferrous metals |

- a safe, sequential ignition system or procedure
- adequate methods of continuous monitoring of the pilot and the main gas flames during the whole time that gas is supplied
- the provision of approved safety control devices
- the use of flame traps where stoichiometric mixtures of air and gas are conveyed by pipework
- for plant operating at temperatures below the incandescent range, that is below 600°C, the provision of approved explosion reliefs.

The design and operation of burner control systems and devices is dealt with in Chapter 4 and flame monitoring systems are discussed in Chapter 5.

Explosion reliefs are required by the Health and Safety Executive (booklet HS(G)16) to be fitted to plants classified as 'ovens'. The recommendations specify the:

- size and design of relief
- materials of construction
- location
- methods of fixing.

An explosion relief is a weak section of the oven which will give way at a low pressure and provide an opening for the gases to vent quickly and safely.

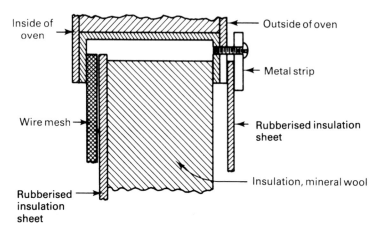

*Fig. 27    Explosion relief*

Box ovens usually have the relief located at the back of the oven and an approved design is shown in Fig. 27. The oven must be spaced at least 380 mm from any wall immediately behind it, and if sited in a corner, at least 600 mm from both side and rear walls. Personnel must be prevented from entering any space into which flames from a relief may be discharged.

Treble cased ovens usually have two reliefs, one in the inner case and one in the outer double case. The two reliefs are at the back of both compartments but do not interconnect.

Conveyor ovens usually have open ends through which some pressure would be dissipated. Ovens where the length is more than six times the diameter may require additional reliefs.

For furnaces operating at temperatures in the incandescent region no explosion reliefs are necessary. Any stoichiometric mixture admitted will burn steadily and not explosively. These furnaces do, however, present other problems.

Many high temperature furnaces work continuously for many months without being shut down. Some would be destroyed if the temperature was to fall suddenly or rapidly. In addition to damaging the furnace, an unexpected shutdown could also spoil the contents of the furnace resulting in an entire batch of rejects. Failure of a furnace can be a very costly business indeed.

For this reason extra precautions are taken. Continuous checking ensures that the equipment is operating satisfactorily. Control systems may be duplicated or triplicated. Flame monitors may be arranged so that shutdown will only occur if two out of the three systems simultaneously call for action.

Complex control systems may be expensive but cost of failure would be many times the cost of the protection system.

Furthermore, before carrying out any work on the design, installation or commissioning of an industrial process plant, reference should be made to current recommendations. These include:

- Gas Safety Regulations
- Appropriate British or European Standards
- British Gas Publications
- HMSO publications on Health and Safety at Work
- Manufacturer's instructions.

# Industrial Gas Burner Systems and their Control

Chapter 4 is based on an original draft by J. R. Cornforth

## BURNERS AND SYSTEMS

### Introduction

Gas burners were dealt with in Vol. 1, Chapter 4 which concentrated on their principles of operation and their application to domestic appliances. This chapter goes on to consider industrial and commercial burner systems.

Inevitably there is some apparent overlap. Some of the burners used in small commercial and low temperature industrial equipment are very similar to some domestic burners. However, burners used in larger equipment and high temperature applications are very different both in the pressure of the air and gas used and in the manner of their mixing.

The varied requirements of commercial and industrial equipment call for a variety of flames which differ in their four main characteristics:

- shape
- size
- aeration
- temperature.

Suitable burners may be selected from a wide range of proprietary types or be developed by the industry for a special application. For any given application there may be more than one way of achieving a satisfactory result.

### Types of Burner Systems

Burners for commercial or industrial equipment may be divided into five main categories as follows:

- diffusion flame or post-aerated burners
- atmospheric or natural draught burners
- air blast premix burners
- nozzle mixing burners
- other burners and special applications.

*Diffusion flame burners*, or neat gas burners, obtain all their air for combustion from the surrounding atmosphere.

*Natural draught burners* use an atmospheric injector assembly, Fig. 1.

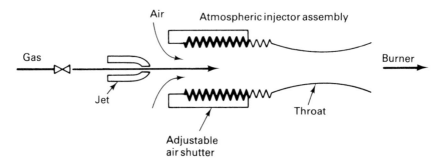

*Fig. 1    Natural draught injector*

Here, the gas mixes with a proportion of the combustion air before entering the burner. The air, at atmospheric pressure, is entrained by gas under pressure issuing from a jet. If the gas is at normal supply pressure, the burner is a 'low pressure' type and the primary aeration is less than stoichiometric. The remaining air is obtained as secondary air from the atmosphere into which the burner is firing. If gas is available at relatively high pressures of about 2 bar, stoichiometric proportions may be obtained by using a high pressure natural draught burner.

*Air blast burners* also achieve stoichiometric air/gas mixtures by using air at high pressure to entrain gas at atmospheric pressure. Because all the air required for combustion is available in the burner nozzle, combustion is much more rapid and a short, intense flame results. The system generally uses fanned air at about 75 mbar, Fig. 2. Both this and the natural draught system are known as 'premix' systems because air and gas are mixed in varying degrees before they reach the burner. The same result could be achieved by a mixing machine, but the initial and running costs would be considerably higher.

*Nozzle mixing burners* are supplied with both air and gas under pressure and there is no prior mixing until they reach the burner

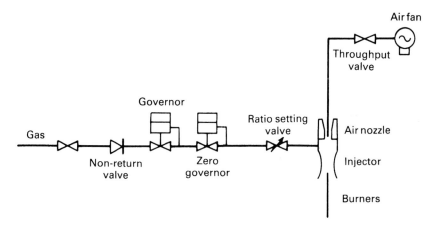

*Fig. 2    Air blast burner system*

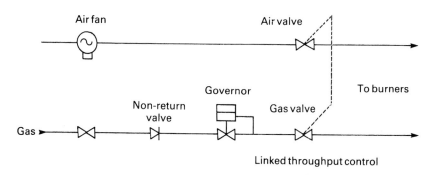

*Fig. 3    Nozzle mixing burner system*

nozzle, Fig. 3. The air and gas are proportioned separately by linked valves or other techniques and fed independently to the burner nozzle. Mixing, by this method, is very positive and mixtures are usually stoichiometric. Intense rates of combustion are developed for high temperature work. The burners are usually very flexible and can be run on either air or gas rich mixtures. They have a greater degree of turn down than a corresponding air blast burner.

*Other burners* and applications include those which are either outside the four main categories or are special applications of them. They include:

- radiant, or surface combustion burners
- catalytic combustion burners
- radiant tube burners
- self recuperative burners

- compact regenerative burners
- packaged burners
- dual fuel burners.

### Diffusion Burners

The diffusion flame often burns with a luminous appearance. This is due to the unburnt gas issuing from the burner being 'cracked' by the heat developed in the burning outer flame. Minute carbon particles are formed which, on reaching the outer zone of the flame, react with oxygen and burn with a yellow luminosity.

Because of the low flame speed and high air requirements of natural gas there are relatively few burners of this type used. The exception to this case are burners for firing large glass tanks. These are almost exclusively diffusion flames from gas flowing at pressures in excess of 100 mbar into a medium velocity stream of hot air at about 800°C. Large power station water tube boilers often inject neat gas into the combustion chamber where there is ample space for combustion.

Air recirculating ovens, used for drying processes, often employ a suction burner, frequently the 'Firecone' shown in Fig. 4. The recirculating fan creates a negative pressure inside the combustion chamber and combustion air is induced naturally. The size of the holes in the

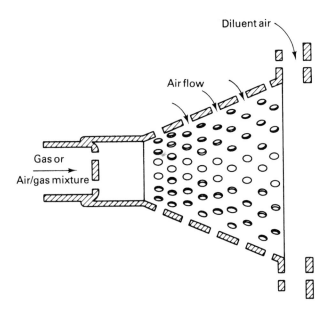

*Fig. 4   Diffusion burner (Firecone)*

cone through which the air flows are carefully graduated with small holes near the gas inlet and larger ones approaching the rim of the cone in order to avoid flame chilling and incomplete combustion.

Whilst many small industrial jets are now aerated, some small diffusion flame burner heads are produced. Figure 5 shows one in which neat gas issues from two opposite holes and impinges on two stainless steel wings to give a fan shaped flame.

Other neat gas burners include pinhole burners, target burners and matrix burners which are described in Vol. 1.

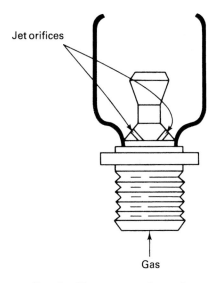

*Fig. 5    Neat gas jet (Drew)*

**Natural Draught Burners**

The majority of these are low pressure natural draught burners, using gas at 15 to 20 mbar. The energy of the gas is used to entrain about 50 to 55% of the air required for combustion in the form of primary air. The burners usually operate in open atmospheric conditions.

*Uses*

Burners of this type form the largest proportion of industrial burners. They are used for relatively low temperature processes in the range of 200 to 400°C. Applications include small box ovens and underfired liquid heating tanks.

*Types*

Probably the most common type is the simple drilled bar burner, Fig. 6. Other examples in general use range from simple single port burners, such as the Bunsen burner, to large assemblies of drilled or ribbon-ported bar or ring burners.

The burner port area for a particular heat input rate is termed the 'burner port loading'. It may be calculated from:

$$\text{burner port loading MW/m}^2 = \frac{\text{heat input rate (MW)}}{\text{total burner port area (m}^2\text{)}}$$

For drilled ring burners with about 50% primary aeration, burner port loadings should be limited to about 10 MW/m$^2$ or lift off will occur. For lower primary aeration, port loadings may be increased to about 25 MW/m$^2$, but care must be taken to avoid yellow tipping and soot formation.

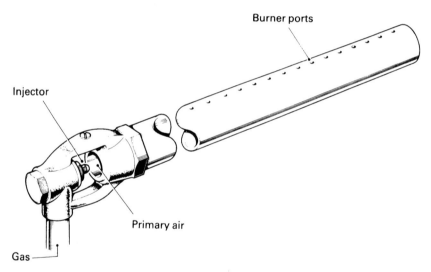

*Fig. 6   Ported bar burner*

*Flame Retention*

If higher burner port loadings are required, some method of flame retention is necessary. With natural draught burners there are three main types:

- sudden enlargement of mixture flow
- wake eddies produced downstream of an obstacle
- stabilisation by auxiliary flames.

*Sudden enlargement* is achieved by welding or rivetting continuous strips along the bar on either side of the main flame ports, Fig. 7. This gives a sudden enlargement of the mixture flow cross-section and generates a recirculation of hot gas products towards the root of the flame, so providing a source of heat for ignition. The strips also shield the flames and reduce the effect of cross draughts. They should be saw cut at intervals along their length to reduce any bending effect caused by unequal expansion of the strips and the bar material.

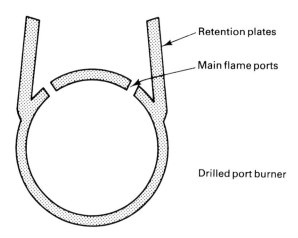

*Fig. 7    Flame retention; sudden enlargement*

*Wake eddies* have the effect of producing a zone in which mixture velocity is reduced, giving a higher probability of the mixture becoming heated for ignition. They also create a recirculation behind the eddy, Fig. 8.

*Auxiliary flames* or retention flames are, in effect, pilot flames, generated by much smaller ports adjacent to the main flame ports. A flame retention plate is welded or rivetted along each side of the main flame ports, Fig. 9. The small pilot drillings act as metering orifices and slow down the rate of mixture flow so that a continuous strip of pilot flame forms between the bar and the retention plate. This flame has the effect of constantly keeping the main flame lit. With this design there may be problems due to the retention strip buckling or dust deposits collecting in the metering orifices.

A form of this burner which overcomes both these problems is shown in Fig. 10. The retention plate is secured in a horizontal position by screws in elongated holes which allow for expansion. The plate prevents dirt from falling into the central metering orifices. The

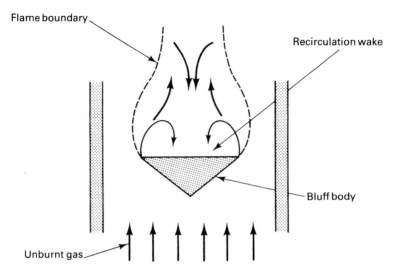

Fig. 8  *Flame retention; wake eddies*

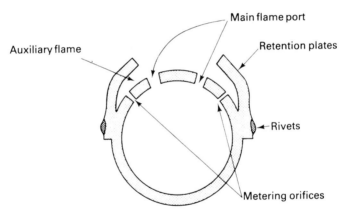

Fig. 9  *Flame retention; auxiliary flames*

burner gives short, well-defined flames at high outputs and is completely stable. This is because secondary air is easily entrained by the roots of the main flame.

The ribbon burner, Fig. 11, also produces auxiliary flames. The ribbon forms a combination of large and small burner ports, the small ports being located at either side of each crest of the corrugations. These burners are very versatile and can be used with high and low pressure natural draught injectors. They may also be used for air blast and premix systems giving stoichiometric aeration. The ribbon

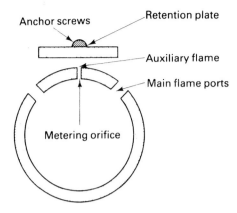

*Fig. 10    Horizontal flame retention plate*

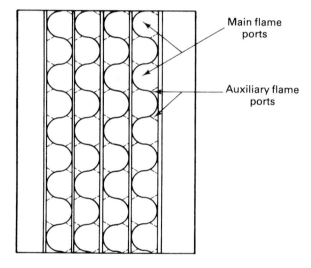

*Fig. 11    Ribbon burner*

must be securely welded or rivetted in position otherwise it may be forced out when the burner gets hot.

A simple method of providing auxiliary flames in fairly high rated bar or ring burners is by the use of the 'Insertajet' (Aeromatic Limited), Fig. 12. Originally produced for conversion, these jets are now available in larger outputs up to 300 W per jet at normal gas pressures. They may also be used at higher pressures. The burner ports are drilled out to 5.9 mm diameter and the 'Insertajet' tapped lightly in until it reaches the shoulder. The jets must be a tight interference fit otherwise they may come loose when the burner heats up. On ring burners it is only necessary for 50% of the ports to be fitted with 'Insertajets'. This is adequate to ensure that the remainder of the flames are continuously lit.

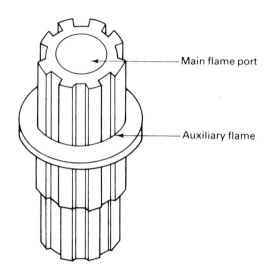

*Fig. 12    Aeromatic 'Insertajet'*

Another group of natural draught burners is the single port burner shown in Fig. 13 and known as cup burners. These are made in the range 3.0 to 150 kW and can be used for a variety of purposes. The small size can be used as a lighting torch head or can be used in groups by screwing them into a manifold or gas pipe to provide a high heat release bar burner.

*Pilot Burners*

When fanned air is not available, natural draught pilots are often used for smaller burner installations. They usually have an adjustable air port to vary the hardness of the flame and they can be associated with either thermoelectric or flame rectification monitoring systems, Fig. 14. The primary function of the pilot burner is always to keep the main burner alight.

Where more than one bar burner is fed from a single control valve, a 'ladder pilot' may be fitted. This is a small bar burner with ports that give a continuous pilot flame from one end to the other, Fig. 15. The ladder pilot is fitted at right angles to the main burners and the flame sensor must be at the opposite end to the ignition device. This should ensure that all main burners are alight when the flame is detected (see Chapter 5).

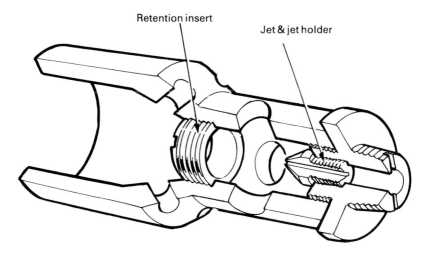

Fig. 13    *Cup burner*

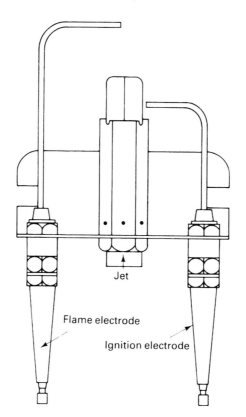

Fig. 14    *Natural draught pilot burner*

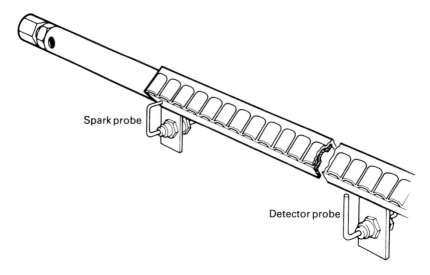

*Fig. 15    Ladder pilot burner*

*Control Systems* (Non-Automatic Burners)

Because the gas throughput on natural draught systems is generally relatively small the controls are kept simple. A typical control train is shown in Fig. 16 for a purely manual system which is to be constantly supervised. The pilot line is separately valved and governed so that the pilot is not starved of gas when the main valve is opened.

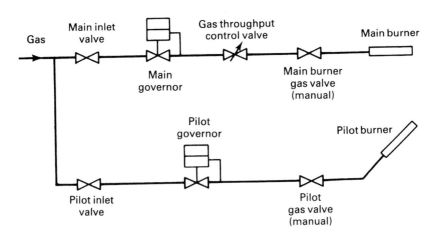

*Fig. 16    Natural draught burner control system*

To ensure safety, a flame protection device should be added (see Chapter 5). Where the heat input rate is not more than 120 kW, for example in a drilled bar burner, thermoelectric flame protection is considered adequate (provided that the closing time following flame failure is not more than 45 seconds). The thermoelectric flame protection system should meet BS 6047 Part 1. For rates above this, either flame rectification or ultra-violet devices should be used in conjunction with at least one certificated safety shut off valve to BS 7461.

## High Pressure Natural Draught Burners

Where the gas supply is available at relatively high pressures, a natural draught injector may be used to provide near stoichiometric mixtures. For example, a gas pressure of 2 bar will give a stoichiometric mixture pressure of 6.25 mbar. This is sufficient pressure for the burner to be stable over a satisfactory turndown range. Alternatively, if this pressure is not available, a typical natural draught system could be employed with the remaining air being supplied as secondary air. If higher gas pressures are required a reciprocating or vane type booster could be used. Air blast systems can offer similar advantages with much simpler equipment.

## Mechanical Mixing Techniques

In some installations natural draught burners may be supplied from a mixing machine. These systems were often used during conversion to reduce the cost. By producing an air/gas mixture of the same Wobbe Number as manufactured gas, the existing burners could still be used. In some cases the whole factory may be supplied by one mixing machine, in others it could be a single item of plant, for example a continuous baker's oven with a hundred or so ribbon burners.

In most mechanical premixing machines, the air and gas are drawn through suitable metering elements into a mixing device and then compressed. The rotary compressor is the most widely used type. It produces a mixture of substantially constant air/gas ratio at pressures up to about 350 mbar. It is important that the volume of air inspired should form a gas-rich, non-explosive mixture which may then be safely distributed to a number of burners with atmospheric injectors to give a final mixture near to stoichiometric. A simplified system is shown in Fig. 17. An essential feature of the system is the ability of the machine to deliver a constant ratio mixture when the back pressure or the number of burner heads in use is varied.

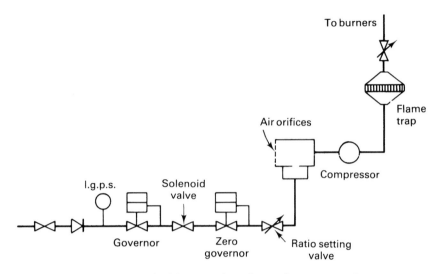

*Fig. 17   Simplified layout of mechanical premix machine*

For safety it is essential to fit a low pressure cut-off device in the gas supply to the compressor. Additionally, a flame trap element should be fitted in the mixture supply as near to the burners as is practicable.

The system described gives only a 'partial premix' requiring some atmospheric air to complete the stoichiometric proportions. A 'full premix' system can be used to provide a stoichiometric mixture. In this case the most stringent safety measures must be taken, particularly with regard to the use of flame traps which may need to incorporate fusible links or over-heat cut-offs.

## Air Blast Burners

This system uses air under pressure to entrain gas at atmospheric pressure, the opposite of the natural draught system. Because the air pressure is commonly about 70 mbar a stoichiometric mixture is easily produced. As all the air required for combustion is delivered to the burner head, this results in intense combustion and a fairly short flame.

### Uses

Air blast burners are generally used for higher temperature work than diffusion flame burners where high heat inputs are required. They are frequently used on all types of heat treatment and annealing furnaces

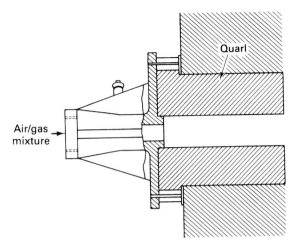

*Fig. 18    Air blast burner with integral quarl*

and also on furnaces for reheating steel prior to forging small articles. With correct furnace design, temperatures up to 1200°C can easily be achieved.

*Types*

These burners may be divided into two distinct groups:

- burners firing into a refractory 'quarl'
- burners without refractory quarls.

The 'quarl' is a refractory burner head or tunnel. Where this is integral with the burner casting it may be sealed into the furnace wall, Fig. 18. In other types the cast iron burners are bolted onto the side of the furnace casing and a suitably flared or parallel tunnel is then cut in the furnace refractory wall, Fig. 19. In these types of burner the flames are stabilised by radiation from the incandescent quarl. It is therefore important that quarls are correctly designed. For a parallel tunnel the relationship between the length and the diameter are as follows:

If the diameter of the air jet nozzle is d
And the diameter of the mixture tube orifice is D
Then:

$$D =   1.6 \, d$$

Tunnel diameter     =   4.8 d
Tunnel length       = 16.0 d

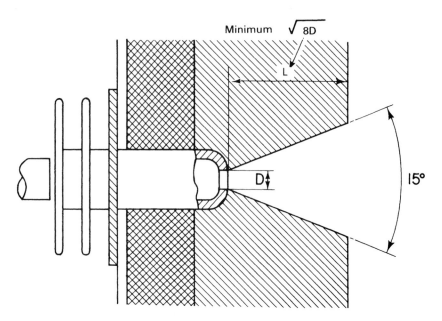

*Fig. 19    Air blast burner firing into flared tunnel*

### Flame Retention

When air blast burners are used without refractory quarls, a flame retention head or non-blow-off burner tip is required. Small auxiliary flame ports are incorporated into the burner head to produce low velocity pilot flames which keep the base of the main flame alight. A typical burner tip is shown in Fig. 20. It has small metering orifices drilled around the inner core. The air/gas mixture then expands between the concentric tubes and forms a low velocity stable flame which keeps the root of the main flame continuously lit. The tip allows the use of high mixture velocities without blow-off.

These tips are usually made from steel and designed for use in the open, to avoid over-heating and to prolong their working life. They may also be used for 'gap-firing'. Here the tips are spaced at a short distance from the furnace wall and fire into a circular port in the wall refractory. With this method, excess atmospheric air is induced into the air/gas mixture stream and it is used where lower furnace temperatures are required. Gap-firing also keeps the tip relatively cool.

### Control Systems

In a typical system air from a fan at about 75 mbar is allowed to expand through the jet of an air blast injector into a venturi mixing

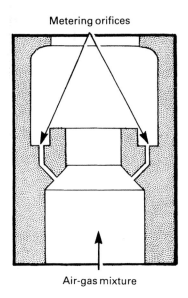

Metering orifices

Air-gas mixture

*Fig. 20    Non-blow-off burner tip*

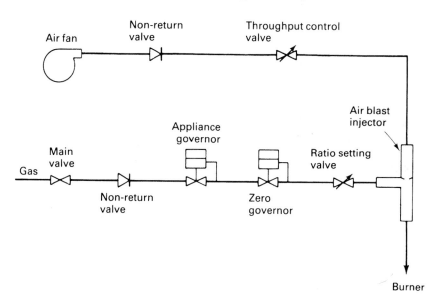

Air fan

Non-return valve

Throughput control valve

Air blast injector

Appliance governor

Main valve

Ratio setting valve

Gas

Non-return valve

Zero governor

Burner

*Fig. 21    Air blast control system*

tube, Fig. 21. The expanding jet of air entrains gas available at zero, or atmospheric pressure, which is admitted to the mixture tube immediately downstream of the air nozzle. The mixture is then supplied directly to the burner. There must be no restriction or valve between the injector and the burner. Providing that the gas is supplied at zero pressure, usually by means of a zero governor, and that the burner is firing into approximately atmospheric pressure, then the quantity of gas entrained is proportional to the air flow through the injector. The gas ratio setting valve is often incorporated into the injector, when it is known as the 'obturator setting screw'.

In the tunnel burner, injection, mixing and combustion occur together in the tunnel. With this and any other system using fanned air, it is a requirement of the Gas Act that a suitable non-return valve be fitted in the gas supply line. This is to prevent any air from entering the gas distribution system. It is also advisable to fit a non-return valve in the air supply line, especially if the fan is situated above the burner, to prevent gas entering the fan housing and escaping to the atmosphere.

## Commissioning

The throughput control valve, which is a single linear flow type valve or quadrant cock in the air supply line, should be set to 'low fire' rate and the burner ignited. The air/gas ratio should be set, while at low fire rate, by adjusting the zero governor tension spring. The gas pressure downstream of the governor should then be substantially at zero gauge pressure.

The system should be turned to high fire rate, by the quadrant cock and the air/gas ratio set by adjusting the obturator screw or a separate ratio setting valve. It may then be necessary to re-check the setting at low fire rate. When set correctly proportional control is achieved by the one valve in the air supply line.

The setting up of the system is made easier if flow meters can be fitted in the air and gas lines or if an orifice plate or insertion meter can be used. In most cases, on the district, the system must be checked by viewing the burner; the correct settings may be determined with experience. Flue gas analyses should be used whenever possible.

## Appliance or Mixture Pressure Back Loading

If the chamber into which the burner fires is not at atmospheric pressure, then to maintain the self-proportioning action, the reference pressure for both the appliance and the zero governors should be the appliance chamber pressure and not atmospheric pressure.

To compensate for changes in appliance back pressure, the top diaphragms of both the appliance governor and the zero governor are back loaded with the furnace chamber pressure. This is usually communicated by small bore copper tube from a position in the appliance adjacent to the burner. Furnace chamber pressures are seldom higher than 5 mbar.

Compensation for changes in mixture pressure is obtained by mixture pressure back loading the two governors from a position in the mixture manifold. Apart from the point of connection, the two systems are identical, and are shown in Fig. 22. With any back loading technique it is necessary to have the inlet pressure to the zero governor at least 0.75 mbar higher than the back loading pressure and the top chambers of both governors airtight.

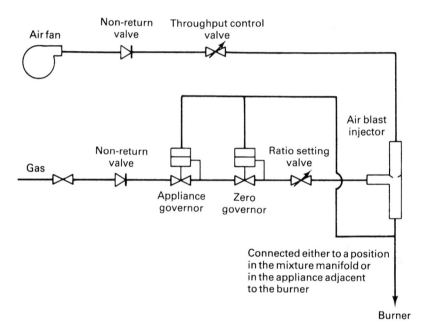

*Fig. 22    Appliance or mixture pressure back loading*

## Nozzle Mixing Burners

In this system, gas and air do not mix until they enter the burner quarl. The burner body merely serves as a distribution box, conveying the air and gas to the quarl. The other difference between this and the previous systems is that both air and gas are under pressure. The burners are generally very versatile. They have a wider turndown range than the equivalent air blast burner and can be used with either excess air or excess gas. Because all the air for combustion is supplied to the burner, relatively rapid mixing results giving maximum heat release rates and short flames.

*Uses*

Nozzle mixing burners are used on all types of high temperature plant, from small heat treatment furnaces to large reheating furnaces and non-ferrous metal melting furnaces. Because of their flexibility and relative ease of commissioning, these burners are generally replacing the equivalent air blast systems.

*Types*

There are many different types of nozzle mixing burners. In each case, flame retention is achieved by the recirculation of hot products in the quarl and also by radiation from its incandescent refractory walls. This enables the burner to operate satisfactorily at low throughputs, both in stoichiometric conditions and with large amounts of excess air. If a more gentle flame is required for less intense local heating, gas rich firing can be employed.

A typical burner is shown in Fig. 23 in schematic form. A cut away section of the nozzle is shown in Fig. 24. Here, both the gas and air flows are axial, with the air annulus surrounding the gas flow. In some designs a small amount of air may be fed into the gas stream before the quarl either axially or tangentially, to assist flame retention.

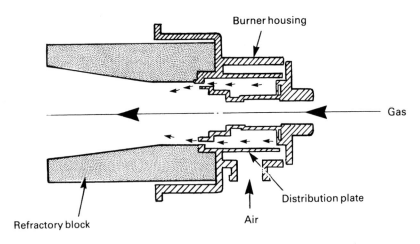

*Fig. 23    Nozzle mixing burner (Schematic)*

To give even more rapid rates of heat release some burners are fitted with swirl vanes in the gas and air sections to increase the speed of mixing.

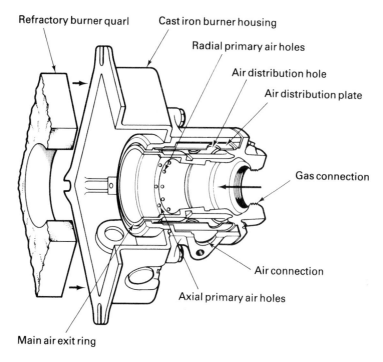

Refractory burner quarl    Cast iron burner housing
Radial primary air holes
Air distribution hole
Air distribution plate
Gas connection
Air connection
Axial primary air holes
Main air exit ring

*Fig. 24    Nozzle mixing burner*

*High velocity burners* are used for processes which rely mainly on forced convection as the method of heat transfer. In this case the main obstacle to efficiency is the thin film of gas on the surface of the stock. Increasing the velocity of the heating gases relieves or removes the insulating film so increasing the heat transfer efficiency. A high mass flow of hot gases, forced to envelope the stock completely, results in high speed heating with very low temperature differentials. Because there are no high temperature gradients, hot spots cannot form and overall uniformity of temperature is improved.

This can be achieved by using recirculating fans or, more simply, by utilising the jet effect of a high velocity burner, Fig. 25. Because combustion is completed within the burner and because the exit quarl is convergent, the products leave the burner with a high velocity at approximately 200 m/s. In a properly designed furnace the thrust from the burner can be used to entrain combustion products and recirculate them around the load. The burner can be operated on high excess air so it is equally suitable for low as well as high temperature applications.

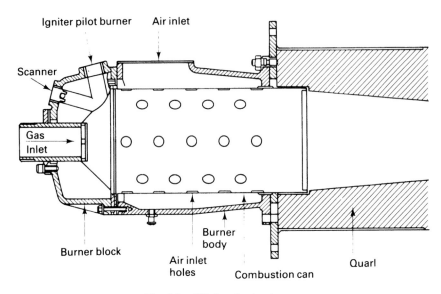

*Fig. 25  High velocity burner*

Several small burners in a furnace could be replaced by one single high velocity burner, so giving even temperature distribution and making the provision of flame protection much simpler.

*Flat flame burners* are used when flame impingement on the stock must be avoided. A typical burner is shown in Fig. 26. It is essential that the outlet of the refractory quarl is in alignment with the furnace wall or roof so that the flame can flow from the burner on to the adjacent surface. The air inlet orifice is arranged to swirl the air so that the combustion products tend to hug the surface of the quarl and move along the surface of the furnace wall with little forward velocity. There is, therefore, no flame impingement on the stock. The successful uses of this burner include galvanising bath heating and crown firing of small beehive brick kilns.

## Control Systems

The two main systems for the control of nozzle mixing burners are:

- the linked valve system
- the pressure divider technique.

*Linked valves* control the separate flows of air and gas to the burner. One valve is situated in the air supply line and the other in the gas supply. They are linked either mechanically or electrically so that the flows of air and gas can be adjusted simultaneously. The mechani-

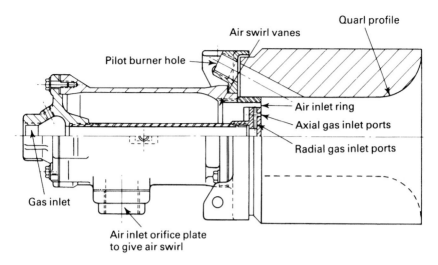

*Fig. 26    Flat flame burner*

cal linkage is usually driven by an electric motor and the electrical link is usually operated electrically or pneumatically. In some simple systems the two valves are adjusted independently of each other, especially if the air/gas ratio is required to vary during the process cycle.

Changes in back pressure are often appreciable compared with the supply pressures and tend to affect the air and gas flows unequally, so causing a deviation from the ratio required. Similarly, if preheated air is used, the air/gas ratio can vary from the desired value due to the change in air density at the burner nozzle. If, however, the differential pressure across the valves is kept substantially constant, a suitable choice of valves can maintain the air/gas ratio to within narrow limits. Control of the differential pressure is achieved by fitting pressure governors upstream of the valves, back loaded from pressure tappings downstream, Fig. 27. By positioning the air valve upstream of any recuperator, the air controlled by the valve is virtually at constant temperature and density.

If the changes in back pressure are relatively small, adequate control may be achieved without back loading. This is done by using constant upstream supply pressures which are large in comparison to the pressures downstream of the control valves.

The valves most commonly used for high/low control are linked butterfly valves. With these it is relatively easy to set the flows to the correct air/gas ratio at both high and low limits. Variations in the ratio may occur at mid-positions during the changes between high

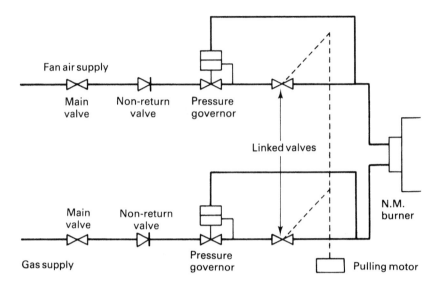

*Fig. 27   Nozzle mixing burner control system, mechanically linked valves*

and low limits. Care must be taken to ensure that these variations do not seriously affect the stability of the burner.

Another system frequently used for high/low control is the weep relay valve system, Fig. 28. This has solenoid valves situated in the weep lines of relay valves in the gas and air supplies. The solenoids are controlled by a thermostat or similar device. The flow rates at high and low positions are set by adjustable stops above the diaphragm and below the valve of the relay valve.

If proportional or modulating control is required it is essential to match the air and gas flows over the whole throughput range. This requires linear flow valves such as the adjustable port valve. The installation would be as in Fig. 27.

*The Pressure Divider Technique* is applicable when there is little or no injection or pressure interaction between gas and air streams in the burner. This situation occurs, for example, when a simple concentric tube burner fires into a large chamber. In this case the burner consists of air and gas orifices with the two streams mixing at a common down-stream pressure. Consequently, if the pressures upstream of the burner are held at a constant ratio over the whole throughput range, a constant air/gas ratio will be maintained. The control layout is shown in Fig. 29.

A zero governor is fitted in the gas supply and is loaded with a fraction of the air supply pressure from pressure dividing orifices. These are a bleed orifice C and a ratio setting valve D. The system has

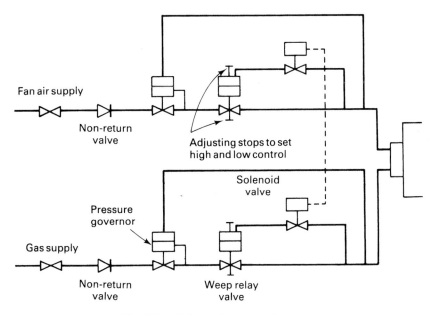

*Fig. 28    Relay valve control system*

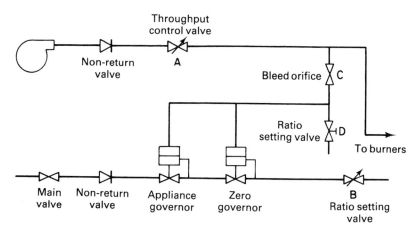

*Fig. 29    Pressure divider control system*

the advantage that throughput may be controlled by a single valve, A, in the air supply line. As the air flow through the valve is increased, the diaphragm of the zero governor is further depressed by the bleed air and the gas flow increases in proportion. The bleed orifice and the ratio setting valve are usually incorporated in one proprietary valve body.

When the gas pressure is less than the air pressure, valve B is not fitted.

### Commissioning Pressure Divider Systems

Set the valve A to low fire rate and adjust the zero governor tension spring to give a downstream pressure just above atmospheric, consistent with a satisfactory flame at the burner.

Turn to high fire rate and adjust by trimming the ratio setting valve D, provided that the correct size bleed orifice C has been fitted. It may then be necessary to readjust the low fire setting. When the system has been set up all variable adjustments should be locked in position. Once correctly set the system will give a constant air/gas ratio over the whole turndown range so, if a linear throughput valve is used, the system will give proportional control.

In an installation in which the gas pressure is equal to or greater than the air pressure, the bleed orifice is removed so that full air pressure is fed to the top of the governor diaphragms. The ratio setting valve is located at B and the pipework at D is blanked off. Low fire is adjusted by means of the zero governor tension spring and high fire rate is adjusted by the ratio setting valve B.


## OTHER BURNERS


### Radiant Burners

These burners were introduced in Vol. 1, Chapter 4 and some types are used on domestic appliances. The burner heads are of porous or perforated refractory material or of wire gauze. They are known as 'surface combustion' burners because combustion usually takes place at the surface of the porous medium or about 1.5 mm below the surface in the perforated medium. The air/gas mixture is supplied either from an air blast injector or an atmospheric injector and produces an area of even radiation as the surface becomes incandescent. Surface temperatures of about 850°C are common although special burners can reach over 1400°C.


### Uses

Radiant burners have been extensively used in overhead radiant space heaters. They are also used for special, high temperature drying operations and as the heat source in large, overfired grills.

*Types*

Burners may be divided into categories based on the type of radiating surface as follows:

- porous refractory
- perforated refractory
- wire gauzes
- catalytic combustion
- high temperature porous medium burners.

*Porous refractory burners* may have either atmospheric or air blast injectors producing an air/gas mixture which is fed through the pores in a refractory plaque.

*Perforated refractory* tiles are used in burners like the Schwank burner, Fig. 30. This has perforations of about 1.4 mm diameter set in individual plaques of about 65 × 45 mm built up to form the radiating area. The air/gas mixture is supplied from an atmospheric injector.

*Fig. 30 Perforated refractory burner (Schwank)*

*Wire gauzes* are used in some domestic grills and in some overhead radiant heaters (Chapter 9). They are suitable for low temperature applications. A metal gauze is fitted about 5 mm above the plaque surface on perforated refractory burners when burning natural gas, to give flame retention.

*Catalytic combustion* has been widely used on the Continent, principally with lpg. The gas reacts with a catalyst to burn at a low temperature, about 450°C, without an apparent flame. Problems with early development catalytic combustion burners were associated with methane slippage. A big advantage with this type of heater is on paint drying plant and heating ovens containing solvents. Work on developing a more effective catalyst for natural gas is continuing.

*High temperature porous medium burners* have been developed by research laboratories and the manufacturers to give operating temperatures up to 1400°C with high radiant efficiency. The Shell P.R. (Porous Radiant) Burner, Fig. 31, uses porous elements made either from:

- zircon, or
- sillimanite.

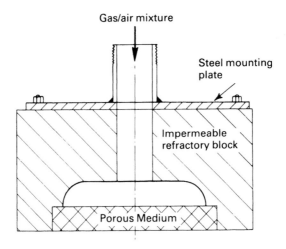

*Fig. 31   Porous radiant burner (Shell)*

The zircon burner can be operated at higher surface temperatures than the sillimanite, but the sillimanite panels offer less resistance to the mixture flow and can be operated on lower mixture pressures.

## Radiant Tube Burners

These were introduced in Chapter 3 and are used where indirect firing is essential. For example, in many heat treatment processes where the work must not come into contact with the products of combustion or the flame.

The tubes can provide a range of temperatures from about 300 to 1100°C. They may be used for a large number of industrial processes. Radiant tubes can be divided into three main categories:

- straight-through
- double pass single-ended
- recirculating.

*Straight-through types* include tubes of the following shapes:

- straight
- curved parabolic
- 'U' shaped
- 'W' shaped.

Each of these tubes may be fitted with external recuperators, if required.

*Double pass single-ended tubes* have the burner and flue combined at one end. They have an internal tube which transmits the hot gases to the closed end of the outer tube where they are diverted back through the annulus in a counter-flow direction. A refinement of this principle incorporates a recuperator and Fig. 32 shows a burner of this type. Thermal efficiency may be about 60 to 70%. Single-ended tubes may easily be replaced in a furnace, unlike the 'U' and 'W' shaped tubes.

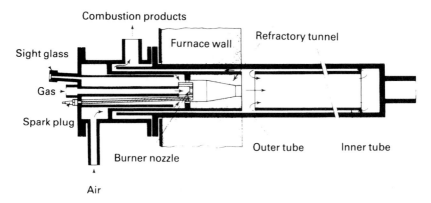

*Fig. 32    Radiant tube burner with recuperator*

*Recirculating types* of radiant tubes have a concentric inner section which further increases the operating efficiency of the tubes by recirculating the hot combustion products. A ceramic tube, of silicon carbide bonded with silicon nitride, of the same type will withstand process temperatures up to 1250°C and is suitable for bright annealing of stainless steel and for special ceramic wares.

## Self Recuperative Burners

Waste heat recovery can be achieved either by an external heat exchanger or by a recuperator integral with the burner. The latter system avoids the heat losses experienced from the pipework associated with external recuperators. There is, however, the problem of air density variation as the burner heats up.

A typical burner was described in Chapter 3. It consists essentially of a high velocity nozzle mixing burner surrounded by a counter flow heat exchanger supplying hot combustion air to the burner nozzle. This can give air preheating up to about 650°C.

Normally, all the combustion products are extracted through the recuperator by means of an air driven eductor mounted on the burner flue outlet. The furnace pressure is kept constant by regulating the eductor air supply.

The burners may be used for a variety of processes including batch furnaces for heat treatment of metals, non-ferrous metal melting, salt baths and pottery kilns.

## Compact Regenerative Burners

The burners are fitted in pairs with only one burner firing at any one time, Fig. 33. Waste heat from the products of combustion is recovered by transferring heat to refractory spheres contained in the flue exit of the burner not firing. When the temperature reaches a certain level, the flow is reversed and the other burner fires, utilising the waste heat accumulated in that regenerator. The products flow through the first burner that was originally firing and reheat the regenerator attached to that burner. The firing cycle for each burner is 2–5 minutes and the flue exit temperature after leaving the regenerator is about 100°C.

Over the temperature range 800–1400°C, 90% of the heat from the flue gases is transferred to the incoming combustion air giving the potential for heating stock up to 80% gross thermal efficiency.

The burner systems are available in size ranges from 300 kW to 3.5 MW.

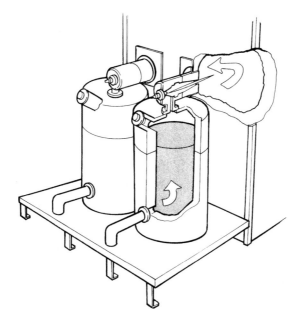

*Fig. 33    Compact regenerative burner*

### Packaged Burners

These are self-contained units for use on industrial or commercial appliances. They are easily installed by bolting on to the combustion chamber of the appliance and connecting to gas and electrical supplies. If the rating of these burners is between 7.5 kW and 60 kW, they should be designed to comply with BS 5885 Automatic Gas Burners Part 2. Above 60 kW the burners should be designed to comply with BS 5885 Automatic Gas Burners Part 1 and the requirements of BS EN 676 Automatic forced draught burners for gaseous fuels.

Packaged burners normally have the following components in a common housing:

- burner head
- combustion air fan
- programming control unit, with ignition and flame detection devices
- two safety shut-off valves
- main gas governor, and pilot gas governor
- pilot solenoid valves
- air proving device
- isolating valves.

A packaged burner is shown in Fig. 34. The programming control unit should meet the appropriate standards as mentioned above and the safety shut-off valves should meet BS 7461 Specification for electrically operated gas shut-off valves with adjusters or gas flow control.

*Fig. 34    Packaged burner (Nuway Ltd)*

**Control System**

A typical control layout for a packaged burner is shown in Fig. 35. The sequence is monitored by the programming control unit and typically consists of the following operations:

1.  The air proving device is proved in the 'no air' position prior to start up, otherwise start up will be prevented or lockout will result.
2.  A pre-purge period of proved air pressure of at least 30 seconds, at the full combustion air rate. Five air volume changes of the combustion chamber must be given.
3.  A safe start check. Flame simulation in the pre-purge period must cause lockout.
4.  A start gas flame ignition period of no longer than 5 seconds. This is the time when the spark and the pilot gas valve are both energised, and when no check is made for the presence of the flame.

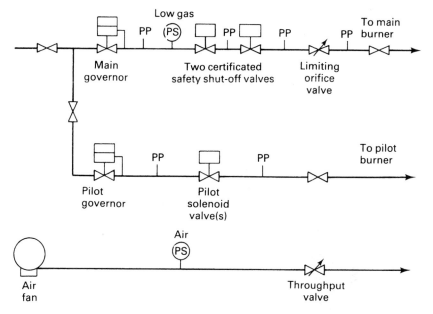

*Fig. 35    Control layout for forced or induced draught packaged burner*

5.  A start gas flame proving period of at least 5 seconds. This follows immediately upon the previous period, and the pilot flame must be sensed throughout this period. During this time the pilot flame is alight without the spark.

6.  The main flame ignition period, of no longer than 5 seconds. This is the time when the pilot burner should be lighting the main burner, and therefore main and pilot valves are open. Again, the flame must be sensed throughout this period.

7.  The main burner run period. The pilot valve is closed at the end of the previous period, and hence the main burner is alight alone. The flame must be sensed throughout this period. In modulating systems, this period may be split into the main flame proving period, after which the burner is switched to full modulation control.

8.  A post-purge period. This is optional, and may come into operation on shut-down or lockout of the burner.

9.  If the flame is lost in either the pilot flame proving or the run position lockout will occur within 1 second. When lockout does occur, this will require manual intervention by an operator to reset the unit to restart the system.

10.  With modulating units:

- the air flow damper for pre-purge must be proved in the high fire position
- the air and gas throughput dampers are proved in the low fire position at start up only.

11.  Loss of air at any time must give either safety shut-down or lockout.

The flame safeguard should be checked at least daily on continuously operated plant by manually shutting the burner down and allowing it to start up again.

The main natural gas safety shut-off valve requirements are based on heat input rates as follows:

- from 60 to 600 kW — one Class 1 and one Class 2 valves
- above 600 up to 1000 kW — two Class 1 valves
- above 1000 up to 3000 kW — two Class 1 valves with a system check
- above 3000 kW — two Class 1 valves with proving.

*Uses*

Package burners are frequently found on small sectional boilers, commercial air heaters and on a multitude of small, low temperature heating applications.

## Large Gas and Dual Fuel Burners

*Large Gas Burners*

Single large gas burners if they are fully automatic are covered by BS EN 676 and BS 5885 standards as for packaged burners. However, they do have additional requirements with regard to the control system and the requirements of the safety shut-off system.

Whilst most large gas burners will have a pilot burner, some will be ignited using an expanded main flame technique. Here part of the main burner which will smoothly ignite the remainder is supplied with the gas from what would have been the pilot line. Thus ignition takes place on the main burner rather than at a pilot burner. This small section of the main flame stays ignited after the remainder of the main flame has ignited.

The safety shut-off system required for single automatic gas burners between 1000 kW and 3000 kW is two Class 1 valves and a system check. The system check on this type of burner is the incorporation within the safety shut-off valves of a closed position indication switch. This is a switch that is activated when the gas valve is within 1 mm of being fully closed or nearly closed so that no more than 10% of full flow of leakage occurs.

Above 3000 kW the requirements are more stringent and no leakage is allowable. This requires proof of closure or valve proving, which must be capable of detecting a leak of 0.05% of maximum rate of the burner or 300 litres/h, whichever is smaller.

The first proving system shown in Fig. 36 uses two approved normally closed safety shut-off valves to close the fuel line and a third normally open valve to vent the space between them to the atmosphere. On opening the safety shut-off system the vent valve will be proved closed before the block valves are energised. On closure all valves are de-energised simultaneously, Fig. 36.

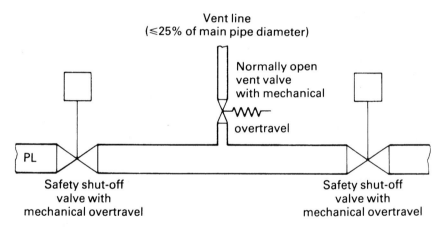

Fig. 36    *Double block and vent system*

Valves with overtravel characteristics are plug, ball or some types of gate valve. They are fitted with proof of closure switches which will initiate lockout if the valves are not in the correct position.

Before the burner start sequence can commence the two normally closed block valves must be proved closed and the normally open vent valve must be proved open.

The vent pipe bore should not be less than 25% of the main pipe diameter or 15 mm, whichever is the greater.

The second system checks the valves for leakage so mechanical overtravel with position proving switches is not required. The systems in common use include:

- vacuum proving
- pressurising with an inert gas such as nitrogen
- opening and closing the valves in sequence using line gas proving pressure.

In the vacuum proving system, Fig. 37, the space between the two block valves and the normally open vent valve which is made to close, is evacuated by a vacuum pump. The required vacuum should be reached in a set time interval. This checks for a large valve leak. The pump is stopped and a proportion of the vacuum must be maintained over a second time interval. This checks for small valve leaks.

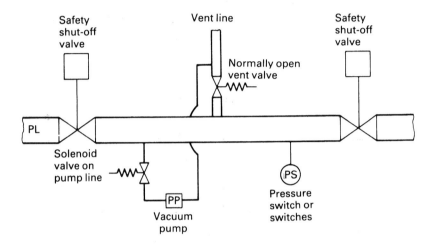

*Fig. 37   Vacuum proving system*

When pressurising with an inert gas the nitrogen, for example, is admitted into the space between the two block valves and the normally open vent valve, which has been made to close. A set pressure must be achieved in a given time interval to prove that the valves are reasonably tight, Fig. 38.

The sequential proving system first closes the vent valve and monitors the pressure in the space between the three valves over a set time interval. If the pressure rises above atmospheric, the upstream valve is leaking and the system will lockout. If this part of the check is satisfactory the upstream block valve (or a bypass) is opened and then closed. This admits line gas pressure into the space between the three valves. The pressure is monitored for a set time interval and, if it drops substantially it indicates that the downstream valve, the vent valve or the valve flanges are leaking, Fig. 39. Vent valves on pressure proving systems should have a port diameter of not less than 6 mm.

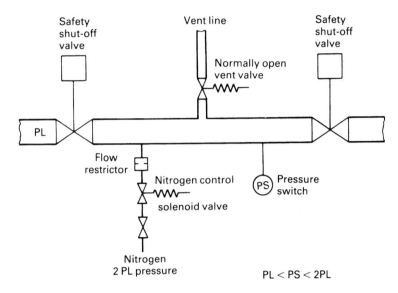

*Fig. 38    Nitrogen proving system*

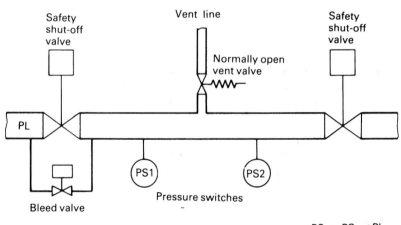

*Fig. 39    Sequential pressure proving system*

Other requirements for large single gas burners are:

1.  The whole of the pre-purge period must be proved to be in the high air flow condition. This is monitored by a position switch on the air damper or by an air pressure switch downstream of the air damper.

2. Before the main flame establishing sequence can begin, the air and gas dampers must be proved to be in the low fire position. This is done either by damper position proving switches or by air and gas pressure switches.

3. High gas pressure protection is required on systems above 3 MW, or where the plant governor inlet pressure exceeds 75 mbar, where the pressure drop across the governor is more than 30% of the normal minimum operating outlet pressure.

   Low gas pressure protection is only required if the burner is unstable under lean gas conditions.

4. It is recommended that self-checking flame safeguards be fitted. Where these are not incorporated the burner shall be shut down at least once per day to check the operation of the detector.

5. If the start gas is greater than 10% of the main gas rate then two Class 1 safety shut-off valves are required on the start gas supply.

Where a group of gas burners is applied to a low temperature plant or where the gas burners are not fully automatic then the installation should comply with IGE/UP/12.

*Dual Fuel Burners*

When the energy requirements of an organisation are large, the customer may have an interruptible tariff agreement. This enables him to purchase gas at an advantageous rate, provided that his plant can operate on an alternative fuel (usually oil), when gas is in great demand, for example, during the winter months.

The burners are principally constructed of components common to both gas and oil systems. These common items are:

- air fan and air manifold
- programming control unit
- ignition transformer and spark plug
- flame detection system
- pilot burner.

The burner heads and fuel control systems are different. Gas burners are usually nozzle mixing, having a metal fabricated burner head.

The oil burner is usually a spinning cup to accept 950 seconds oil, Fig. 40. Pressure jet burners are used when the standby fuel is light oil of 28 or 35 seconds viscosity.

The standby oil system should be adequate for at least one month's running. Light oil is generally used but heavy oil may be found if it was in use before gas firing was introduced.

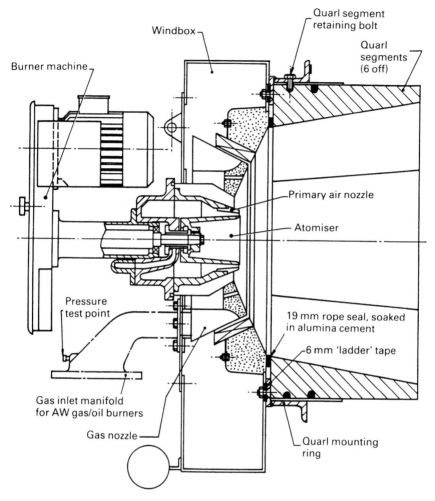

*Fig. 40    Dual fuel burner (Hamworthy)*

When the system is switched from one fuel to another, one of the following sequences is employed.

1.  *Full restart switchover*. This entails the complete interruption of firing on the one fuel and the full start up sequence on the second fuel, including the pre-purge.
2.  *Piloted switchover*. This requires establishment of the pilot, the interruption of the main flame firing and the restart of the main flame on the second fuel for each burner in turn. Purging is dispensed with because of proved continuity of the flame.

3. *Dual valve switchover*. This system uses a pair of linked valves, one in each fuel line, separate from the main throughput control valves. The valves are arranged so that, as one opens, the other closes. Both valves are set at a fixed plant load to give a constant total thermal input.

4. *On load switchover*. Switchover is carried out while maintaining the plant under full process control. This also involves the simultaneous reduction of one fuel input rate with a corresponding increase in the other fuel rate while maintaining the total required thermal input.

5. *Sequential shut-down switchover*. This entails the complete interruption of firing on one fuel and restarting on the second fuel, excluding the purge, for each burner in turn. The system should only be used on multi-burner plant.

A typical control train for large burners is shown in Fig. 41.

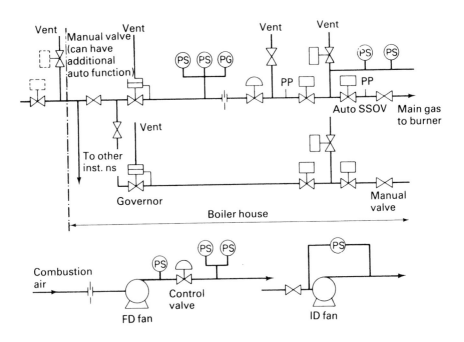

*Fig. 41   Control layout for burner system over 3 MW*

*Uses*

The most popular use for these burners is on large boilers, for example, Economic, Lancashire and water tube boilers. They are also applied where large heat input rates are required on low and medium temperature plant.

## Low NO$_x$ Burners

The formation of oxides of nitrogen (NO$_x$) is discussed in Chapter 13. The importance of NO$_x$ as a pollutant has been widely recognised by the burner manufacturers and this has led to a new breed of low NO$_x$ natural gas burners for cold air and warm air firing. The hot air burners are used extensively as self-recuperating burners or regenerative burners.

The techniques used to achieve low NO$_x$ include partially premixing the natural gas and the air and staging the combustion. The production of thermal NO$_x$ depends upon the temperature and oxygen levels. Fig. 42 shows the effect of temperature and Fig. 43 shows how the concentration of oxygen affects the production of NO$_x$. It is important to remember that the general higher temperature levels reached in industrial furnaces is 1,500 – 1,600°C when studying these two Figures.

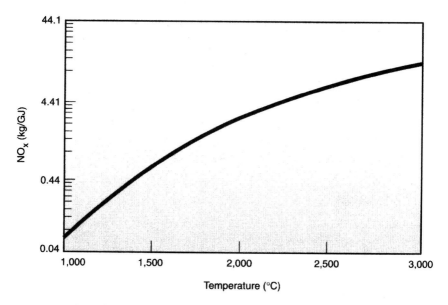

*Fig. 42    The effects of temperature on NO$_x$ formation (DETR Best Practice Programme Guide 252)*

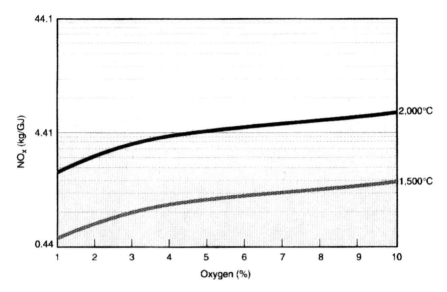

*Fig. 43    The effects of oxygen levels on NO$_x$ formation (DETR Best Practice Programme Guide 252)*

Burner manufacturers use staged combustion and sometimes flue gas recirculation to lower NO$_x$ levels. Also if nitrogen is reduced or eliminated from the combustion air, for example, by using oxygen instead of air, the NO$_x$ values will be reduced.

Reducing NO$_x$ after combustion is much more difficult and expensive. Two methods used are Selective Catalytic Reduction (SCR) and Selective Non-catalytic Reduction (SNCR). SCR is mainly used on clean industrial applications where ammonia is injected over a catalyst bed and the chemical reaction reduces NO$_x$ to nitrogen. Reduction rates up to 80 – 90% can be achieved. SNCR techniques can reduce NO$_x$ by 60 – 90% and requires higher temperatures than SCR, but no catalyst.

## CONTROL DEVICES

### Introduction

To ensure safe and efficient combustion, gas and air supplies must be adequately and accurately controlled. The devices used on industrial and commercial equipment include:

- cocks and valves
- governors

- non-return or back-pressure valves
- low pressure cut-off valves
- relay valves
- solenoid and safety shut-off valves
- pressure switches
- flame traps
- fusible links.

The list excludes flame protection equipment which is dealt with in the next chapter. Time switches and thermostats may also be used.

A number of the devices listed have domestic applications and were described in Vol. 1, Chapter 11.

## Cocks and Valves

IGE/UP/2 publication is helpful in selecting manual valves for various applications.

Manually operated valves used on industrial equipment include:

- plug valves
- butterfly valves
- gate valves
- ball valves
- diaphragm valves
- disc-on-seat valves
- needle valves
- proportioning valves.

### Plug Valves

These are used in the form of quadrant cocks and thumb cocks. They are used for isolation as well as the control of flow. Heavy duty lubricated plug valves will withstand pressures up to 7 bar. The flow through the valve is linear, that is, the flow is directly proportional to the angle of opening.

A cross-section of one type of plug valve is shown in Fig. 44.

### Butterfly Valves

These valves are generally used to control low pressure air or gas flows. They have large port openings and so allow high rates of flow. Where complete shut-off is required it is essential that a valve with a soft, neoprene-lined seating is used, Fig. 45. This valve can withstand pressure up to 7 bar and may be inserted between flanges in a pipe run.

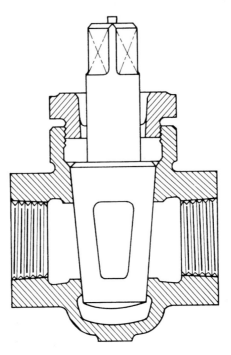

*Fig. 44    Plug valve*

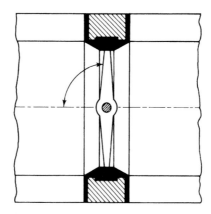

*Fig. 45    Butterfly valve; soft seating*

*Gate Valves*

This type of valve is usually used for isolating purposes. Because the gate of the valve is lifted completely out of the pipe run when fully open, the valve gives full flow with minimal pressure loss. A parallel slide valve is shown in Fig. 46. Gate valves are slow acting and require several turns of the wheel to close the valve completely.

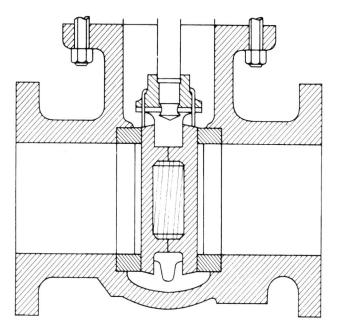

*Fig. 46    Gate valve; parallel slide*

## Ball Valves

These valves, Fig. 47, may be used either for isolation or for through-put control. With nylon seals the valves can withstand pressures up to about 7 bar. Like the plug valve, the ball valve has mechanical overtravel. That is, it moves beyond the pneumatically sealed position to a mechanically closed position. When operated automatically it can be used for such controls as the double block and vent system.

## Diaphragm Valve

The valve is closed by depressing the flexible diaphragm until it makes contact with the seating. Further compression of the diaphragm ensures complete closure, even when grit or small particles in the fluid lodge under the diaphragm. These valves offer very good control of gas flow with their fine progressive adjustment, Fig. 48.

Because the moving parts of the valve are above the diaphragm, they do not come into contact with the fluid. So the valves may be

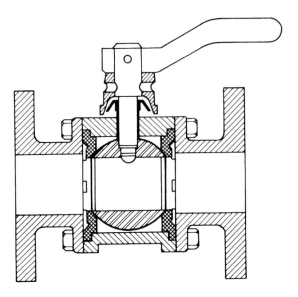

*Fig. 47    Ball valve*

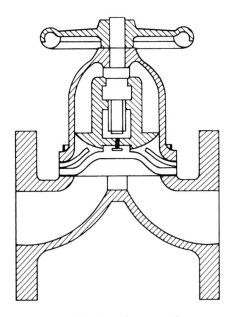

*Fig. 48    Diaphragm valve*

used to control corrosive liquids, provided that these are compatible with the diaphragm and the valve body.

## Disc-on-seat Valves

As their name suggests, these valves have a non-metallic disc which closes on to a knife-edged seating. The seat can be either in the vertical or horizontal position. A typical valve is shown in Fig. 49. These valves are used for isolating purposes only and the same disc and seat arrangement is found on automatic valves such as the solenoid, safety shut-off and slam shut valves.

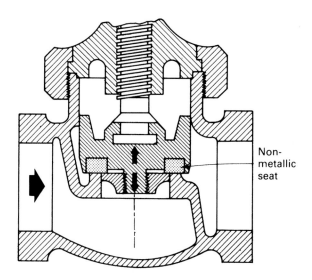

Non-metallic seat

*Fig. 49    Disc-on-seat valve; vertical seat*

## Needle Valves

Needle valves, Fig. 50, are used where fine adjustments of low gas rates are required. The handwheel withdraws the tapered 'needle' out of the similarly tapered seating to increase the flow rate. The valves can be obtained to withstand pressures of up to 28 bar.

## Proportioning Valves

There are two main types of proportioning valves:

- adjustable port valves
- adjustable flow valves.

*Adjustable port valves*, Fig. 51, consist of mechanically linked plug valves on a common spindle or separate valves with connected valve levers. Rotation of the valve spindle controls the throughput. Adjust-

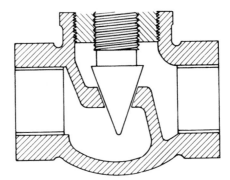

*Fig. 50    Needle valve*

ment of the valve slides controls the proportioning by altering the height of the rectangular ports to vary the relative port area ratio.

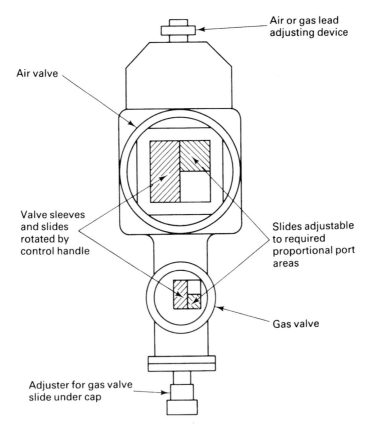

*Fig. 51    Adjustable port valve*

Provision is made for adjusting the position of the spindles at which each valve opens or closes.

*Adjustable flow valves* are illustrated in Figs. 52 and 53. On both the valves shown the flow rate is varied by the rotation of a cam which alters the amount of port opening. The cam is in the form of a spring which can be adjusted to the required profile by means of a series of screws. In Fig. 52 the cam is rotated to move the valve spindle, whilst in Fig. 53 the cam is stationary and the rocker arm is traversed around it.

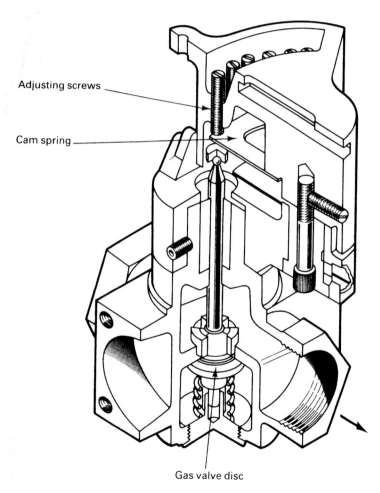

*Fig. 52    Adjustable flow valve; micro-ratio*

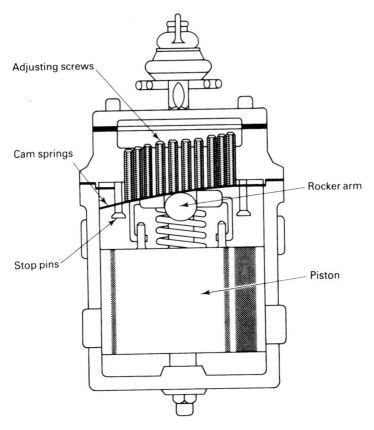

*Fig. 53   Adjustable flow valve (Stordy)*

The cylindrical piston in Fig. 53 has a rectangular port which is moved up or down by the cam as the piston is rotated, so varying the opening.

## Governors

*Low Pressure Governors*

Gas supplies to industrial or commercial equipment are normally controlled by low pressure governors. These were described in Vol. 1, Chapter 7.

The type used is the double diaphragm compensated constant pressure governor. It usually has neoprene main and auxiliary diaphragms, a valve covered with nitrile rubber and spring loading.

The governor is usually the first device in the control train downstream from the isolation valve. A small governor should also be included in the pilot gas supply line.

When volumetric control is required the constant pressure governor is used with back loading from downstream of a valve or orifice. The governor responds to the differential pressure across the orifice and maintains a constant volume.

*Zero Governors*

When gas is to be entrained by a stream of air under pressure, its pressure must be reduced to zero gauge pressure, that is, atmospheric pressure. The device used is a zero governor.

This is similar to the constant pressure governor but it has a thin, flexible tension spring in place of the normal stout compression spring, Fig. 54. The tension spring supports the weight of the moving parts, that is, the diaphragms, valve and valve spindle. Small changes in the outlet gas pressure may be obtained by adjusting a nut at the top of the spring support.

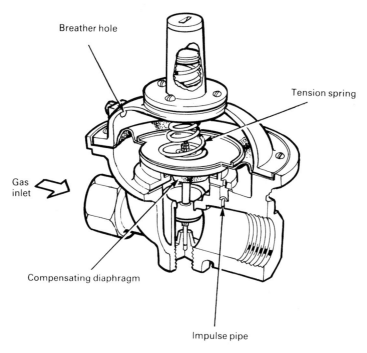

Breather hole

Tension spring

Gas inlet

Compensating diaphragm

Impulse pipe

*Fig. 54   Zero governor*

Because the top of the main diaphragm is exposed to the atmosphere, via the breather hole, and the spring supports the moving parts, then the downward force acting on the diaphragm is atmospheric pressure. This is balanced by the upward force of the outlet pressure below the diaphragm. So any increase above atmospheric pressure downstream of the valve will cause it to close, whilst any fall in outlet pressure will result in the valve being opened by the pressure of the atmosphere on the main diaphragm.

Zero governors are used in air blast systems and should be installed downstream of a constant pressure governor. They are also used in the pressure divider system of control for nozzle mixing burners. In this case the outlet pressure is not zero but the same pressure as in the impulse line.

### Non-return Valves

A non-return valve, sometimes called a back-pressure valve, is designed to allow fluids to pass in one direction only. Any reversal of flow closes the valve instantly.

It is a requirement of the Gas Act 1986 that, where air from a fan or compressor, or any other type of gas, is used in conjunction with a gas burner, a non-return valve must be fitted in the main gas supply. This is to prevent air or any other extraneous gases being admitted into the gas service pipe.

The maximum permissible reverse leakage rates are shown below in Table 1.

The valves must be capable of withstanding a minimum reverse pressure as follows:

- 7 bar on 25 mm valves
- 2 bar on valves 25 to 150 mm
- 1 bar on valves above 150 mm.

TABLE 1 Maximum Permissible Reverse Leakage Rates For Non-return Valves

| Size (mm) | Leakage rate ($dm^3/h$) |
|:---:|:---:|
| 25 | 1 |
| 40 | 2 |
| 50 | 3 |
| 65 | 4 |
| 80 | 6 |
| 100 | 10 |
| 150 | 20 |
| 200 | 48 |
| 250 | 60 |
| 300 | 85 |

Where the air fan is situated above the burner, it is good practice to insert a non-return valve in the air line to prevent gas from entering the air fan casing.

Typical preferred non-return valves are shown in Fig. 55. In the valve at (a), gas entering lifts the two leather diaphragms off their seats on the spring loaded valve head so allowing gas to flow to the outlet. In the event of reverse pressure the diaphragms return to their seating, preventing any return flow past the valve. If the return pressure is increased, the complete metal valve is forced down on to its seating against the spring pressure. This provides an additional seal to withstand the higher back pressure.

The valve must be mounted horizontally.

The valve at (b) is a simple, flap type, disc-on-seat valve with a cast iron body and a knife edge valve support pivot. The valve seat is self-aligning.

The valve must be mounted horizontally, with the dome uppermost.

The valve at (c) has a pair of nitrile rubber diaphragms which touch around the periphery. In forward flow, gas forces the diaphragms apart but under reverse pressure the lips of the diaphragms are forced together so preventing gas flow through the valve.

The valve can be mounted in any position.

If the electronic flame protection system is employed and the safety shut-off valve is able to withstand twice the maximum pressure of the extraneous gas or air, applied in the reverse direction, this would meet the requirements of the non-return valve.

Non-return valves should be periodically checked for correct operation and pressure points are incorporated on the inlet and outlet for this purpose. Leather diaphragms and valve seats should be regularly dressed with an appropriate oil to keep them supple.

British Gas plc Publication IM/14 Standard for Non Return Valves gives additional information about the more detailed requirements of non-return valves.

## Low Gas Pressure Cut-off Systems

The object of these systems is to ensure that, when the gas pressure falls to a predetermined value above atmospheric pressure, a valve closes and cannot be reopened until:

- all downstream burner valves are closed
- the system is restarted manually, for example by depressing a button.

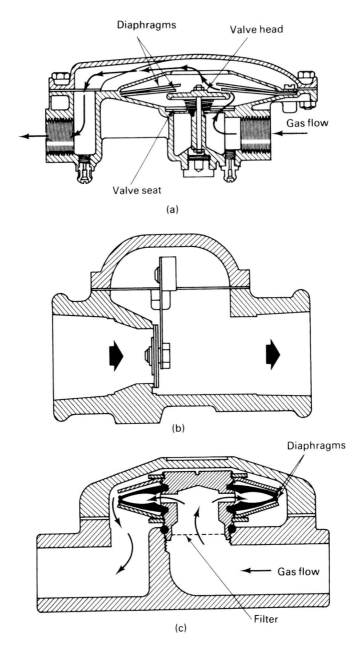

*Fig. 55  Non-return valve: (a) double diaphragm; (b) roll check valve (Donkin); (c) Adaptogas*

*Low Pressure Cut-off Valves*

The valve originally used for this purpose was the low pressure cut-off valve described in Vol. 1, Chapter 11. A typical valve is shown in Fig. 56. This has an auxiliary diaphragm so that inlet pressure does not exert any downward force on the valve. It is reset by a weep of gas through the by-pass orifice when the pressure reset plunger is depressed. The size of the orifice is a compromise. It must be small enough to detect a downstream burner valve left open and large enough so that the operator does not have to keep the plunger depressed for too long. These valves are fairly slow in closing and should therefore only be used on small installations. They are now seldom fitted and are being superseded by low pressure switches in conjunction with certificated safety shut-off valves or, in some cases, by flame protection systems.

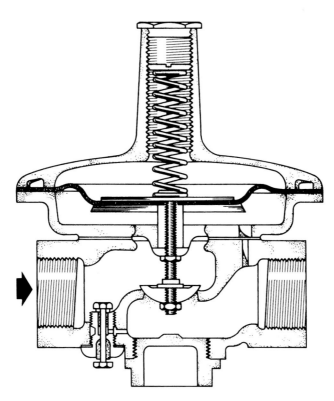

*Fig. 56    Low pressure cut-off valve*

*Low Pressure Cut-off Switches*

A diaphragm operated low pressure cut-off switch is illustrated in Fig. 57. This has a micro switch which is held closed by gas pressure acting under the diaphragm against the set tension of a spring. An earlier type uses a mercury tilt switch, Fig. 58.

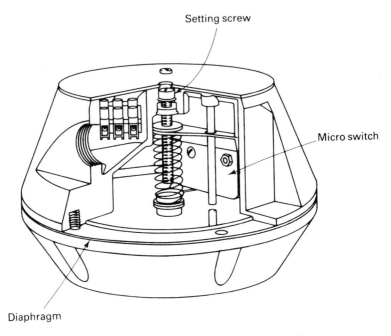

Setting screw

Micro switch

Diaphragm

*Fig. 57    Diaphragm type pressure switch*

Pressure operated switches may be used for high gas pressure protection by arranging for the switch contacts to remain closed until the set pressure is exceeded. They may also be used for low air pressure protection.

*Anti-suction Valves*

Another form of cut-off which has been used to protect gas meters from implosion due to suction from a compressor is shown in Fig. 59. The valve closes when the inlet pressure falls to a set amount above atmospheric pressure. This prevents a suction or reduced pressure reaching the inlet side. The valve is fitted in the gas supply on the inlet of the compressor.

Like other diaphragm operated gas valves, these are being superseded by pressure operated switches controlling safety shut-off valves.

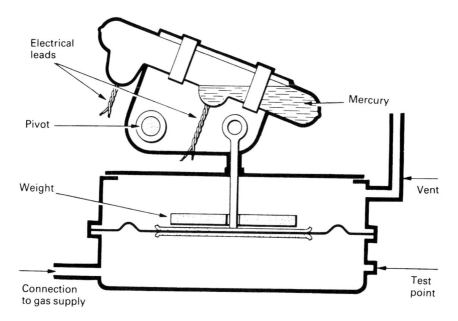

*Fig. 58    Mercury tilt switch*

## Air Flow Failure Devices

The purpose of these devices is to cut off the gas supply if the combustion air supply fan fails.

The most common method of achieving this is to install a pressure switch in the air supply which upon detecting an unacceptable fall in air pressure, then causes a safety shut-off valve to close in the gas supply. It is also acceptable to use a centrifugal switch connected directly to the air fan impeller to prove air. This should only be used on burner systems having a thermal input of less than 2 MW, Fig. 60.

## Combined Gas and Air Pressure Failure Systems

It is common practice to combine gas and air failure protection in one system.

Pressure operated switches in the gas and air lines are wired in series with the safety shut-off valve in the gas line. If any switch opens, the safety shut-off valve will close. A switch will open in the event of:

- low air pressure or
- low gas pressure or
- high gas pressure.

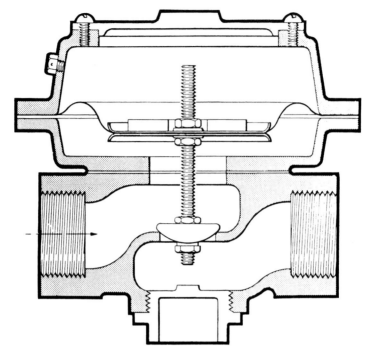

*Fig. 59    Anti-suction valve*

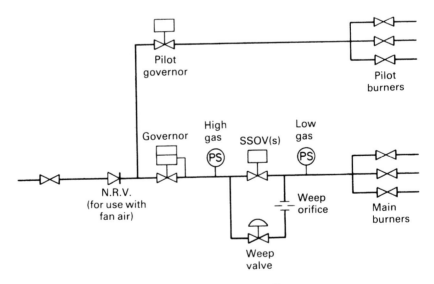

*Fig. 60    Low pressure cut-off system*

The gas supply can only be restored when the air fan is operated and by closing all gas burner valves and depressing the weep valve to actuate the low gas pressure switch. Guidance on the calculation of the size of the orifice used in the weep system is contained in British Gas plc Publication IM/20 Weep By-Pass Pressure Proving System.

On high temperature plant operating at temperatures of 750°C or above, a low gas pressure protection system should be used if a flame safeguard is not fitted. This is to comply with the requirements of the IGE/UP/12 The safe use of gas-fired process plant.

## Relay Control Valves

The operation of relay valves was described in Vol. 1 and a typical industrial type is shown in Fig. 61. This may be used to control either air or gas flow. Various shut-off valves may be fitted in the weep line, the most common on smaller installations are indirect acting thermostats. Solenoid valves are also used, frequently in conjunction with a thermostat switch or a time clock.

There are three methods of operation:

- on/off
- high/low
- modulating.

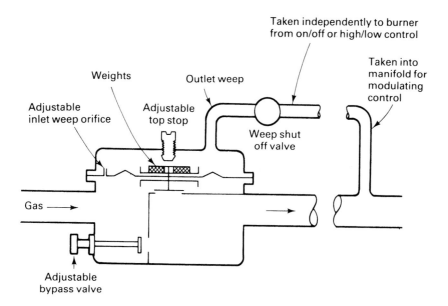

*Fig. 61    Relay control valve*

*On/off operation* takes place with the adjustable bypass valve closed. The weep gas is burned independently adjacent to the pilot burner. The bypass valve may take the form shown or be an adjustable stop below the valve to limit its closing.

*High/low operation* uses the bypass valve to adjust the low-fire rate. If an adjustable top stop is fitted this is used to adjust the high-fire rate. The weep gas is again burned independently adjacent to the pilot burner.

*Modulating control* is provided with the bypass valve closed and the weep gas returned to the gas manifold downstream.

Relay valves provide a very cheap method of proportional control on small burner systems.

## Solenoid Valves

Solenoid valves were introduced in Vol. 1, Chapter 9. They are commonly used for electrically operated flow control and in certain circumstances as safety shut-off valves. Solenoid valves open or close immediately the current is switched on or off.

When used as safety shut-off valves, fast opening, direct acting solenoid valves should satisfy the requirements of BS EN 676 and BS 5885. The valves should be certified by a recognised Test Laboratory to the CE requirements for valves.

The compression springs on large solenoid valves can exert only a low closing force. This makes them unsatisfactory for use where positive closing must be ensured. Also, under certain conditions their fast opening may be undesirable. A sudden surge of gas may extinguish pilot burners.

Because of these characteristics, solenoid valves are more frequently used to control small burner systems or pilot lines on larger burners.

## Safety Shut-off Valves

One of the requirements of BS 5885 'Automatic Gas Burners' is that these valves should be relatively slow opening to ensure smooth reliable ignition of the main flame from the start gas, which has a proving period of less than 5 seconds. There are two types of slow opening valves:

- electro-hydraulic
- electro-mechanical

Both these systems enable the valve to open slowly against the closing force of a strong valve spring to ensure positive shut-off. The valves must close in 1 second or less and have the appropriate CE approval.

## Electro-hydraulic Valve

This valve is described in Vol. 1. An electric motor drives a pump to raise the pressure of oil above a diaphragm which moves to open the gas valve. The valve is held shut by a powerful compression spring. When the current is switched off an electromagnetic relief valve opens allowing the oil to flow quickly back into the reservoir and the valve closes.

## Electro-mechanical Valves

One type has an electric motor driving a rack and pinion through a magnetic clutch. The rack forces the valve open against the tension of a strong spring. When de-energised, the clutch is disengaged and the valve is closed by the spring.

In another type, Fig. 62, the electric motor turns a cam. The cam follower is connected, by a toggle plate and a lever arm, to the valve spindle. A solenoid is incorporated which when energised, provides a fulcrum for the lever arm on top of the latch pedestal. When the valve is energised, the latch pedestal is vertical and the lever arm opens the gate valve against the tension of the compression spring. A limit switch controls the amount of valve travel.

When the current is switched off, the solenoid is de-energised, the latch pedestal moves away, allowing the lever arm to drop and the spring to close the gas valve.

## Multifunctional Controls

These are defined as valves embodying two or more discrete flow control devices including at least one safety shut-off valve, normally Class 1, within a common housing. They are described more fully in Vol. 1, Chapter 11. They may incorporate a governor with one or two safety shut-off valves, which themselves may be fast opening or slow opening or one of each. Some multifunctional controls have by-passes around the downstream safety shut-off valve to enable a start gas supply facility.

Others have an air/gas ratio controller incorporated with the safety shut-off valves within the common housing, Fig. 63.

Figure 63 shows a diagrammatic view of an air/gas ratio controller integrated with a safety shut-off valve to form a multifunctional control. It controls the gas pressure as a function of the pressure of the combustion air so that the gas/air ratio remains constant over the entire output range of the burner. Hence changes in the air volume due to voltage fluctuations on the air fan have no effect. In addition

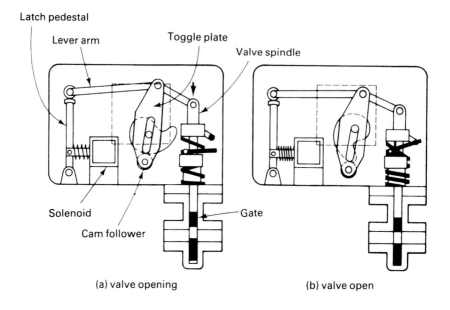

(a) valve opening

(b) valve open

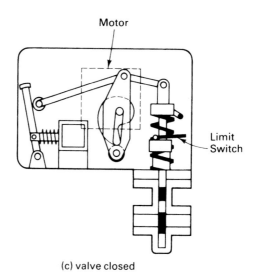

(c) valve closed

*Fig. 62   Safety shut-off valve*

deviations from the correct air/gas ratio can be eliminated from
varying combustion chamber pressure by use of a special impulse
pipe connection.

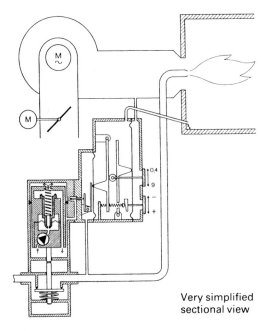

Very simplified
sectional view

*Fig. 63    Multifunctional control with air/gas ratio control (Landis Gyr)*

The electro-hydraulic actuator consists of a cylinder filled with oil
and an electric oscillating pump with piston and relief valve. The
relief valve is open when de-energised and fitted in the bypass
between the inlet and outlet of the oscillating pump. The cylinder
carries a seal which hydraulically separates the inlet from the outlet
side of the pump, and serves, at the same time, as a guidance for the
piston. At the bottom the piston is guided by a rod rigidly connected
to the piston. The rod transfers the travel of the piston direct to the
valve stem.

When the gas valve is closed, only the pressure of the air supplied
by the fan acts on the controller. It causes the air diaphragm to move
to the left and, via the lever system, the gas diaphragm to move to the
left, thus closing the bypass valve in the actuator of the gas valve. The
actuator can therefore open the gas valve if given the appropriate
signal from the flame safeguard. When the gas valve opens, the
pressure downstream of the valve increases immediately and thus the
pressure at the gas diaphragm. As soon as the forces acting on both
diaphragms are in balance (taking the lever ratio into account) the
bypass valve in the actuator is opened to such an extent that the
return flow through the bypass valve and the flow supplied by the

pump are identical. This means that the piston of the actuator and thus the disc of the valve remains in the position reached.

If the heat demand increases and the air damper opens further, or the speed of the fan increases, the controller closes again the bypass valve due to the greater pressure on the air diaphragm so that the actuator opens the gas valve further until the forces acting on the diaphragms are in balance again.

The air to gas pressure ratio and thus the air/gas volume ratio remain constant over the entire output range.

With this valve a separate gas governor is not required and because there is no additional pressure losses a smaller valve size than usual may be used. The gas train is thus shorter and in some respects easier for the installer to fit.

A photograph of this valve is shown in Fig. 64.

*Fig. 64  Safety shut-off valve with air/gas ratio control (Landis Gyr)*

## Flame Traps

Where premixed air and gas is used in gas-fired equipment a flame trap should be fitted to prevent a flame from passing back through the pipework to the mixer. A satisfactory flame trap should do two things:

- stop the flame passing through the pipe system
- cut off the fuel supply if a flame reaches the flame trap.

A typical flame trap element, Fig. 65, is built up in a similar way to the ribbon burner from alternate strips of flat and corrugated steel ribbon, wound in a spiral. It has the effect of breaking up the flame over the large surface area and quenching it in the very small passages between the corrugations.

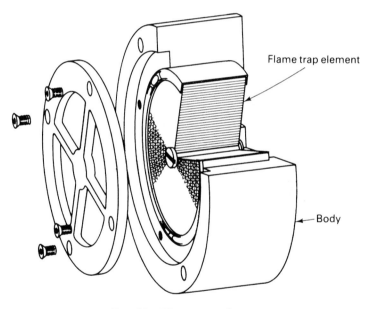

Fig. 65    Flame trap element

Flame traps should be situated as near as possible to the gas burner. This is so that the flame does not have a long pipe run in which it might accelerate to such a speed as to form a detonation wave and make the trap useless.

If a flame was allowed to continue burning against the element it might heat the trap until the mixture ignited on the upstream side. To avoid this thermocouples can be placed on the burner side which will sense the presence of a flame and close a gas valve via temperature detector contacts.

Flame traps are used in other situations including:

- pressure relief systems and governor vents
- purge outlet pipes.

**Fusible Links**

These are a form of limit control which may be used to:

- prevent accidental overheating
- protect premix systems against lighting back
- control a process by shutting off the gas when a given temperature is reached.

The devices may be either mechanical or electrical. Mechanical devices may employ the link as part of a chain supporting a weight which, when released, closes a gas valve. Another type holds a gas valve open against the tension of a spring.

Electrical control may be effected by fitting a fusible link as a thermocouple interrupter. When the link melts the circuit is broken.

Because softening occurs before the melting point is reached the links are liable to cause premature shut-down and have largely been superseded by other more accurate temperature controls.

Fusible plugs are often found above oil-fired burners on boilers and on some dual fuel burners as relates to BS 799. There is an obvious location for a fusible link above an oil burner since if oil spillage occurs and a fire results, this is likely to be at the burner, whereas any leakage of gas from the system could be anywhere in the pipework system, not necessarily at the burner.

When changing over a solid fuel boiler to gas firing any fusible plug should be removed and a low water alarm fitted (Chapter 12).

**Servicing**

To maintain a gas-fired plant in operation at its optimum performance and to ensure that all the safety features in its control system are always effective, planned, regular servicing is essential. A routine programme should be designed so that equipment may be serviced when it is least likely to be needed for production.

Where equipment has been in use for several years it should be updated to incorporate the appropriate requirements of relevant Standards and Codes of Practice during periodic servicing.

The items of plant to be serviced include:

- burners
- fans and boosters
- manual valves
- safety shut-off valves
- non-return valves
- governors
- throughput and mixture controls
- flue gas controls
- flame protection equipment
- electrical equipment.

## Burners

For burners where the air/gas ratio has been correctly maintained, very little servicing is normally required.

On natural draught systems the injectors and burner ports should be checked to ensure that they are not clogged with dust which would reduce the primary aeration. Auxiliary flame ports on bar or ring burners should be free from deposits or scale which could render them ineffective.

On forced draught systems, make sure that burner quarls are in good condition and that no loose refractory material obstructs the entrance to the combustion chamber. Check all flames visually for stability and correct flame profile.

## Fans and Boosters

These items of equipment should be checked in accordance with manufacturer's instructions.

Lubrication should only be carried out if specified and with the correct oil or grease. Check for excessive noise or vibration and ensure that the mountings are secure and in good condition. Check that the air fan inlets are not obstructed by dust or dirt and clean out if necessary. Ensure that all glands and seals on gas boosters are sound and that there is no smell of gas in the area. Ensure that warning notices are correctly displayed.

## Manual Valves

All manual isolating valves should be pressure tested annually to ensure that they are leak tight. This is done with a gauge fitted between closed valves. If there is a perceptible rise in pressure the faulty valve should be exchanged or repaired.

The servicing requirements of the various types of valve are as follows:

*Plug Valves*

The low pressure gas and air quadrant cocks and thumb cocks should have the plugs lightly greased to ease their operation. If the valves leak the plug should be lapped in with a fine paste, greased and retested.

Lubricated taper plug valves for use at higher pressures should be filled with a slug of grease periodically, depending on the frequency of their use.

*Ball Valves*

The majority of these valves have nylon ring seals which take up any wear. The servicing required is minimal.

*Gate Valves*

These valves have a tendency to jam in the open position if not frequently used. It is therefore important that they should be checked at six month intervals to ensure that they open and close freely. The gland seal should be tested with leak detection fluid to ensure that there are no leaks around the valve spindle.

*Diaphragm Valves*

These should be checked annually for correct operation. The valves should be dismantled about every three years so that the diaphragm may be examined for wear and replaced if necessary.

*Safety Shut-off Valves*

All certificated safety shut-off valves and normally open vent valves must be pressure tested at least annually or more frequently, depending on how much the plant is used. Single valves should be tested against a downstream manual valve. The space between the valves is vented, then any perceptible rise in gauge pressure between the valves over two minutes indicates the need for further investigation. On some of the solenoid valves, removal of the bottom cover reveals the valve and seating. Any foreign matter should be carefully removed, the valve seat cleaned with a non-fluffy material, the cover replaced and the valve retested. On the electro-hydraulic valves access to the valve and seating is gained by completely removing the actuator, leaving the valve body in the line.

If the actuator fails to open the valve it should generally be returned to the manufacturers for repair. So spare actuators or valves should always be available to maintain the continuity of the production process. It is absolutely essential that identical valves are used for replacement. Often just one digit different in the valve designation will mean that the replacement is wrong.

With two safety shut-off valves in series, as for example on automatic packaged burners, the space between the valves should be vented through the test point and then checked for pressure by a gauge connected to the test point. If the pressure rises above atmospheric, the upstream valve is faulty.

If the test is satisfactory, the space between the valves is then pressurised either by an air pump or with line pressure via a temporary flexible tube connection. When the pressure is established there should be no rise in pressure between the second safety shut-off valve and the final manual isolating valve. If this pressure does rise, the second safety shut-off valve is leaking.

On large burner systems a normally open vent valve is fitted between the two safety shut-off block valves. To check the vent valve, first check the two block valves then energise the vent valve by connecting it to an external electrical supply and pressurise the space between the three valves. If pressure is lost over a period of two minutes, the valve must be repaired or replaced. Finally reconnect the vent valve to its electrical supply and check that it operates in the correct sequence.

*Non-return Valves*

The most widely used valve is the double diaphragm type shown in Fig. 55(a). Because the diaphragms and the valve seat are still made of leather they should be dressed with a light oil of a type recommended by the manufacturer to keep them from drying out in the presence of natural gas. This is carried out by removing the top cover to expose the diaphragms and by removing the large plug beneath the valve to dress the valve seat.

The disc-on-seat flap valve, Fig. 55(b), should have the seat cleaned of any foreign material whilst the rubber diaphragm type, Fig. 55(c), should require very little servicing.

Annually the effectiveness of all non-return valves should be checked. If possible remove the valve from the line and carry out a reverse flow check against a closed valve using a bubble test meter. If there are more than one or two bubbles per minute the valve must be replaced.

*Air and Flue Dampers*

Some sectional boilers with natural draught burners are fitted with safety shut-off valves with a lever which operates the air inlet damper. When the valve is closed the air inlet louvres should be in the closed position. Check that when the valve is open the louvres are in the fully open position. Reset and tighten the air linkage mechanism as necessary.

Check the operation of any flue dampers. Ensure that the damper flap is in good condition and not corroded away inside the flue. Check that the damper can be correctly adjusted and that it is free from dirt and fluff.

*Governors*

Governors normally require minimal servicing. The large governors fitted upstream of the primary meter are the province of specialist staff. Individual appliance governors should be examined to ensure that diaphragms are sound and in good condition. Valves and seatings should be cleaned and any rust or dust deposits removed. Check that impulse pipes are clear, sound and secure. The outlet pressure should be reasonably constant between low and high fire rates.

*Throughput and Mixture Control Systems*

Many low temperature, low thermal input appliances such as vats and tanks, air heaters and sectional boilers are controlled by relay valves. These may give on/off, high/low or modulating control. They should be checked for operation by adjusting the thermostat and noting the response of the burner. Ensure that weeps and orifices are clear, valves clean and diaphragms sound. Check that top and bottom stops for high and low fire rates are correctly set and tightened.

The mixture control on many industrial furnaces is the premix air blast system. Before commencing work it is important to know whether the air/gas mixture is correct at high fire and low fire. This may be done by measuring the oxygen in the mixture, by using a special test burner, by measuring the $CO_2$ from the resultant flame, or it can be judged by flame appearance. If the air/gas ratio needs adjustment then carry out the following procedure. Check that at low fire rate, the zero governor outlet pressure is atmospheric. Adjust the tension spring or service the governor as necessary. Check at high fire rate and adjust the injector obturator screw as necessary to give the required flame. Retighten all locking nuts after adjustments have been made.

Nozzle mix burner control is commonly linked by butterfly valves. Check the linkage between the valves and examine the flame at low and high fire. Adjustment of the top and bottom stops of the valves should be made to give the correct aeration. If a pressure loading system is employed, adjust the zero governor tension spring and the trimming valves to obtain correct aeration at low and high fire rates.

Larger burners with full modulating control use either adjustable port valves or adjustable cam profile valves. In the first system check the port settings and any linkages to ensure that a stoichiometric air/gas mixture is maintained over the turndown range. In the second system, check the tightness of the cam profile, adjusting screws and any wear in the Bowden cable or other linkage between the valves.

Unless the air and gas supplies are metered, the normal check on throughput control equipment is by recognition of the sight and sound of a correctly aerated flame. This ability may be developed as the result of experience.

A more objective test of the completeness of combustion and the efficiency of the appliance is by carrying out a flue gas analysis.

Whenever possible this should be taken at the various burner settings. With natural draught installations the products of combustion measured at points across the flue should contain at least 4% $CO_2$ when measured by an electronic $CO_2$ meter which often incorporates temperature measurement and hence an efficiency figure. Draeger and Fyrite analysers are also used for simple tests. On forced draught systems there should be a minimum of 9% $CO_2$ at all settings with a maximum of 6% $O_2$ which may be checked with an electronic instrument based on the electrochemical principle or paramagnetic principle. For less accurate results a Fyrite instrument can be used. In all cases there should be no trace or only minimal traces of CO when checked by a Draeger tube.

On boilers the maximum permissible CO concentration is 100 p.p.m. for non-shell boilers and 200 p.p.m. for packaged shell boilers measured at the back end of the combustion chamber before the water tubes with a water cooled probe.

*Air/Gas Ratio Control*

It is important to control the air/gas ratio on burners to avoid gas rich firing, the build up of carbon monoxide and the possibility of explosive mixtures developing. One method of achieving this aim is to use a multiple diaphragm differential pressure controller or ratio controller in conjunction with an air orifice valve and a gas orifice valve, Fig. 66.

The differential pressures generated across each orifice is proportional to the square of the flow rate. The air differential is applied

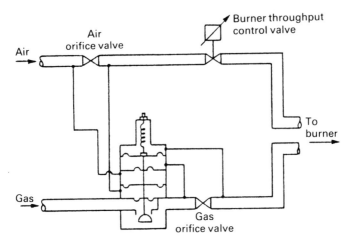

Fig. 66   *Differential pressure ratio control system using multiple diaphragm controller*

across the lower diaphragm producing a downward force to open the valve. The gas differential is applied accross the upper diaphragm producing an upward force to close the valve. The controller reaches equilibrium where these differential pressures are equalised and hence the air and gas flows are maintained in proportion. The controller can also be used in other ways, particularly on larger installations.

Another method of maintaining air/gas ratio is to use an electronic ratio controller, Fig. 67.

In this device, orifices in the air and gas supply lines divert a small fraction of the mainflow around separate by-pass lines. A thermistor in each by-pass line measures the flow, which is compared electronically and any deviation from the preset ratio causes the motorised ratio control valve to restore the set ratio. They can control accurately over 10:1 burner turndown ratios.

*Flame Protection Equipment*

This equipment and its servicing is dealt with in the next chapter.

*Electrical Control Equipment*

All electrical equipment should be checked annually to ensure that all switches and interlocks function correctly. Gas and air flow and pressure switches are particularly vulnerable. Pressure switch diaphragms may rupture, pressure settings vibrate out of adjustment and microswitch contacts weld together. Check that normally closed contacts are closed when the system is at rest. Artificially reduce the

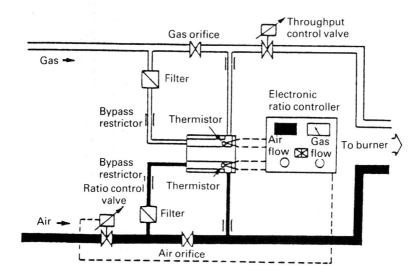

*Fig. 67    Electronic ratio control system*

air or gas pressure being monitored by the switch and check with a gauge that, when the pressure falls below the setting level, the appropriate interlock opens and shuts the system down.

Position proving switches on valves, doors and dampers should also be checked to ensure that, if the device is not in the correct position, the switch will be either open or closed as required by the circuit to provide protection. Similarly, on boilers and water heaters, high, low or extra-low water level switches, steam over-pressure or temperature over-heat limit switches should be checked. A multimeter should be used to establish that contacts are in their correct position and function satisfactorily.

Time clocks should be manually rotated to ensure that the process or heating system switches on and off at the appropriate times. Temperature and over-heat settings should be checked to ensure that the systems operate at the desired temperatures. No system should be run at a temperature higher than that absolutely necessary for the purpose, otherwise fuel will be wasted.

Semi-automatic and fully automatic programming control units should be checked annually by running the system through its sequence and noting the time intervals for each stage of the cycle. Simulate loss of flame and loss of air and gas pressures. The unit must go to lockout and shut down, closing the safety shut-off valves immediately.

If a system fails to start up it is usually an interlock which is at fault, rather than the control unit itself. Check all interlocks first and when these have been proved to be satisfactory, isolate the equipment electrically and remove the plug-in control unit from the fixed base. Thoroughly clean the spring loaded connection contacts between the base and the control unit with fine emery paper if necessary. On most units the chassis pins are self-cleaning and removing and replacing the unit cleans the contacts. If, when replaced, the unit still fails to operate, a new plug-in control unit should be fitted to the base. No attempt should be made to repair the faulty unit which should be returned to the manufacturer for servicing. To ensure continuity of the heating process, a spare control unit should be kept by the customer.

Any flexible metal protection sheathing which is fitted to shield interlocks such as position or pressure proving switches must be made secure. The earth continuity bonding of all controls, interlocks, motors and switchgear should be checked annually, preferably with a Megger.

### General Notes

Since the introduction of natural gas, many processes have been changed over to its use. New burner systems are continuously being developed and new Standards and Codes of Practice devised to ensure that plant is operated as efficiently and safely as possible. A list of current standards and codes associated with industrial gas burners and controls is given below. It is likely that European standards will continually be introduced in the near future and certain British Gas plc Codes will be replaced by IGE documents.

### List of Relevant British Standards and Codes of Practice Relating to Industrial Gas Burners and Controls

| | |
|---|---|
| IM/1 | Guidance Notes on the Use of Oxygen in Industrial Gas Fired Plant and Working Flame Burners |
| IM/9 | Code of Practice for the Use of Gas in Atmosphere Gas Generators and Associated Plant (1977) Parts 1, 2 and 3 |
| IM/10 | Technical Notes on Changeover to Gas of Central Heating and Hot Water Boilers for Non Domestic Applications (1989) |
| IM/11 | Flues for Commercial and Industrial Gas Fired Boilers and Air Heaters (1989) |
| IM/13 | Specification for Pressure Switches for Industrial and Commercial Gas Fired Plant (1980) |
| IM/14 | Standard for Non Return Valves (1989) |

| | |
|---|---|
| IM/15 | Manual Valves – A Guide to Selection for Industrial and Commercial Gas Installations (1989) |
| IM/19 | Automatic Flue Dampers for Use with Gas Fired Space Heating and Water Heating Appliances (1982) |
| IM/20 | Weep By-Pass Pressure Proving Systems (1982) |
| IM/21 | Guidance Notes for Architects, Builders etc., on the Gas Safety (Installation and Use) Regulations (1985) |
| IM/22 | Installation Guide for High Efficiency (Condensing) Boilers (1986) |
| IM/23 | Guidance Notes on the Connection of Electrical Equipment to Gas Meters (1987) |
| IM/24 | Guidance Notes on the Installation of Industrial Gas Turbines, Associated Gas Compressors and Supplementary Firing Burners (1989) |
| IGE/UP/1 Edition 2 | Soundness testing and purging of industrial and commercial gas installations |
| IGE/UP/2 Edition 2 | Gas installation pipework, boosters and compressors on industrial and commercial premises |
| IGE/UP/3 | Gas fuelled spark ignition and dual fuel engines |
| IGE/UP/4 | Commissioning of gas fired plant on industrial and commercial premises |
| IGE/UP/6 | Application of positive displacement compressors to natural gas systems |
| IGE/UP/9 | Gas turbines |
| IGE/UP/10 | Flueing and ventilation for larger gas appliances |
| IGE/UP/11 | Gas in educational establishments |
| IGE/UP/12 | The safe use of gas-fired process plant |
| Report 770/71 | Overheat Protection for Steam Tube Ovens (1972) Health and Safety at Work etc. Act 1974 Gas Safety (Installation and Use) Regulations 1984 Health and Safety At Work Series HS(G)16 Evaporating and Other Ovens Health and Safety At Work Series HS(G)11 Flame Arrestors and Explosion Reliefs |
| BS EN 676 | Automatic forced draught burners for gaseous fuels |
| BS EN 746 | Industrial thermal processing equipment |
| BS 779 | Cast Iron Central Heating Boilers 44 kW and above |
| BS 799 | Oil burning equipment |
| BS 5440 | Code of Practice for Flues and Air Supply for Appliances not exceeding 60 kW |
| BS 5864 | Code of Practice for the Installation of Gas Ducted Air Heaters not exceeding 60 kW |
| BS 5978 | Safety and Performance of Gas Fired Hot Water Boilers (60 kW to 2 MW) |

| | |
|---|---|
| BS 5986 | Specification for Electrical Safety and Performance of Gas Fired Space Heating Appliances (60 kW to 2 MW) |
| BS 5990 | Specification for Direct Gas Fired Forced Convection Air Heaters for Space Heating (60 kW to 2 MW) |
| BS 5991 | Specification for Indirect Gas Fired Forced Convection Air Heaters for Space Heating (60 kW to 2 MW) |
| BS 6047 | Flame Supervision Devices for Domestic, Commercial and Catering Gas Appliances |
| BS 6230 | Specification for installation of Gas fired forced convection air heaters for commercial and industrial space heating, exceeding 60 kW |
| BS 7461 | Specification for electrically operated gas shut-off valves with adjusters or flow controls. |

# Flame Protection Systems

Chapter 5 is based on an original draft by W. A. Pidcock

## Introduction

The purpose of flame protection systems is to safeguard gas-fired equipment from hazard during any phase of its operation. These phases are:

- start up
- normal run
- shut down.

The most critical phase is that of starting up, so, as well as giving protection from flame failure at any time, the system must be linked to the ignition procedure.

Industrial and commercial gas-fired equipment and plant use a variety of flame protection devices. These range from domestic models on the small appliances to industrial models on large equipment. Flame protection systems should possess a number of indispensable characteristics and should generally not be used on industrial equipment unless they comply with the following requirements.

Flame protection systems should:

- ensure that the appropriate lighting up procedure is correctly applied before the burner will light
- either prevent any gas being supplied to the main burner until the pilot flame is established
- or prevent the full gas rate being supplied to the main burner until a flame at a low rate has been established and sensed for a trial period
- stop all gas being supplied to the burners after flame failure and require manual resetting; unprotected pilots should not be used with devices for industrial applications

- be adjusted to sense only that part of the flame which will ignite the main burner and not respond to any other flame or a flame simulating condition
- in the case of electronic devices, be provided with a 'safe start check' to prevent energising the gas valves and the ignition if a 'flame on' condition exists before ignition
- be of fail safe design
- thermoelectric types should ensure that the main gas valve is manually isolated until the pilot is established
- be mechanically and electrically sound and readily serviced
- when correctly installed, be free of any tendency to fail to danger.

In addition to these essential characteristics there are a number of other features which are desirable.

Flame protection systems should, if possible:

- be protected against interference by unauthorised persons
- operate satisfactorily under all changing conditions of

  - throughput
  - flue draught
  - mixture ratio
  - gas characteristics

- operate satisfactorily within the ambient temperature range to be expected on the appliance
- where powered by mains electricity, operate satisfactorily within supply voltage variations of +10% and −15% of the nominal rating
- be unaffected in its operation by foreign matter
- tolerate reasonable vibration and shock
- where it includes a gas carrying component, pass the required volume of gas within the permitted pressure loss.

## Ignition Devices

Because flame protection systems are linked to the appropriate lighting procedures, it is useful to review the methods by which industrial and commercial appliances may be ignited.

These are:

- manual ignition       – by match or taper
                        – by lighting torch
- pilot flame

● electric ignition             – piezo-electric spark
                                     – mains transformer spark
                                     – electronic pulse spark.

Ignition and ignition devices were dealt with in Vol. 1, Chapters 2 and 11 respectively.

*Manual Ignition*

Hand-held matches or tapers are used on domestic appliances and may be used on some of the smaller commercial appliances. For industrial appliances the lighting torch is recommended. This is typically a length of 6 mm steel pipe connected to a piece of flexible tubing. The length of the torch must enable the operator to ignite the burner while within easy reach of the burner control valve and whilst viewing the main burner. The gas rate of the torch must not be more than 7.3 kW or 3% of the main burner gas rate, whichever is the greater.

*Pilot Flames*

Pilot flames may be:

● continuous burning
● intermittent
● interrupted.

A continuously burning pilot remains alight while the appliance is in operation, whether the main burner is on or not. Some pilots are increased in size when required to light the main burner, but this is limited to commercial equipment of low thermal input. They may be reduced after ignition or when the main flame is shut off.

An intermittent pilot is lit prior to ignition of the main flame and is shut off simultaneously with it.

An interrupted pilot is lit prior to ignition of the main flame and is shut off when the main flame is established. This is the type of pilot that is nearly always used with industrial burners of the fully automatic type.

All pilot flames should:

● be positioned to provide immediate and smooth ignition of the main flame at any normal gas rates
● be situated to avoid interference from draughts or combustion products
● be provided with access for ignition and facility for viewing.

Where flame protection is provided, the pilot connection should be taken from the main supply downstream of the main inlet valve.

The pilot burner must be securely mounted to prevent movement affecting its operation. The pilot supply should be rigidly fixed to prevent vibration occurring which could extinguish the flame.

A start-gas flame may be established either at the main burner or at a separate pilot burner. The rate for natural gas shall not exceed 25% of the stoichiometric gas rate of the burner to be fired. The energy release shall not be more than 53 kJ/m$^3$ of combustion chamber volume for every 100 mbar pressure rise the combustion chamber can withstand.

### Electric Ignition

*Piezo-electric* igniters are used on some commercial appliances but are generally not used industrially.

*Mains transformer* ignition may be produced by a step-up transformer. This may either have the electrodes connected to each end of the secondary winding and have a central earth tapping or, more commonly, have one end of the winding connected directly to earth and the other connected to the spark electrode. This is shown in Fig. 1. System (a) is generally used because it only requires one electrode.

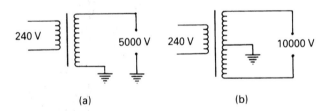

(a)     (b)

*Fig. 1   Mains ignition transformer: (a) electrode connected to one end of secondary winding; (b) electrode connected across both ends of secondary winding*

It is not possible to obtain sparks simultaneously at two gaps, using the centre earth tapped transformer with each electrode sparking to earth. Sparks will be obtained at the shorter gap or at random at both gaps.

Ignition transformers should be mounted near to the burners but protected from heat or damage. The high tension lead should be insulated from contact with other components or people and not run near any flame detector leads. Electrical pick up can cause signals which simulate a flame.

Spark gaps may be from 2.4 to 9.5 mm.

Power input is normally 100 to 200 W.

*Electronic pulse igniters* produce sparks by injecting a high speed pulse of electrical energy into a transformer. This may often be an auto-transformer of the car ignition coil type. The input pulse may be obtained from a solid state switching circuit controlled by a thyristor which triggers repeatedly to discharge a capacitor into the pulse transformer. A single pulse transformer is shown in Fig. 2, but, with the appropriate circuitry, up to 100 transformers can be operated from one pulse generator.

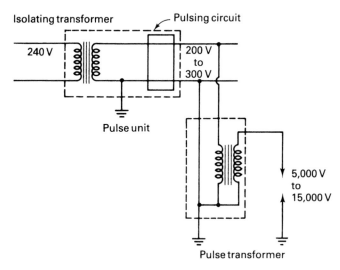

*Fig. 2    Electronic pulse ignition circuit*

Power input for a single spark unit is about 40 W and for multiple units about 2 to 5 W per spark.

Whilst this type of system has been fitted to many low input burners, for example, on industrial continuous baking ovens, the method is not recommended. The British Gas plc Code of Practice on Low Temperature Plant recommends that flame safeguards are fitted to each burner rated above about 7.3 kW.

## FLAME PROTECTION DEVICES

### Types

Flame protection devices may be divided into four main categories, as follows:

- thermal expansion
- thermoelectric
- flame ionisation
- photosensitive.

Of these, the devices in the first category are not used for industrial equipment and generally only the liquid expansion and vapour pressure types are used commercially. The other three categories are used where appropriate for commercial or industrial applications.

**Thermoelectric Devices**

The simple direct acting gas valve type was described in detail in Vol. 1, Chapter 11. These devices are generally more reliable than the thermal expansion types but are much slower to react than the electronic devices. They are relatively cheap and those which operate the gas valve do not require an electrical supply. They are used extensively for low temperature applications and commercial appliances.

When starting up, the push button should not need to be depressed for more than 30 to 40 seconds. When the flame fails, all gas to the pilot and the main burner should be shut off within about 45 seconds. In operation this is usually within 20 seconds. Generally thermoelectric devices should meet British Standard 6047 Flame Supervision Devices for Domestic, Commercial and Catering Gas Appliances and should only be fitted to natural draught burners below 150 kW. However, the 45 seconds drop out time may be extended to 60 seconds for natural draught boilers conforming to BS 5978: Part 2 and air heaters with natural draught burners to BS 5991.

There are two types of thermoelectric devices:

1.  The gas valve is attached directly to the armature of the electric magnet. Thermoelectric valves are fairly slow in operation and, in the 45 seconds allowed between loss of flame and closure of the valve, gas is passing into the combustion chamber. These devices should not be placed adjacent to surfaces with high thermal inertia as this would keep the thermocouple warm for a longer period. They should only be used on low throughput, non-automatic natural draught appliances.

2.  The magnet is used to operate an electric switch which controls a solenoid or safety shut-off valve. This type requires an electrical supply. The second type is less commonly used, electronic devices are usually preferred.

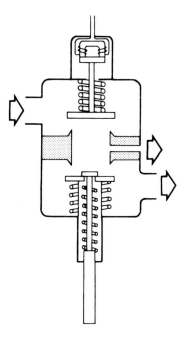

*Fig. 3(a)    Thermoelectric flame protection device; gas valve type*

*Fig. 3(b)    Thermoelectric gas valves (Teknigas)*

## Devices and Systems

The direct gas valve type was described in Vol. 1 and is shown in Fig. 3(a). Fig. 3(b) shows actual valves. Its application to protect a single bar burner is given in Fig. 4.

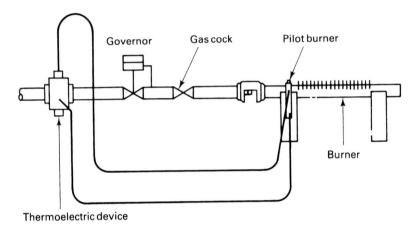

*Fig. 4    Direct protection to a bar burner*

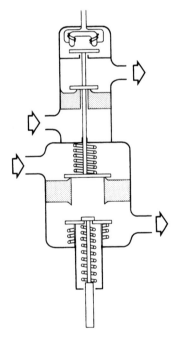

*Fig. 5    Device with separate pilot valve*

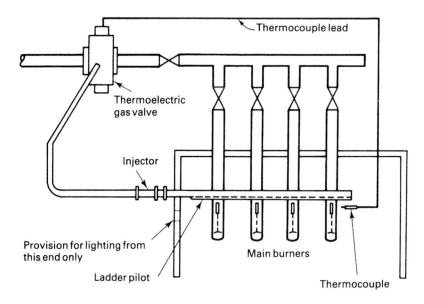

*Fig. 6    Ladder pilot on multiple burners*

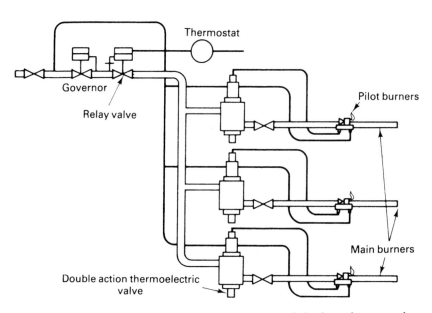

*Fig. 7    Protection of multiple burners by separately fed pilot valve on each burner*

A similar device is illustrated in Fig. 5. This has a separate pilot gas supply controlled by a separate pilot valve independently housed and mounted on the same spindle as the main gas valve and the armature.

Multiple burner systems have always presented problems and two methods of providing flame protection are shown in Figs. 6 and 7.

The first is a ladder pilot system using one direct acting device to supply the main burners and the ladder pilot. This system has the disadvantage that it can fail to danger if the ladder burner becomes blocked over part of its length.

The second system uses individual devices of the separately fed pilot type for each main burner and is a more positive system.

### Installation

When installing thermoelectric devices the following points should be noted:

- the device containing the gas valve should be located downstream of the main gas manual isolation valve and upstream of any controls using a weep line
- the pilot supply should be taken from a point upstream of the main governor
- the device should be located so that an operator can depress the start button and easily light and observe the flame
- it should be fitted so that the gas valve moves vertically downwards when closing, assisted by gravity
- the pilot and thermocouple should be located in accordance with the manufacturer's instructions and shielded from draughts
- the thermocouple should be protected from radiation from sources other than the flame, the bracket should be attached so as to conduct heat away from the thermocouple
- care must be taken to ensure that the correct temperature difference is maintained between the hot and cold junctions, the pilot and thermocouple assembly should not reach a temperature of more than 300°C to avoid overheating the cold junction
- the gas valve and electromagnet assembly should be rigidly mounted to avoid being affected by shock or vibration.

### Commissioning and Servicing

After installation and again when servicing the system, it must be checked to ensure that it will operate satisfactorily at all normal conditions of draught, throughput and mixture ratio change. A pilot turn down test should be carried out to ensure that, when the pilot is turned down to a point at which it will just hold up the main gas valve, it will still light the main burner satisfactorily.

Procedure for the test is as follows:

- turn off the burner valve
- turn down the pilot flame to a point at which it will just hold up the gas valve; each setting should be held for at least 3 minutes, to reach thermal equilibrium
- check visually that the igniting flame is long enough to light the main burner
- turn on the burner valve and check that ignition is smooth and satisfactory
- repeat the test at the extremes of draught throughput and mixture ratio which might normally occur.

A shut down test should be carried out to ensure that the device is left operating satisfactorily. It should also be carried out periodically and might, with advantage, be adopted as part of the normal closing down routine. The procedure is as follows:

- run the appliance up to normal temperature
- turn off the main isolation valve
- note the time taken for the gas valve to shut off by listening or feeling; this should not exceed 45 seconds
- record the time taken when commissioning and check on subsequent tests that the time taken remains substantially the same
- complete the shut down routine by closing burner valves.

Instead of listening for the 'click' when the valve closes it is possible to obtain a visual indication by turning the isolating valve down until only small beads of flame remain on the burner. At this point the pilot will have shortened so that it does not materially heat the thermocouple. Note the time taken for the beads of flame to start to go out, indicating that the valve has closed. Complete the shut down by closing the burner valves and the isolation valve.

Further information on servicing thermoelectric devices will be found in Vol. 1, Chapter 11. Testing of thermocouples is in Vol. 2, Chapter 14.

## Flame Conduction and Rectification Devices

These devices require a 240 V a.c. mains electrical supply. They employ a flame electrode as a sensor and have an electronic amplifier to make it possible for the small current involved to operate a relay. Their advantages over thermal expansion or thermoelectric devices are:

- they detect only a flame and are not actuated by heat
- they have a quick response to the presence of a flame or to flame failure
- they can be positioned to prove the presence of a pilot flame at a point where the main flame must be ignited
- the electrode has a longer working life.

Both systems rely on the ability of a flame to conduct electric current. During combustion, large numbers of free electrons and ions are present in the flame, so the flame acts as an electrolyte in which a current can flow. The ions and electrons are attracted to suitably charged electrodes and currents of about 10 to 12 μA d.c. may be conducted.

*Flame Conduction*

This method used a steady d.c. potential applied between the flame electrode and the earthed burner. The small current conducted was then amplified directly to operate a relay.

The method had a major disadvantage. A build up of dust or condensation on the insulator of the flame electrode can provide a resistive path to earth which results in a current flow indistinguishable from that produced by the flame. Refinements were devised to overcome the problem but it is electronically simpler to use the ability of a flame to rectify a current. Flame conduction systems are not therefore recommended.

*Flame Rectification*

This method is not liable to flame simulation and has superseded the flame conduction device.

It was discovered that if an alternating potential is applied between the flame electrode and the earthed burner nozzle arrangement then an alternating current flow through the flame will result. However, if the earthed portion of the burner nozzle in contact with the flame is larger in area than the flame electrode, then more positive ions will strike the earthed electrode when it is negatively charged than will strike the flame electrode when it becomes negatively charged. The net result is that more current flows when the burner is negative than when it is positive. This causes a bias of current flow in one direction and it is termed rectification. The rectifying action effectively causes a direct current flow that can be amplified and used to operate a power relay.

Any dust build up does not affect the rectification and cannot simulate a flame.

The rectification depends on an adequate area of burner nozzle in contact with the flame, not less than 4 times the area of the flame electrode in the flame. Where this is insufficient the area may be increased by fitting extension rods, a wire spiral or fins to the burner nozzle, Fig. 8. The need for an increase in area is shown by a low microamp reading when checking the flame signal as instructed by the manufacturer.

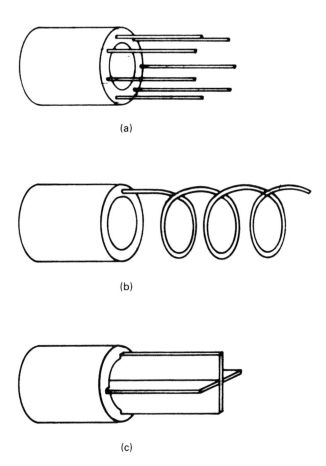

(a)

(b)

(c)

*Fig. 8    Extensions to earthed electrode: (a) extension rods; (b) wire spiral; (c) fin assembly*

*Installation*

The successful application of rectification devices is dependent on:

- the correct location of the flame electrode within a suitable flame (see Fig. 9)
- the provision of an adequate earth surface within the flame.

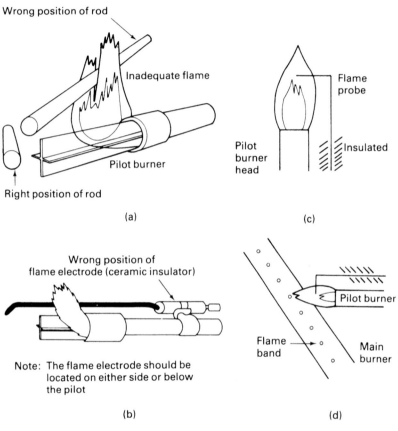

Fig. 9  *Positioning of flame rectification detectors*

The flame electrode should:

- be situated so that, to be detected, the pilot flame must always be of sufficient size to ignite the main burner under all normal working conditions
- be located so that sparks from the igniter cannot arc to the electrode and damage the electronic components
- be made of a material suitable for use at the operating temperature required

- preferably be a short, straight rod, mounted above and to one side of the main burner to avoid problems from the electrode drooping or sagging
- have the insulator located so that it will not reach a temperature above 350°C
- where the lead is subject to high temperatures, employ suitable materials as insulators (glass fibre, P.T.F.E. and silicone rubber may be used up to 200°C).

The installation should be in accordance with current standards and the manufacturer's instructions.

## Commissioning

The installation should be checked to ensure that it meets the recommended requirements. Tests should be made to ensure that the electrode is correctly positioned and that the system operates satisfactorily. When closing down the appliance it is usual to check the operation of the controls by carrying out a shut down test as follows:

- run the appliance up to temperature
- turn off the main isolating valve
- check that the safety shut off valve closes within 1 second or that the flame relay opens within 1 second
- complete the shut down procedure.

## Servicing

Regular servicing is essential to avoid breakdown and ensure safety. Manufacturers' recommendations should be followed. Generally, the following operations should be carried out.

## Weekly

Carry out a shut down test. Also, with the burner firing normally, disconnect the electrode lead and check that shut down occurs within 1 second.

## Monthly

Ensure that the flame electrode is properly located with respect to the main burner and the pilot burner.

## Six-Monthly

Clean the flame electrode insulator and check that the lead is in good condition. Check that the burner nozzle earth electrode material, if fitted, is in good condition. Replace any defective parts. Check that flame relay contacts are clean and operating satisfactorily.

*Annually*

Renew flame and earth electrodes and electrode leads.

## Photosensitive Devices

These devices require a 240 V a.c. main electrical supply. They employ a photoelectric flame sensor, sometimes called a 'head' or 'scanner' in conjunction with an electronic amplifier.

They have the same advantage over thermal devices as flame ionisation types and are more easily applied because:

- flame contact is not required and positioning of the sensor is therefore simplified
- no electrodes are required
- there is no deterioration at high temperatures.

The operation of the sensor depends on it receiving radiations from the flame. These initiate a flow of electrons which produces a signal current that may reach 100 µA. The current is amplified to operate a relay.

Although flames generally radiate over wavelengths from infra-red, through the visual range to ultraviolet, gas flames only produce weak visual radiations and flame sensors must therefore be sensitive to infra-red or ultraviolet rays. Fig. 10 indicates the spread of wavelengths over the relative spectrum.

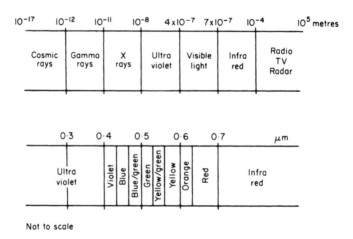

*Fig. 10 Electromagnetic wavelengths*

## Infra-red Detectors

The wavelength of the radiation sensed by this detector is usually from 1 to 3 μm. Because all heated objects emit infra-red radiation, a simple detector cannot differentiate between a flame and a hot refractory surface. However, the radiation from a refractory is relatively steady, whilst that from a flame modulates or 'flickers'. The detector is therefore designed to respond to a modulating radiation and not to a steady output. A lead sulphide cell is commonly used.

Oil flames have strong infra-red emissions and weak UV emissions therefore infra-red detectors are useful on dual fuel applications.

## Ultraviolet Detectors

These detectors respond to radiations with wavelengths of 0.19 to 0.3 μm. At these wavelengths the energy emitted from a flame is normally much higher than that from a hot or incandescent surface which radiates at 0.7 μm or above. Sensors which operate in this narrow wave-band are therefore readily able to discriminate between the flame and surrounding refractory surfaces.

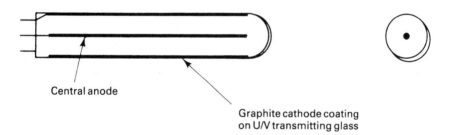

Central anode

Graphite cathode coating
on U/V transmitting glass

*Fig. 11    Co-axial electrode (Geiger-Müller)*

Early detectors used the Geiger-Müller tube whilst the type in common use is the symmetrical diode. Both types use a gas-filled ultraviolet transmitting quartz glass envelope. The G-M tube is shown in Fig. 11 and has a co-axial electrode structure consisting of a cathode of graphite material deposited on the inner surface of the glass and a central rod-like anode.

The symmetrical diode, Fig. 12, has two wire electrodes placed parallel to each other with a narrow space between. An alternating voltage from 220 to 900 V, depending on the type of unit, is applied across the electrodes. When ultraviolet radiations within the appropriate wave-band strike the electrode which is in the negative half-cycle, an electron is emitted. As this electron accelerates towards the positive electrode it strikes atoms of the inert gas which fills the tube, so

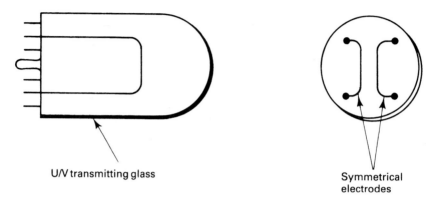

U/V transmitting glass

Symmetrical
electrodes

*Fig. 12    Symmetrical diode*

ionising the gas. This results in an increased discharge between the electrodes and current flows. The current dies away as the voltage drops to zero but the discharge will be repeated when the voltage rises again, if UV radiation is still being received.

The a.c. produced by the tube may be rectified and amplified to operate a control relay.

UV cells are excellent flame detectors but have one basic flaw – they can age. This may happen after 10,000 – 30,000 hours of operation. An aged cell may 'strike' at any time with or without UV radiation. An aged or self-striking cell will continue to signal a flame long after the flame has failed. Hence a safe start check is made by applying a test voltage, considerably higher than normal operating voltage, to the detector during start up and prepurge periods. If the detector does not discharge, the ignition sequence continues. If a discharge occurs, the detector is faulty and the system goes to lock out. Fig. 13 shows UV detector heads.

UV detectors can be made self-checking by:

- automatic shutters to cut off the radiation
- electronic cut-outs.

An automatically controlled shutter is periodically interposed about 3 times per second between the detector and the source of radiation to halt the discharge. If the current does not stop flowing, the burner is shut down and the system locks out.

Electronic cut-outs also periodically halt the discharge automatically. Both self-checking devices check for a false flame signal during the operation of the burner.

This type of flame safeguard is frequently fitted to large burners and also used in applications where high security levels are required.

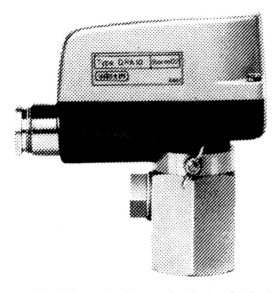

*Fig. 13    Ultraviolet detector head (Landis Gyr)*

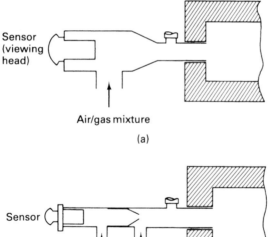

Sensor
(viewing
head)

Air/gas mixture

(a)

Sensor

Air    Gas

(b)

*Fig. 14    Location of ultraviolet flame sensor: (a) in air/gas mixture; (b) in combustion air (viewing in line of flame **not** recommended)*

## Installation

The satisfactory operation of photo-sensitive detectors depends on:

- the correct mounting and wiring of the sensor
- the correct sighting of the sensor with respect to the pilot and main burner flames.

*Mounting* should be carried out in accordance with the manufacturer's instructions. The tube should not reach temperatures of more than 50 to 75°C depending on the type.

Where necessary the sensor can be located so that it is cooled by the combustion air or by the air/gas mixture, Fig. 14. Because fuel gases can absorb UV radiation from a flame the sensor should never normally be mounted to sight through a gas supply.

*Wiring* should be carefully installed with all joints sealed against moisture and leads from the sensor to the control box run separately and not in the same conduit as other wiring to avoid induced signals.

The length of the lead should normally not exceed 15 m and with some types should be considerably shorter. Some of the early symmetrical diode detectors had the flame relay mounted with the sensor in the viewing head.

*Sighting* is critical and the following points are important:

- the line of sight should ensure that the minimum pilot flame detected is of such a size and in such a position that it will always ignite the main burner (Fig. 15)

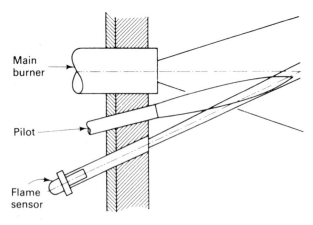

*Fig. 15   Location of ultraviolet sensor (preferred method)*

- UV detectors are sensitive to ignition sparks and a check should be made with all gas valves closed to ensure that the sparks are not detected
- although the UV sensor should never be sighted through fuel gas, it may view the flame through a stoichiometric air/gas mixture up to a distance of about 1 m
- for both infra-red and UV sensors the viewing tube should preferably be of black steel and not stainless or galvanised steel
- the line of sight should be such that the flames are viewed under all normal operating conditions; this may be checked by reading the voltage across the coil of the flame relay in accordance with the manufacturer's instructions.

*Commissioning*

After carrying out normal checks a pilot turn down test should be applied as described earlier. If the main burner fails to ignite within 2 seconds, shut down immediately and resight the viewing head.

When checking shut down, the safety shut off valve should close within 1 second, or if a self-checking flame safeguard is used, the delay may be 2 or 3 seconds depending upon the frequency of the automatic check.

*Servicing*

This should be carried out regularly and frequent checks should be made to ensure that the system will operate satisfactorily and go to lockout if a hazardous situation occurs.

A UV cell usually has a maximum life of 10,000 hours or one year at 50°C. Although the most common fault is for a cell to fail to detect a flame which is present the cell can 'go soft' and detect a flame which is not present. This is a failure to danger, but it should be detected when the cell is checked each time the burner is started up but will not be detected during normal running. Hence the UV cell should be changed after 1 year's operation. Care should be taken to replace the cell with an identical replacement of the same sensitivity.

Automatic programming control units incorporate a safe start check and if the cell senses a flame in the pre-purge period the system goes to lockout.

Semi-automatic units prevent the pilot valve and ignition transformer being energised if the cell has gone soft.

The more frequently units are started up, the more often a check is made on the UV cell. If not shut down daily by a time clock, the system must be shut down manually at least once per day.

## Automatic and Semi-automatic Controls

The basic differences between manual, semi-automatic and automatic control are shown pictorially and by means of a table in Fig. 16. The comments on operator dependency are important. With manual control and to some extent semi-automatic control, there will be an operator present who can detect any problems and take corrective action. With automatic plant, this is not the case and the integrity of the primary controls – flame safeguard unit and the safety shut off valves is extremely important. It is for this reason that British Gas plc has tested and certificated these controls to a high standard.

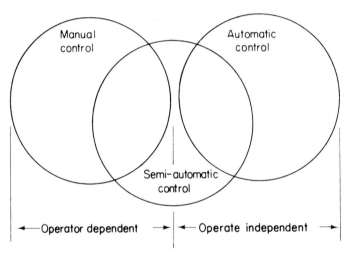

| Stages | Time limits for each stage | | |
|---|---|---|---|
| | Manual | Semi-automatic | Automatic |
| Pilot ignition | — | Unlimited (or preset) | $3\frac{1}{2} \pm 1\frac{1}{2}$ S |
| Pilot proving | — | None | > 5 S |
| Main flame establishment | Unlimited | Unlimited (or preset) | < 5S |
| Flame supervision | Varied | Continuous on pilot (and main flame) | Continuous on pilot and main flame |

*Fig. 16   Comparison of manual, semi-automatic and automatic control*

Definitions of automatic and semi-automatic controls are as follows:

*Automatic Control*

An automatic control system is one which can establish a main flame from the completely shut down condition without any manual intervention. So the main burner may be controlled automatically by a thermostat, timer or similar switching device. Fig. 17 shows fully automatic flame safeguards.

The flame on an automatic forced draught burner is established and controlled by the programming control unit (the fully automatic flame safeguard), in stages as follows:

- pre-purge          – combustion air fan on to clear any combustibles which may have accumulated in the combustion chamber
- pilot ignition     – pilot gas and ignition spark both on
- pilot proving     – detection of stable pilot by flame safeguard with ignition spark off
- main flame ignition – pilot gas and main gas supply on
- main flame proving – detection of stable main flame with pilot gas off (if interrupted pilot is used).

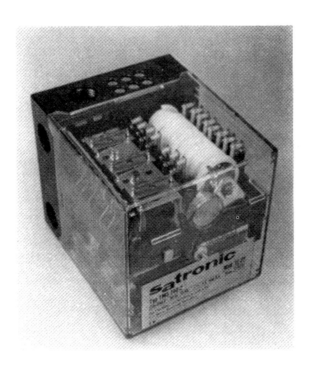

*Fig. 17(a)    Fully automatic flame safeguard (Satronic)*

*Fig. 17(b)    Fully automatic flame safeguard (Landis Gyr)*

| | Pre-purge | Pilot flame establishment | | Main flame establish | Run period |
|---|---|---|---|---|---|
| | | Pilot flame ignition | Pilot flame proving | | |
| Time limits, in seconds | 30 | 2 to 5 | 5+ | < 5 | |
| Thermostat contacts closed | | | | | |
| Air fan on | | | | | |
| Ignition transformer on | | | | | |
| Pilot gas valve open | | | | | |
| Main gas valve open | | | | | |

*Fig. 18    Automatic control unit sequence diagram (on-off operation)*

On shut down, which could be due to the thermostat contacts opening, the programmer returns to the beginning of the sequence in readiness for the thermostat contacts to close again. Fig. 18 shows the time interval for each stage of the sequence.

Most flame safeguards are electromechanical devices and have been used for many years. Because of this experience, the failure modes are well known and various designs have been of the fail safe type.

In recent years, the micro-processor has made enormous advances and is used most extensively on process plant in industry. For safety reasons where flame safeguards have been required, these have been of the electro-mechanical type interfaced with the micro-processor. This is because the failure modes of micro-processors are unpredictable and hence have not been used.

However, there are micro-processor based flame safeguards which have built in features that interrogate themselves, checking various functions and giving diagnostic messages, as well as operation state messages on a screen contained on the flame safeguard, see Fig. 19.

*Fig. 19   Microprocessor flame safeguard (with range of modules)
( Allen-Bradley)*

There are several of this type of flame safeguard available that have been thoroughly tested by British Gas plc and are certificated.

Also annunciator type attachments are available which do not jeopardise the safety or function of the flame safeguard but enable remote readings to be given.

The technology is also available to receive these messages on computers remote from the plant and be operated remotely (except for lockout). Current Codes of Practice require manual intervention at burner lockout.

Another useful feature of the computer connection to the flame safeguard is that it is possible to retrieve, for example, the last, say, six lockout conditions, e.g. flame signal at the time of lockout etc., and analyse thoroughly fault conditions causing these events.

Advice on the use of programmable electronic systems in safety related applications in the gas industry are covered in the publication IGE/SR/15.

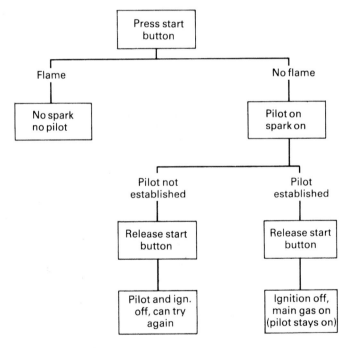

*Fig. 20   Typical light-up sequence for semi-automatic control unit*

## Semi-automatic Control

The opposite to an automatic control is a manual system where the operator carries out each stage of the sequence. The semi-automatic system incorporates elements of the two extremes. It usually provides spark ignition and flame detection but always requires some manual action to bring on the main burner from the shut down condition. This may simply involve releasing a spring loaded start button after the pilot has ignited. The pre-purge may be very short on such units or may rely on the purge being established by some other means.

A typical semi-automatic control sequence is:

- pilot ignition            – operator presses start button, pilot gas and ignition spark on
- pilot proving/main        – operator releases start button, pilot gas main flame ignition    and main gas on, provided that pilot flame has been detected.

The sequence is shown as a chart in Fig. 20.

With both automatic and semi-automatic control, combustion safeguards, such as ultraviolet and flame rectification detectors and supply safeguards, for example pressure switches, are incorporated. These devices give shut down or lockout requiring manual resetting

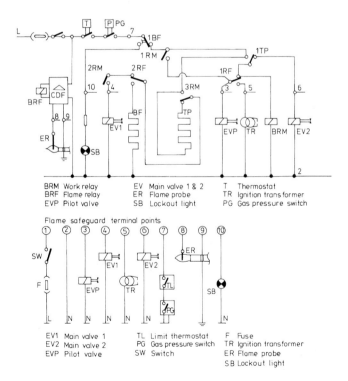

*Fig. 21   Flame rectification wiring diagram*

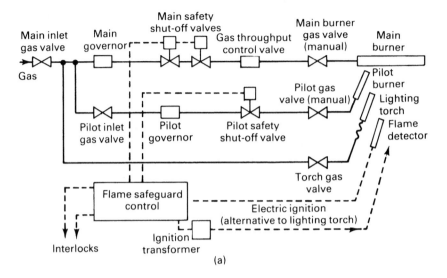

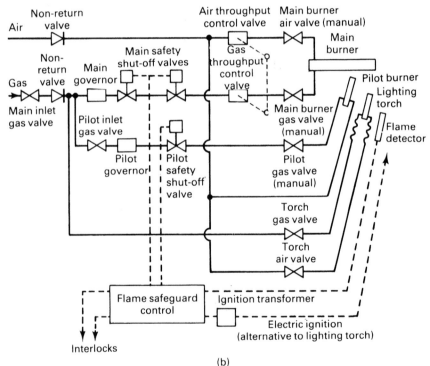

Fig. 22   Controls, including safeguards, on: (a) natural draught; and
(b) forced draught burners

when a fault condition occurs. Fig. 21 shows the connections that may be necessary to connect a semi-automatic flame safeguard to a burner system. Fig. 22 shows layouts of controls, including flame safeguard, for natural draught and forced draught burners.

Figure 23(a), (b) and (c) shows three semi-automatic flame safeguards produced by different manufacturers.

*Fig. 23(a)    Semi-automatic flame safeguard (Pactrol)*

*Fig. 23(b)    Semi-automatic flame safeguard (Teknigas)*

*Fig. 23(c)    Semi-automatic flame safeguard (Saflame)*

CHAPTER 6

# Commercial Catering

---

Chapter 6 is based on an original draft by T. Fox

---

## Introduction

There are many different catering operations carried out in a wide variety of establishments. Food is served in pubs and clubs, hotels and hospitals, shops and schools, cafés and canteens. The increasing demand for good quality catering at outdoor events such as sporting tournaments, exhibitions and festivals has led to an increase in the use of mobile catering units fuelled by Liquid Petroleum Gas (LPG). This is a specialised area that will be covered later in the chapter.

Nearly half of all meals served in a year by catering establishments are produced in hotels and guest houses, cafés, snack bars and restaurants, public houses and clubs, fish and chip shops and take-aways.

About one quarter are produced in schools and colleges, while hospitals, office and industrial canteens together produce about one tenth.

In addition to the establishments which produce meals or snacks for consumption on the premises or to take away, there are others which process food for sale in shops. These are bakers, confectioners and the makers of pies, biscuits, frozen and chilled foods. Where these operations are carried out by the large combined organisations, the plant is mechanised and the kitchen becomes a factory.

This chapter deals generally with the smaller scale unautomated commercial cooking operations.

## Cooking Processes

The basic cooking processes and the domestic appliances on which they may be carried out were described in Vol. 2, Chapter 6. The information given applies equally to those catering appliances which are simply larger and more robust versions of domestic appliances.

For large-scale cooking operations specialised equipment is used and the cooking process may be modified.

*Baking*

Baking is cooking food dry in an oven, principally by convection with some radiation from the oven walls. Cooking times may be reduced by using forced convection. Bread is best baked on a falling temperature from 260°C down to 230°C and steam may be introduced into the oven for a period to give a crisp brown crust to the products (see Chapter 8).

*Blanching*

This is carried out to remove skins from fruit or nuts. Vegetables may be blanched before being frozen for storage to check the action of the enzymes which cause deterioration.

Blanching is also carried out to whiten food or to remove strong flavours. It is usually done by placing the food in a wire basket and immersing it in boiling water for 2 to 5 minutes. This is immediately followed by rapid cooling in cold water.

*Boiling*

This is heating food by immersing it completely in a liquid, essentially water, at 100°C. Water is boiling when bubbles continually rise to the surface. Because some evaporation takes place more liquid may need to be added during boiling.

Meat, poultry and green vegetables are added to boiling water which is quickly brought back to the boil.

Root vegetables, except new potatoes, are placed in cold water and then brought to the boil.

Any scum arising from boiling must be removed. With the exception of green vegetables, food may be covered during cooking.

*Braising*

Braising is slow cooking in a tightly covered container with some liquid or sauce, usually in an oven. Meat and poultry are browned by frying in a little fat before being braised. They are then laid on a bed of sliced vegetables and half covered with liquid. Fish and vegetables may also be braised. In the case of meat and fish, the braising liquid is used to make the accompanying sauce.

*Frying*

There are various methods of frying:

- deep frying
- shallow frying
- dry frying.

Deep frying is done by totally submerging the food in hot fat or oil. With the exception of potatoes, the items are usually coated with batter, seasoned flour or egg and breadcrumbs before frying. Potatoes are dried thoroughly. Normal frying temperatures are 185 to 195°C.

Shallow frying is cooking in a small amount of fat in a shallow pan. The food is cooked on the best side first and turned so that the fat seals the whole of the outside. The items should be well drained after cooking. 'Sauté' is a term given to small items of food which are tossed while being shallow fried. 'Meunière' is the shallow frying of fish, usually in butter.

Dry frying is carried out on a lightly greased flat plate and is used for cooking eggs, bacon, liver, hamburgers and pancakes. The plate or griddle is tilted to allow any fat to run to a drain so that the food stays reasonably dry.

## Grilling

This is cooking by radiant heat. The food is placed on bars and in the original grills, was heated from below. Grills may now be 'underfired' or 'overfired'. Overfired grills or salamanders have their heat source above the food while underfired or 'flare' grills have theirs below. In the latter, the liquid fat from the meat falls on the hot fuel and flares up. The smoke and flames give the meat a characteristic flavour and appearance.

## Poaching

Poaching is heating food totally submerged in water just below boiling point at a temperature of about 93 to 95°C. This is as close to boiling point as possible without there being any movement in the water. Poaching is used for cooking fish, fruit and eggs.

## Poeler

This is cooking meat or poultry in a covered container with butter, in an oven. The meat may be laid on a bed of sliced vegetables and there must be sufficient butter for basting. The lid is removed to complete the cooking and the liquid is used to make the accompanying sauce.

## Roasting

Roasting may be carried out in an oven or on a rotating spit. The food should be cooked with a little fat and must be basted occasionally to keep it moist. Joints of meat and poultry should be seasoned before cooking and should preferably be raised off the bottom of the roasting tin. Roasting should begin at a high temperature to seal in the meat juices. The temperature may then be reduced. Vegetables should be partially cooked before being put in the oven.

### Simmering

Simmering is very gentle boiling or cooking at a temperature just below boiling point but above poaching temperature. The liquid should just show slight movement. Soups and stock are simmered gently. The surface must be skimmed periodically to remove any impurities.

### Steaming

This is cooking food by enclosing it in a compartment filled with steam from boiling water. Steaming may be carried out at atmospheric pressure or at higher pressures from 35 mbar to 1 bar. At the higher pressures, temperatures are raised and cooking times are reduced.

The method is used for puddings and vegetables which can be cooked slowly without losing their colour or flavour.

### Stewing

This is the slow cooking of meat and vegetables in enough stock or sauce to just cover the food. Gentle simmering concentrates the flavour and tenderises the tougher cuts of meat. Casseroles and hot-pots are forms of stew which are baked in the oven.

## Equipment

Gas-fired catering equipment is designed to comply with the recommendations of BS EN 203–1: 1993.

The type of equipment used in an establishment depends on the following factors:

- the standard of catering
- the menu offered, number of courses, choice of dishes
- type of service, cafeteria, table service either plated or waiter served
- number of meals provided
- type of food used, fresh, frozen, chilled or convenience.

Small cafés and boarding houses may be able to manage with only a single-oven range. The large five-star hotels with specialist chefs for meat, fish, vegetables, soups, pastries and so on will use the complete range of specialised equipment which follows.

## Ranges

A range consists of a boiling table or hotplate mounted on a general purpose oven. An eye level grill may be incorporated into the pot rack.

Ranges may be classified as either medium or heavy duty. Medium-duty ranges usually have open-top hotplates and internally heated ovens with single or double doors, hinged at the side. They may incorporate a grill and they are, in essence, Fig. 1, a larger and more robust version of the domestic cooker.

*Fig. 1   Medium-duty range*

Heavy-duty ranges usually have solid tops and semi-externally heated ovens with drop down doors. They are designed to withstand continuous use with heavy cooking pans, Fig. 2.

*Fig. 2    Heavy-duty range*

## *Hotplates*

Open top hotplates, Fig. 3, may have various designs of ring burners giving immediate flame contact with the cooking utensil.

Solid top hotplates, Fig. 4, are cast iron and are ribbed on the underside. They are usually heated by a central burner surrounded by refractory brick. The 'bull's-eye' and its surrounding rings may be removed to allow flame contact if required. Temperatures on the top range from 540°C at the centre to 150°C near the edge.

Ignition is usually by permanent pilot or spark igniter. All burners must have flame protection devices.

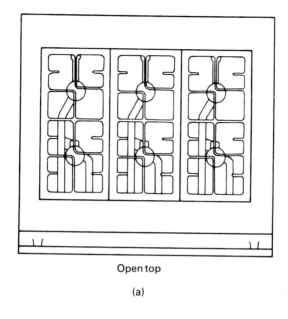

Open top

(a)

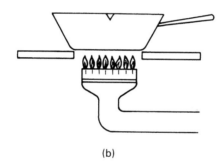

(b)

*Fig. 3    Open top hotplate: (a) plan view; (b) burner*

Gas rates may be as follows:

- open top 3 to 6 kW
- solid top 12 to 15 kW.

*Ovens*

The three basic types of oven are:

- internally heated or direct, Fig. 5
- semi-externally heated, Fig. 6
- externally heated or indirect, Fig. 7.

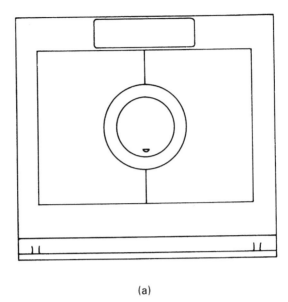

(a)

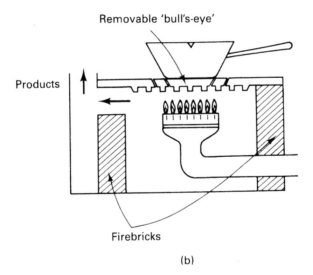

(b)

*Fig. 4   Solid top hotplate: (a) plan view; (b) burner*

Internally heated ovens have a temperature gradient which allows different foods to be cooked at the same time, while externally heated types have an even temperature throughout.

Ranges usually have one of the first two types and the fully indirect design is commonly used as an independent baking oven.

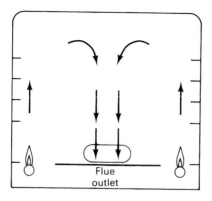

*Fig. 5    Internally heated oven*

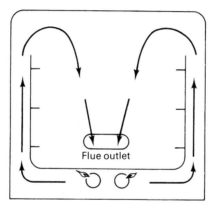

*Fig. 6    Semi-externally heated oven*

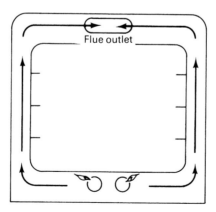

*Fig. 7    Externally heated oven*

Ignition may be by spark or permanent pilot and flame protection devices are incorporated. Most ovens have thermostats.

Gas rates are from 7 to 9 kW.

## Boiling Tables and Stockpot Stoves

*Boiling Tables*

These are hotplates mounted on low tables with a working height of 760 to 860 mm. They may have open or solid tops. Like ranges they are designed for either medium or heavy duty.

Open-top boiling tables have various types of burners and often include double or treble concentric burners. The solid tops are heated by single or multiple ring burners or jet burners.

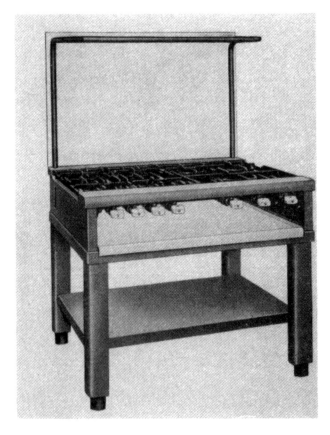

*Fig. 8   Boiling table*

Boiling tables are usually fitted with a steel storage shelf and may have a splash plate and pot racks, Fig. 8. Ignition is by spark or permanent pilot and solid tops have flame failure devices.

Gas rates may be from 4 to 7 kW for open top burners and up to 19 kW for solid tops.

*Stockpot Stoves*

A stockpot stand or stove is a low level boiling table designed to heat one large heavy utensil. The working height is usually 450 to 600 mm. It may be heated by a single ring burner or by concentric rings, Fig. 9. Gas rates are from 8 to 12 kW.

*Fig. 9   Stockpot stoves: (a) single burner; (b) double concentric ring burner*

**Ovens**

Ovens may be classified as follows:

- general purpose and roasting ovens
- forced convection ovens
- convection steamers
- pastry and pizza ovens
- proving ovens
- large-scale baking ovens.

## General Purpose and Roasting Ovens

These may be part of a range or separate units, either as single ovens or tiered, Fig. 10. They are generally either internally or semi-externally heated and heavy duty ovens usually have drop down doors.

*Fig. 10    General purpose oven*

Ignition is by spark or pilot with flame protection. Thermostats may be bimetal or liquid expansion and could be direct, or indirect operating through a relay valve.

Gas rates may be:

- general purpose ovens, 7 to 10 kW
- roasting ovens, 15 to 16 kW.

## Forced Convection Ovens

These ovens have the heated air circulated throughout the interior by means of an electric fan. Figure 11 shows a two-tier model. They may be either externally or semi-externally heated and can be used for all normal roasting and baking operations. The ovens have an even heat

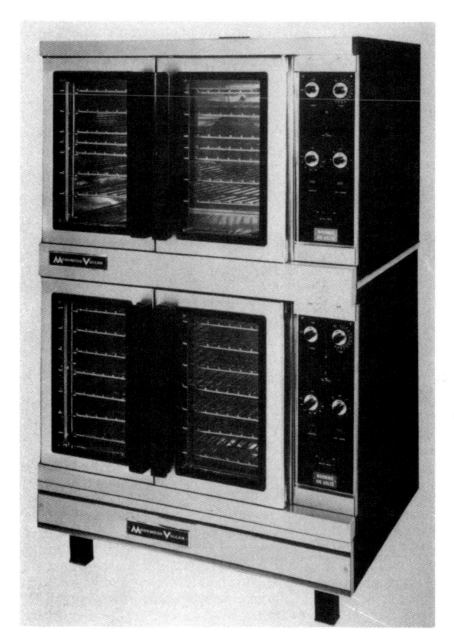

*Fig. 11    Forced convection oven*

distribution throughout and heat up quickly, giving a high output.
They are particularly suitable for reheating or end cooking, chilled

and frozen foods. Some have simplified control systems with preset oven temperatures and push button controls specifically for use as regeneration ovens.

Semi-externally heated ovens, Fig. 12, have some of the combustion products mixed with the circulating air. Air from the fan is deflected upwards through venturi shaped channels, into the side chambers. The reduced pressure at the throat draws in the combustion products. These pass with the hot air through the perforated side panels and around the oven. Some hot gases escape through the flue outlet and the remainder are drawn through the perforated rear panel downwards to re-enter the fan. Water can be injected into some forced convection ovens to provide a more moist atmosphere.

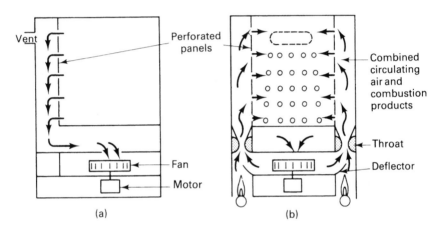

*Fig. 12   Semi-externally heated forced convection oven*

Externally heated ovens, Fig. 13, heat the hot air indirectly and the combustion products pass around the oven to the flue outlet.

Ignition is usually by an automatic mains operated spark igniter with mercury vapour flame failure device.

Other controls may include:

- thermostat
- door switch to shut off fan and main gas when doors are opened
- auxiliary fan switch for rapid cooling
- load control to adjust the heat input to suit the number of racks in use.

Timers are usually included.

Gas rates vary considerably from the small counter top models at about 5 kW, up to 32 kW for a large, free-standing oven.

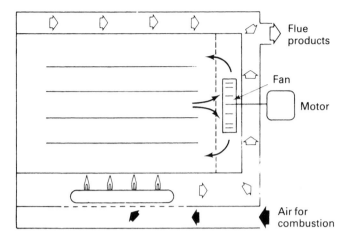

*Fig. 13    Externally heated forced convection oven*

**Convection Steamers**

These appliances are sometimes known as combination steam-convection ovens. The designs are based on forced convection ovens but they have the addition of a steaming facility. They can be used as convection ovens or steamers or can be operated in combination. These appliances can handle a variety of cooking techniques including roasting, poaching, braising, steaming, proving, baking and the regeneration of chilled and frozen foods.

There are generally separate burners for the convection and steaming sections each with gas rates similar to those for forced convection ovens. For the convection mode, controls are similar to those used in forced convection ovens. When steaming, additional sensors are provided to regulate the flow of steam to the oven. Some convection steamers have sequential controls as shown in Fig. 14 to allow changes in the cooking method to be preset. Sometimes a low temperature steaming programme (40°C – 98°C) is offered for use in sous-vide cookery. This is a specialised process in which food is cooked in sealed vacuumised plastic pouches.

*Pastry and Pizza Ovens*

These are externally heated, low height ovens designed to have an even temperature distribution. They are used for baking cakes, pastries and tarts of various kinds. The ovens are from 125 to 300 mm in height and cooking is usually done on the sole plate in those below 200 mm. Larger ovens may have a shelf. Some manufacturers offer ovens of this type for cooking pizzas.

*Fig. 14    Convection steamer*

Because of their low height, pastry ovens are usually grouped in tiers of two, three or four ovens. Each deck may be heated separately or all decks may have a common burner system, Fig. 15. In the second case, separate thermometers indicate the differences in temperatures between the ovens.

The atmosphere in the ovens can be controlled by a vent in the door. This gives a steamy atmosphere when closed and a drier heat when opened.

Ignition is usually by permanent pilot and flame failure devices are fitted. Thermostats are generally indirect, with relay valves. Thermometers are often included.

Gas rates may be from 9 to 14.5 kW.

*Proving Ovens*

When yeast is mixed with sugar and water in a warm atmosphere it ferments and gives off carbon dioxide. Mixed with dough the yeast

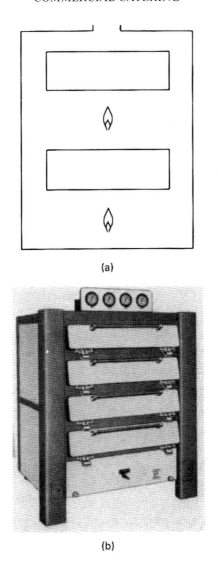

(a)

(b)

*Fig. 15   Pastry oven: (a) decks heated by individual burners; (b) single burner system*

produces little bubbles which aerate the mixture. This fermentation is called 'proving' and it must be carried out in a warm, moist atmosphere to make the dough rise before cooking. Proving ovens are used for this purpose.

The ovens are large, low temperature ovens with water pans to produce vapour and with water feed tanks. They maintain temperatures of 26 to 32°C by means of a thermostat.

Gas rates are about 2 kW.

*Baking Ovens*

There are various types of baking ovens still in use including:

- steam tube ovens
- forced convection peel ovens
- rack ovens
- reel ovens
- continuous hearth conveyor ovens.

Steam tube ovens are heated by heavy gauge, small bore wrought iron tubes running through either the roof or the base or 'sole' of each deck. The tubes are sealed and contain a small quantity of distilled water which turns into high pressure steam when the tube ends are heated, Fig. 16.

*Fig. 16   Steam tube oven*

In addition to the usual gas controls, the oven must have a manually reset overheat cut off.

Forced circulation peel ovens are named after the tool which is used to load them. A 'peel' is a long wooden pole with a flat, spade shaped end used for charging and discharging any fixed oven. The ovens are heated externally by forced circulation of the combustion products giving even heat distribution and quick heating up.

Rack ovens are forced convection ovens, usually externally heated. The bread or other goods to be baked are loaded on to trays which are placed in a rack which is pushed into the oven, Fig. 17. There are various methods by which the racks may be rotated or moved round the oven. Rack ovens are ideal for batch production.

*Fig. 17    Rack oven*

Reel ovens are a form of conveyor oven in which the products are placed on trays suspended from a large rotating reel, similar to a ferris wheel at a fairground, Fig. 18. The oven is usually internally heated and the reel is geared so the articles return to the oven door every $2\frac{1}{2}$ minutes. This enables different products to be cooked at the same time and removed when desired. Reel ovens may also be used for roasting.

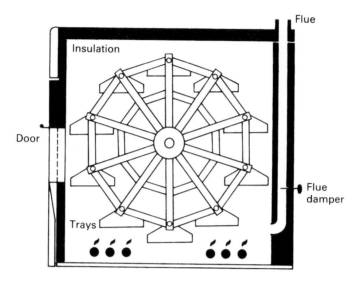

*Fig. 18    Reel oven*

Ignition is by mains operated spark and the temperature is controlled by an electrical thermostat, solenoid valve and relay valve. Flame failure devices are included.

Gas rates vary widely from about 13.5 to 95 kW. Continuous hearth conveyor ovens were described in Chapter 3. They are used in the major plant bakeries and are run at high temperatures with controlled steam injection. This produces a moist loaf lacking the flavour achieved in a smaller oven by the local baker.


**Grillers**

Underfired, or flare grills are commonly used for steaks and chops and are often in public view. The source of heat is below the food and consists of a burner firing upwards on to a bed of refractory material or pumice, Fig. 19(a). Overfired grills, or salamanders, can, in addition, be used for making toast and salamandering. They have the heat source above the food, Fig. 19(b). This may comprise sets of burners firing below refractory or metal frets, or surface combustion plaques. A 'branding plate' is usually supplied. This is a fluted, solid aluminium alloy plate which absorbs heat so that food placed on it, below the frets, is cooked on both sides at the same time. Figure 20 shows an overfired grill with a branding plate.

Ignition on grillers is usually manual or by permanent pilot.

Gas rates range from 5 to 16 kW for overfired grills and up to 30 kW for underfired models.

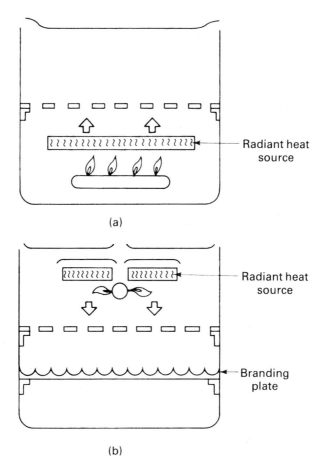

*Fig. 19   Griller: (a) underfired; (b) overfired*

## Griddle Plates

Griddles are solid metal plates heated from below by burners to a temperature of 200 to 300°C, Fig. 21(a) & (b). The plate may have a channel to drain off the fat from cooking. They are used for dry frying eggs, bacon and similar foods or for pancakes or drop scones. Sizes vary from small counter top units to larger free-standing models. They are commonly used for 'call order' cooking in snack bars.

Ignition is usually manual and gas rates vary from 5 to 15 kW.

## Fryers

There are two main types of fryer:

*Fig. 20    Overfired grill with branding plate*

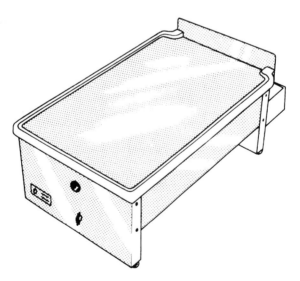

*Fig. 21(a)    Griddle plate, counter model*

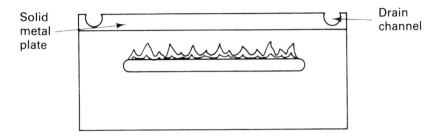

Solid metal plate

Drain channel

*Fig. 21(b)    Griddle plate, cross-section*

- deep fat fryers
- tilting fryers or brat pans.

## Deep Fat Fryers

These are heated pans containing oil or fat. They are used for cooking chips, fish, poultry, sausages, onion rings, fritters and doughnuts, Fig. 22. There are three types of deep fat fryer:

*Fig. 22    Deep fat fryer*

- flat bottomed
- 'V' pan
- immersion tube.

Flat bottomed fryers, Fig. 23, have a flat pan, finned or studded below and heated directly by a burner. This type is used in fish and chip shops.

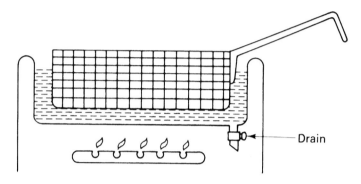

*Fig. 23    Flat-bottom fryer*

'V' pan fryers have a vee-shaped pan, heated above the bottom of the vee, Fig. 24. Immersion tube fryers are heated by burners firing through tubes which pass through the pan from one side to the other, Fig. 25.

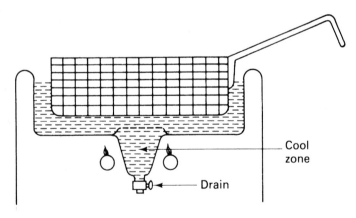

*Fig. 24    'V' pan fryer*

In both the latter types there is a cool zone below the burners into which food particles can sink and remain without charring. This

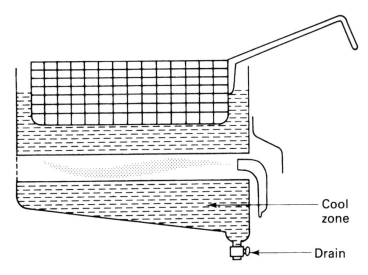

*Fig. 25    Immersion tube fryer*

prevents particles transmitting flavours from previous food to the new food being cooked. In flat pans contamination can only be avoided by straining out the solid particles after every frying.

Frying temperatures are critical and must be maintained with close limits. Different fats and oils have different maximum temperatures or 'smoke points'. At or above the smoke point the flavour of the food will be spoiled and above the smoke point spontaneous combustion can occur. Table 1 gives the smoke points of common cooking fats.

Ignition is by manually lit pilot with flame protection. Thermostatic control is usually by an indirect thermostat and relay valve. The thermostat is adjustable to a maximum temperature of 190°C. Hospital management boards require an additional thermostat or over-heat cut-off set to shut down at 215°C. Timers, computerised controls and automatic basket lifts, may be included.

Gas rates vary from about 8 kW for small counter top models up to 35 kW for large free-standing models.

*Tilting Fryers or Brat Pans*

These are shallow, flat bottomed pans capable of being tilted forward by a hand wheel or lever, Fig. 26. They can be used for shallow or dry frying and also for boiling, poaching or stewing.

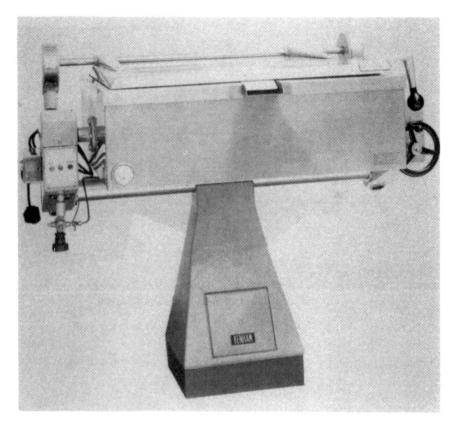

*Fig. 26    Tilting fryer or brat pan*

**TABLE 1  Smoke Point of Common Fats and Oils**

| Cooking Medium | Smoke Point °C |
|---|---|
| Coconut oil | 138 |
| Ground nut oil | 149–243 |
| Dripping | 163 |
| Olive oil | 169 |
| Maize oil | 221 |
| Lard | 190–221 |
| Butter fat | 208 |
| Cotton seed oil | 233 |

Ignition is usually by permanent pilot with flame protection device. A control can prevent the pan from being tilted while the main gas is on.

Gas rates are 12 to 18 kW.

## Steaming Ovens

Steaming ovens may be either atmospheric or pressure types. Atmospheric types are designed for either light or heavy-duty. Figure 27 shows a light duty atmospheric model. Steaming ovens are used for cooking root vegetables, fish or puddings.

*Fig. 27    Light-duty atmospheric steaming oven*

The ovens are heated by steam from a water trough in the base which has gas burners below it.

Atmospheric steamers have their water supplied by an external cistern and steam is allowed to escape through a vent at the top of the oven, Fig. 28. Ignition may be manual or by a permanent pilot with flame protection. Temperature is controlled by an indirect thermostat and relay valve.

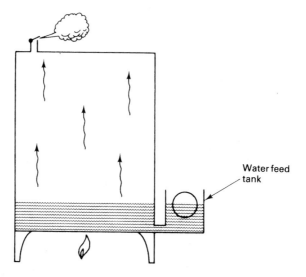

Fig. 28    *Atmospheric steaming oven*

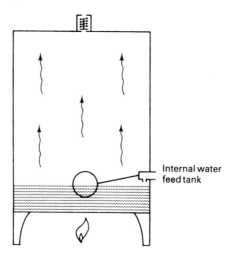

Fig. 29    *Pressure steaming oven*

Pressure steamers have an internal ball valve and a spring-loaded pressure relief valve in place of the vent, Fig. 29. Ignition is usually by protected permanent pilot and control may be by a pressurestat. Most pressure types operate at 35 mbar but models working at pressures of about 1 bar are available. The high pressure models are used for cooking frozen foods.

Gas rates vary and may be from 13 to 20 kW.

*Fig. 30   Jacketed boiling pan*

## Boiling Pans

Boiling pans are large-capacity vessels holding from 45 to 180 litres and used for boiling food in bulk. There are three main types:

- single pan
- jacketed pan
- dual purpose.

A jacketed pan is shown in Fig. 30.

### Single Pan

The pan is heated directly by the burner, Fig. 31. A large bore draw off tap is fitted at the base of the pan so that the liquid may be drained off. Pans are fitted with lids which may be pivoted and counterbalanced on large models. The food is sometimes held in wire baskets to assist handling and prevent it sticking to the sides of the pan. Single pans are used for cooking vegetables or meat and making soup.

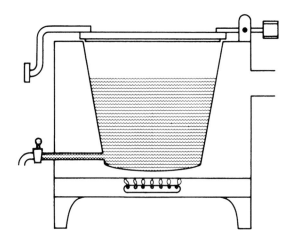

*Fig. 31    Single boiling pan, section*

### Jacketed Pans

The pan is heated indirectly by hot water or steam in an outer pan, Fig. 32. This gives an even heat distribution and prevents local overheating and food sticking or burning. Jacketed pans are used for cooking thick soups, porridge, custard or milk puddings.

A water cock with a swivel arm is usually fitted which supplies water directly to the inner pan and, on some models, through a funnel to the outer pan. There is usually a thermally operated air vent between the funnel and the pan. This is open when cold to allow the pan to fill up and closes when heated to pressurise the pan to about 500 mbar. These pans are fitted with a safety valve and a water gauge.

The pan shown in Fig. 32 has the outer pan supplied by an external cistern with a ball valve.

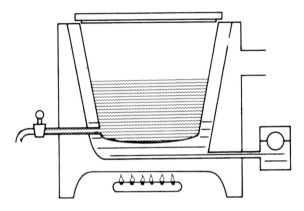

*Fig. 32    Jacketed pan, section*

Some models have a completely sealed water jacket, Fig. 33. They are fitted with a bursting disc to relieve excessive pressure.

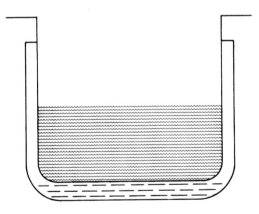

*Fig. 33    Pan with sealed water jacket*

### Dual Purpose Pans

There are single pans which have an inner removable pan or porringer and can be used either as a single pan or as a jacketed pan, Fig. 34.

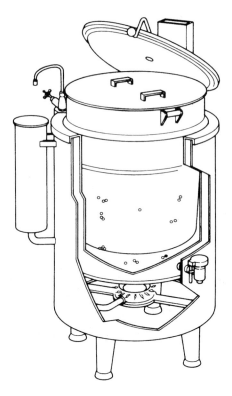

*Fig. 34    Dual purpose boiling pan*

Ignition is usually by permanent pilot, sometimes with flash tube. Flame failure devices are generally fitted. On sealed jacket pans the controls may include pressurestat, relay valve, safety valve, air vent, pressure gauge and multi-functional controls. The safety valve operates at about 750 mbar and gas rates vary from 9 to 35 kW.

## Hot Cupboards

These are insulated cabinets which are used to heat plates or to keep cooked food hot, Fig. 35. There are two main types:

- directly heated
- indirectly heated.

Direct types obtain their heat from a burner situated below a baffle plate, Fig. 36. They are generally used for plate warming at about 60°C.

Indirectly heated models are heated either by hot gases circulated around the cabinet through channels, or by steam generated in a

*Fig. 35    Hot cupboard*

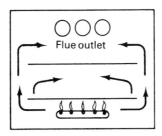

*Fig. 36    Directly heated hot cupboard, section*

water trough in the base, Fig. 37. They have a more humid atmosphere and are suitable for keeping food hot at about 82°C.

Ignition is usually manual and hot cupboards are usually thermostatically controlled to pre-set temperatures. Gas rates range from about 2 to 7 kW.

## Bains Marie

'Bain' is the French word for 'bath' and bains marie are appliances used to keep food hot before or during serving. In their simplest form

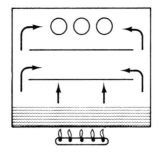

*Fig. 37    Indirectly heated hot cupboard, section*

they consist of an open trough containing water heated from below by a gas burner, Fig. 38. The pans or vessels containing food are stood on a rack in the water.

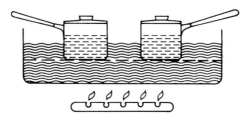

*Fig. 38    Open trough bain marie*

Fitted models are available in which the water is covered by a 'filler plate' into which are set containers of various shapes and sizes to hold different kinds of food, Fig. 39. Some fitted models are directly heated by a burner below a baffle plate, Fig. 40. The temperature of the food is maintained between 75 and 85°C.

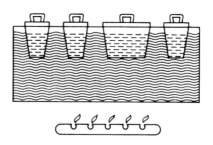

*Fig. 39    Wet heat fitted bain marie*

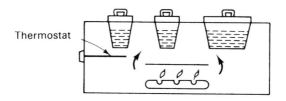

*Fig. 40    Dry heat fitted bain marie*

Ignition is usually manual with interlocking main and pilot taps, thermostats are normally fitted to directly heated types.

Gas rates vary from 5 to 11 kW.

## Dish Washing Equipment

This comprises:

- sterilising sinks
- dish washing machines.

### Sterilising Sinks

There are stainless steel sinks heated from below by a gas burner, Fig. 41. They operate at 82 to 88°C. After washing up, the crockery and cutlery is loaded into racks and immersed in the sink. The utensils retain sufficient heat to dry out rapidly when removed, so eliminating

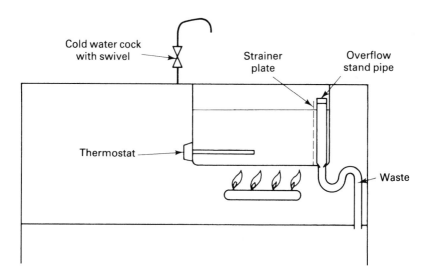

*Fig. 41    Sterilising sink*

the need for drying cloths. The sinks may be part of a washing-up unit, Fig. 42. The units are used in small kitchens as the sole means of washing-up or as stand-by units in larger establishments. Balanced flue models are available.

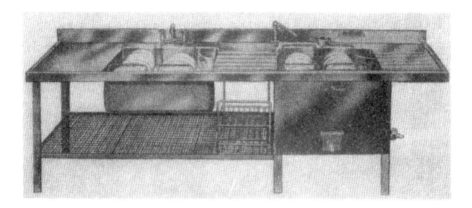

*Fig. 42    Washing-up unit with sterilising sink*

Ignition is by pilot with interlocking taps or flame failure device. Temperature is controlled by an indirect thermostat and a relay valve. Gas rates may be 8 to 16 kW.

*Dish Washing Machines*

There are basically three main types of dishwasher:

- front loading, under counter models
- pass through or hood type
- conveyor type.

The front loading models are similar to the domestic washers with drop down doors and pull-out racks. In the pass through type, utensils are loaded into a rack or basket similar to that used for front loading machines. The rack is then placed in the machine, a hood is pulled down over it and the cycle commences. Tabling is often fitted either side of the machine to take racks before and after washing. The conveyor washers may be manual, semi-automatic or fully automatic. Most machines have time washing cycles and automatic detergent dispensers. They are designed to deal with cutlery, crockery and glass ware. Special machines are available for pot washing.

Dishwashers are usually plumbed into a hot water supply from a separate boiler. Some have integral burners to raise the water temperature to about 82°C for the final rinse.

## Bulk Water Boilers

Bulk water boilers are cylindrical containers heated by gas burners. They are generally constructed of:

- heavy gauge copper, tinned internally
- stainless steel with a tinned copper, Pyrex or stainless steel liner.

There are three main types:

- urns
- jacketed urns
- large boilers.

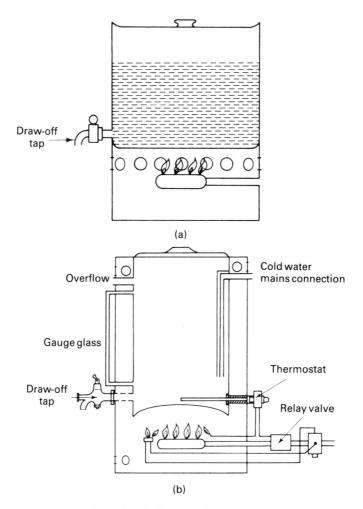

*Fig. 43   Bulk water heating urns*

*Urns*

These are simple containers with a lid. They have a draw off tap near to the base and may have a thermometer to indicate the temperature and a gauge glass to show the liquid level, Fig. 43. Urns are used to produce boiling water for tea or coffee making in quantities from 9 to 45 litres.

*Jacketed Urns*

Heated indirectly by a water jacket to prevent local overheating, these are used for milk or similar liquids, Fig. 44.

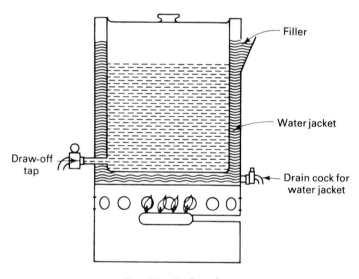

*Fig. 44    Jacketed urn*

*Large Boilers*

These are similar to the urns but are fitted with cold water mains and overflow connections, Fig. 45. They are used particularly where large quantities have to be drawn off quickly. Capacities may be up to 136 litres.

Ignition of the simple models is manual. Larger boilers may have automatic ignition with a flame failure device and temperature control by thermostat and relay valve.

Gas rates range from 8 to 17 kW.

*Fig. 45   Large water boiler*

## Café Boilers

These are automatically operated water boilers which give continuous outputs of boiling water for use within about four minutes of lighting up. Side urns may be fitted for heating and storing milk and for making and storing coffee. A combined appliance with side urns is called a 'café set'.

There are two main types of boiler:

- expansion boilers
- pressure boilers.

### Expansion Boilers

These rely for their operation on the fact that water expands when heated. They are designed with the draw-off tap slightly higher than the water level in the boiler. Only water which is forced up into the expansion chamber by boiling can be drawn off, so only freshly boiled water is delivered, Fig. 46. The water supply is fed automatically from an external cistern with a ball valve, located at the rear of the boiler, Fig. 47.

Ignition is usually by swing-in pilot and the simplest boilers have interlocking taps and a gas tap linked with the draw-off tap. More complex models have automatic control devices and may store a reserve of hot water for any sudden demand.

Gas rates may be from 12 to 36 kW.

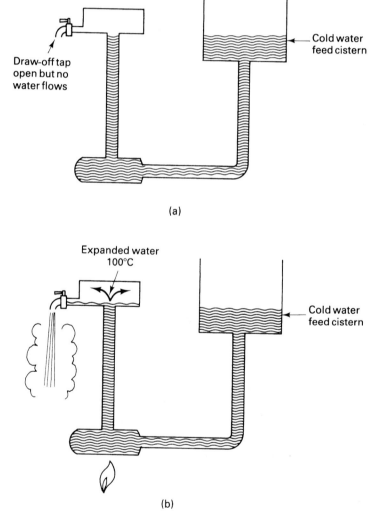

Fig. 46    *Principle of expansion café boiler: (a) cold; (b) boiling*

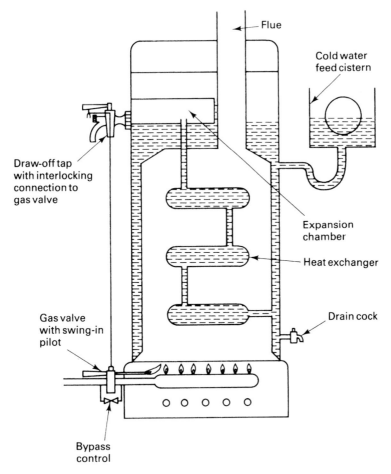

*Fig. 47  Expansion type café boiler, section*

*Pressure Boilers*

Pressure boilers are usually housed below a counter with a draw-off mounted above, Fig. 48. Boiling water is forced up to the draw-off by the pressure of the steam produced, so water can only be drawn off when it is boiling, Fig. 49. Some models include a steam draw off or injector for heating liquids almost instantly. The boilers operate at pressures of 300 to 520 mbar and the temperature of the stored water is about 105°C.

Ignition is by permanent pilot with flame protection.

Other controls are usually:

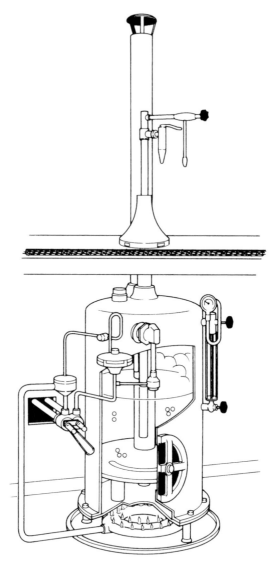

*Fig. 48   Pressure type café boiler*

- pressurestat
- water regulator
- safety valve
- excess temperature cut-off.

The pressurestat, Fig. 50, is a direct acting type. It consists of a bellows A, mounted on a diaphragm B, with its bottom face forming

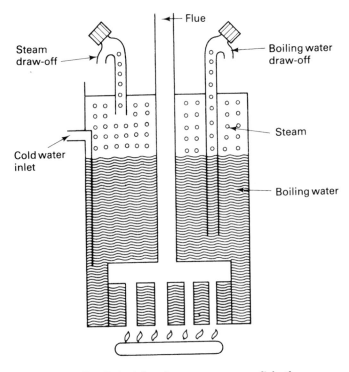

*Fig. 49   Principle of pressure-type café boiler*

a gas valve C. The bellows is connected to the steam pressure and operates against the force of a spring D. The tension of the spring determines the pressure at which the boiler will shut down to a bypass rate.

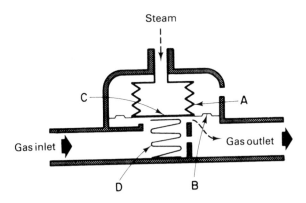

*Fig. 50   Pressurestat*

The water regulator, Fig. 51, ensures that incoming cold water does not reduce the steam pressure below about 170 mbar. It consists of a stout rubber diaphragm A, to which is attached a water valve B. Steam pressure on the diaphragm can open the valve against the force of the spring C and allow water to pass to the ball valve. The spring tension is adjusted to close the valve when the steam pressure falls below 170 mbar. The valve may be opened by screwing in the knurled screw at the top in order to fill the boiler initially. The screw must then be unscrewed before operating the boiler.

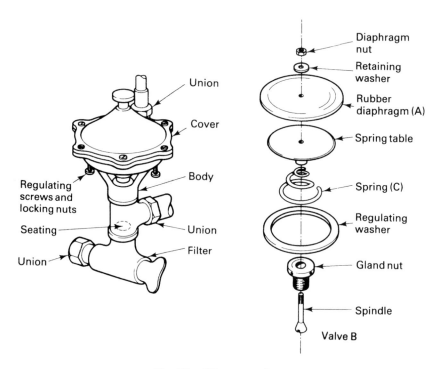

*Fig. 51   Water regulator*

The safety valve, Fig. 52, consists of a valve held against its seating by a spring. Steam pressure on the face of the valve will cause it to open against the tension of the spring and allow steam to be released. The degree of tension determines the relief pressure. On pre-set types the tension nut is screwed down on to a spacing washer of predetermined thickness.

An excess temperature cut-off, similar to that used on some central heating boilers, may be fitted to low water capacity boilers, Fig. 53. It

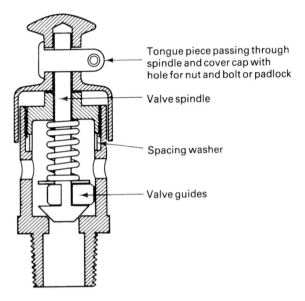

*Fig. 52    Spring-loaded safety valve*

is essentially a heat operated switch which shuts off the main gas and pilot if gas is lit when the boiler is empty.

Gas rates of pressure boilers vary from 8 to 32 kW.

Café sets produce coffee by percolating it in an infuser, Fig. 54. Freshly ground coffee is placed on a filter paper in the infuser, which is then clamped between the inlet and outlet supplies. Turning the operating cock allows boiling water to flow into the infuser and coffee

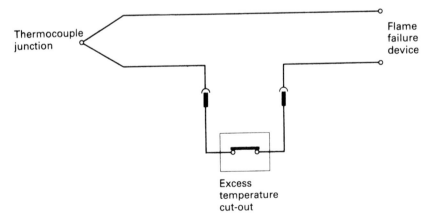

*Fig. 53    Excess temperature cut-out*

to flow into the steam-jacketed side urn. When the required amount has been made, the cock is turned off, isolating the infuser from the boiler and from the side urn.

### Back Bar Equipment

This is the term used to describe the range of small, counter-mounted appliances used in snack bars, cafés and similar small establishments. It embraces griddle plates, fryers, boiling tables, bains-marie and flare grills or salamanders. It could include any appliance required to produce a particular dish.

The appliances are miniature versions of those already described, with similar ignition and control devices. Gas rates are smaller in proportion. One particular piece of equipment which is finding increasing use in pubs and restaurants is the combined microwave and gas convection oven, Fig. 55. This retains the advantage of high speed which can be achieved with microwaves but has the added benefit of

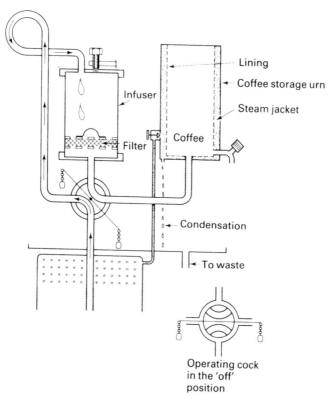

*Fig. 54   Coffee infuser*

being able to brown food such as pastry using gas convection. This appliance can be used in microwave, convection or combination modes.

*Fig. 55    Combined microwave and gas convection oven*

## Installation of Equipment

*General*

All installation work must conform to the requirements of the Gas Safety Regulations and the Building Regulations. It should also comply with the recommendations of the relevant British Standard Codes of Practice including BS 6173 and the appliance manufacturer's instructions.

Appliances should be level, protected from draughts and in a good light. There should be easy access for cleaning and for servicing or adjusting components or controls.

Wall or floor temperatures must not exceed 65°C and fire-resistant sheets or bases may be necessary with some appliances. Adequate ventilation must be provided.

In small kitchens the range is the most frequently used appliance and it should be sited in the most convenient position.

## Gas Supply

The gas meter and the internal pipework must be adequate to meet the maximum demand. Each appliance should be fitted with a gas control cock so that it may be isolated for servicing. Union cocks are generally provided to allow the appliance to be easily disconnected. Appliances may be connected by means of flexible hoses and fittings to BS 669 Part 2. These hoses are made of stainless steel and are covered with a white outer cover of non-chlorinated synthetic material. The use of flexible tubing with plug-in connectors enables appliances to be withdrawn for cleaning. Appliances should be fitted with restraining chains of a length so as to prevent strain being applied to the flex on withdrawal. The chains must be securely anchored to the appliance and the adjacent wall, but under no account may the chains be anchored to the gas pipe.

## Electricity Supply

The electrical installation should comply with the I.E.E.'s 'Regulations for the Electrical Equipment of Buildings' and must comply with the Electricity at Work Regulations 1989.

Supplies for electrical control devices are generally single-phase, but single-phase motors are usually rated below 1 kW and very rarely up to 5 kW. Many of the motors found on large-scale catering equipment are three-phase.

A three-phase supply can be identified from the:

- manufacturer's badge on the appliance
- main isolation device
- motor starter
- motor connections.

Most electrical equipment has a badge stating the type of supply it requires and the voltage at which it operates.

A three-phase main isolation device can be identified by labelling or markings on the front of the unit provided by the manufacturer. The main isolation device may also incorporate three fuses. The main isolation device can normally only be opened when in the off position.

Three-phase motors use a 'starter' as a means of connecting the supply to the motor. This may be controlled manually or by thermostat, timer or other device. The manual type has two shrouded push-buttons usually towards the top of the box. The push buttons are normally marked 'START' and 'STOP' and 'RESET'. Alternatively they may be marked with equivalent international symbols conveying the same meaning. Starters include an overload device which switches the supply off if the motor is overloaded.

Motor connections are not made to any set pattern or colour code. Reference should be made to the manufacturer's connection diagram for the motor which may be fitted to the underside of the terminal cover. Resistance readings between the three winding connections should be equal.

The colour code of three-phase supplies may be either –
*Non Flexible Cables for Fixed Wiring:* the three phases are Red, Blue and Yellow, Neutral is Black or –
*Flexible Cables and Flexible Cords:* the phases are Black or Brown and the terminals are marked L1, L2, L3, Neutral is Blue.

## Water Supply

Generally water supplies should be taken from the mains and a stop-cock fitted in the line. Where supplies are connected to a cistern, for appliances such as pressure steamers or café boilers, there must be sufficient head of water to operate the appliance.

## Waste Pipes

Provision must be made for the removal of waste water and condensation from appliances. Drains may need to incorporate grease traps to prevent blockage.

## Ventilation

It is necessary to ensure an adequate supply of fresh air to a kitchen in order to provide air for combustion and to limit the effects of heat and humidity caused by the cooking.

It is also important that products of combustion and cooking vapours are removed at source and not allowed to disperse throughout the kitchen. This is usually achieved by a combination of hoods, ducts and extraction fans. Fresh make-up air must be fed in to take the place of the foul air extracted. The design and installation of ventilation in commercial kitchens is a specialised field and until recently very little guidance has been available as to the provision of adequate ventilation in all types of catering establishment. In an attempt to help in this important area, the HSE has published a guidance document 'Ventilation of kitchens in catering establishments' that goes some way to giving information to installers. A new document produced by the Heating and Ventilation Contractors Association 'Standard for Kitchen Ventilation Systems' DW/171 is a must for any serious installer. The following information can only be treated as a general guide.

Kitchens require 20 to 40 air changes per hour. Ventilation systems are shown in Fig. 56.

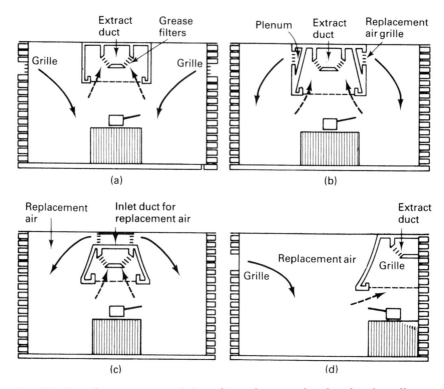

*Fig. 56 Ventilating systems: (a) traditional extract hood with side wall air inlet grilles; (b) extract hood with air inlet plenum; (c) inlet and outlet ducts in parallel over extract hood; (d) side wall hood with cross ventilation of inlet air*

Most catering appliances should be positioned under a hood or canopy which should be sized to extend 200 to 300 mm beyond the equipment. The lower edge of the hood should be about 2 m from the floor. It should have a condensation channel around the inside which may be fitted with a drain. Hoods should be made of anodised, aluminium, reinforced glass, galvanised steel or stainless steel. The structure must be rigid and stable and preferably be suspended from the ceiling.

Frying, grilling and roasting produce greasy vapours and cleanable grease filters should be provided in the flues. They should be cleaned at least once a month.

All extract ducts should be made of sheet metal and of a gauge strong enough to withstand frequent scraping and cleaning. Some grease vapour will pass through the filters and, since its dewpoint is about 152°C, will be deposited on the duct walls. The ducts must have adequate and accessible access doors or panels for cleaning purposes.

Ducts should be as short and as straight as possible and with an upward slope of not less than 1 in 12 towards the termination point.

Fans should be mounted on anti-vibration pads and be joined to the duct by flexible connections in order to minimise noise.

The duct termination outside the building should be selected to avoid causing any nuisance.

The amount of ventilation air required for various appliances is given in Table 2.

**TABLE 2 Ventilation Requirements for Catering Appliances**

| Appliance | Ventilation Rate $m^3/min$ |
|---|---|
| Bain Marie | 11 |
| Café boiler | 14 |
| Sterilising sink | 14 |
| Boiling pan | 17 |
| Griller – salamander | 17 |
| Pastry oven | 17 |
| Range – single oven | 17 |
| Steaming oven | 17 |
| Fryer – deep fat | 25 |
| Grill – underfired | 25 |

## Commissioning

After the installation has been completed, test the gas and water supplies for soundness and purge. Check each appliance in turn for the following points.

### Position

Check against any kitchen plans or layout diagrams that the appliance is fitted in the correct location. Ensure that the appliance is level and correctly situated under any extract hoods or over any gulleys or drains.

### Assembly

Check against the manufacturer's instructions that:

- the appliance is assembled correctly
- doors are properly aligned and door seals effective (recheck door seals after heating up)
- correct furniture has been supplied
- mechanical devices operate satisfactorily.

*Function*

Turn on the water supply and fill those appliances which require water. Completely fill café boilers, steamers, bains marie, sterilising sinks and any pan jackets to the required level. Adjust floats or ball valves as necessary.

Boiling pans, bulk water boilers or urns should have sufficient water to cover the base and the draw-off connection.

Deep fat fryers should be filled to the correct level with oil. The thermostat shield must be covered.

Turn on the gas supply and light the appliance following the recommended ignition procedure.

Check the following points:

- ignition device operates
- pilot flame stable, correct size, correctly positioned relative to flame failure device
- flame failure device operates correctly
- relight the appliance and check the burner pressure, adjust if necessary
- check flame picture and, if necessary, check gas rate by meter test dial
- check appliance taps and control devices
- thermostats and relay valves shut down, bypass rates correct
- pressurestats operate at desired pressures
- overheat cut-offs and thermal switches are satisfactory
- water regulators maintain correct steam pressure.

Check the ventilation system: air inlets clear and adequate, extract fans running smoothly at correct speed and correct direction of rotation, all filters in position and access doors to ducts properly sealed.

Finally, hand over the equipment to the user and ensure that the operating instructions are understood.

## Servicing

In most kitchens catering equipment has long periods of heavy use, so regular cleaning and servicing is essential.

Servicing procedures should follow those already established in Vol. 2 for comparable domestic appliances. However, because of the degree of use, more attention must be given to those components which are subject to spillage and wear. Items to be checked include the following.

*Gas Taps*

Check for smooth operation. If necessary, clean out internal gas ways and regrease.

*Draw-off and Drain Taps*

Dismantle, clean and regrease the plug. Check that the connecting pipes are clear.

*Float Valves and Ball Valves*

Examine moving parts. Check washer and seating for wear and renew or reseat if necessary. Check operation, confirm water level and adjust if necessary.

*Sight Glasses*

Dismantle and clean gauge glass tube. Reassemble and renew packing glands. Check water level.

*Safety Valves*

These may be padlocked and are often ignored. If left for too long they may become scaled up and inoperative. Periodically they should be dismantled, cleaned and have the valve seating ground in. Check that the valve is set to the correct pressure.

*Control Devices*

Check operation and ensure that gas is cut off at the correct temperature or pressure. Check that bypasses, weep jets and orifices are clear. Renew diaphragms as necessary and repack stuffing glands.

*Burners*

Clean burners thoroughly. If spillage has carbonised in the burner parts it may be necessary to use a twist drill of the correct diameter or to soak the burner in a caustic solution. After soaking, rinse and dry thoroughly.

Injectors and jets should be cleaned with a brush. A choked orifice may be cleared by using a piece of soft fuse wire.

### Flueways

Both internal and external flueways must be kept clear. Internal sections in particular can become encrusted with carbonised grease and must be scraped out. Check that flues are functioning and that flame combustion is satisfactory.

### Door Seals

Test door seals and adjust catches or hinges as necessary. Renew worn door gaskets.

### Descaling

In hard water areas and if no water softener is used, water boilers will require periodic descaling. Chemical descaling is normally carried out after removing the appliance to a workshop, but most boilers incorporate access facilities for mechanical descaling.

Soft scale can be removed by wire brushing and washing away by water. Hard deposits will require scraping or chipping which must be done carefully with a blunt tool to avoid causing damage.

Manufacturers often offer a descaling service for their appliances.

### Ventilation

Check that the ventilating system is operating correctly. Clean filters if necessary.

## Fault Diagnosis and Remedy

Most of the faults which occur on catering equipment have already been dealt with in the two previous volumes. Additional points which should be noted are listed in the table opposite.

The Electricity at Work Regulations 1989 impose requirements as to 'live testing' and 'competency'. As a general rule it should always be the ideal not to carry out live testing and that all work on electrical equipment should be carried out by competent personnel. Prior to carrying out any work on electrical equipment, isolate the equipment, remove any fuses and where possible lock the isolation device in the off position using a personnel padlock. Place suitable notices on the main isolating device to the effect 'Danger do not use. Work in progress'. With a suitable voltage potential indicator ensure that no voltage potential exists on the appliance side of the main isolation device between:

1. each phase conductor and neutral
2. phase conductors
3. neutral and earth conductors.

All the above must indicate zero voltage potential. (Remember to test the voltage potential indicator prior to carrying out any tests.) If a voltage potential is indicated in any of the above tests then do not touch the equipment and immediately consult a competent electrician. If the above tests prove that the equipment is in fact isolated from the electricity supply then investigative work can commence.

Check that all motors are free to rotate and check continuity of all appliances and main fuses. If suitable instruments are available motor windings can be tested for open and short circuits. The windings of the motor need to be identified from the motor manufacturer's connection diagram. Once the windings have been identified a resistance measurement can be taken, a zero resistance reading indicates a winding short circuit whilst an infinitely high resistance reading indicates a winding open circuit. Both results indicate a winding fault. It is also useful to check each winding terminal resistance to earth.

The test instrument should give an infinitely high resistance indicating that there is no connection between winding and earth. It is recommended that the above winding resistance tests are carried out with the motor disconnected from the motor starter.

If a three phase motor rotates in the wrong direction then one phase is reversed. Isolate the equipment and confirm isolation as before and reverse any two phase connections on the motor terminal block. The motor will then rotate in the required direction.

In all instances the person carrying out the work should be sure that he is competent to do the work. If he has any doubts then he should refer the work to a suitably competent person.

| Symptom | Action |
|---|---|
| **DEEP FAT FRYERS** <br> Pilot lights up but goes out when press-button is released | *Check:* <br> • *thermocouple correctly connected* <br> • *overheat thermostat connection satisfactory* <br> • *overheat thermostat correctly calibrated and not faulty* |
| All gas shuts off during normal working | *Check oil level and top up if necessary* <br> *Check overheat thermostat correctly set and not faulty* |
| **PRESSURE TYPE CAFE BOILER** <br> Water level too high | *Check float valve:* <br> • *float watertight and not coated with scale* <br> • *valve seating and jumper clean, renew washer if required* |
| Water level too low | *Check water supply:* <br> • *filter clear* <br> • *regulator set at correct steam pressure and admitting water* |
| Steam only obtainable from water draw-off | *Check water level, check that regulator is admitting water* <br> *Check syphon tube not damaged or missing* |
| Safety valve blows | *Check pressurestat setting correct* <br> *Check safety valve clean and seating properly, valve set to correct pressure* |

**Kitchen Planning**

The first step in designing the layout of a catering establishment is to allocate areas for the cooking activities and for dining. Calculations are based on the requirements which include:

- kitchen – types of food to be cooked
  - storage facilities required
  - type of service provided
  - number and size of appliances
  - staff facilities, toilets
- dining room – standard of catering
  - sizes of tables
  - number of diners.

Provisional estimates can be based on the figures in Table 3.

**TABLE 3  Approximate Kitchen and Dining Areas**

Kitchen areas include storage and staff accommodation.
Dining areas include gangways.

| Numbers of Persons Catered for | Kitchen Area per Person |
|---|---|
| 100 | 0.5 to 0.75 m² |
| 400 | 0.3 to 0.5 m² |
| 1000 and above | 0.25 to 0.4 m² |

| Number of Persons per Table | Dining Room Area per Person |
|---|---|
| 6 to 8 | 0.75 to 0.9 m² |
| 4 | 0.9 to 1 m² |

The areas allocated must conform to legislation regarding working space, toilet facilities, fire regulations and other safety measures.

Kitchen areas are usually about 30% of the dining area.

## Kitchen Layout

The kitchen should be designed to enable the work activities to proceed in the proper sequence and as smoothly as possible. The work flow should follow the following pattern:

- reception
- storage
- preparation
- cooking
- serving
- washing up.

A plan of a typical kitchen is shown in Fig. 57 and a key to the symbols used in Fig. 58.

### Reception

Some area is required where incoming goods can be weighed, checked, counted and examined. In small establishments a table may be sufficient.

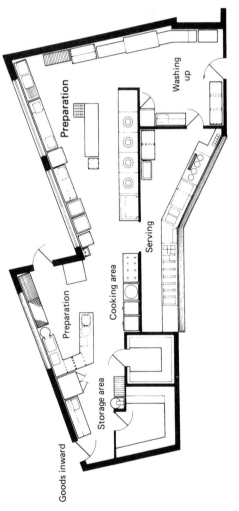

*Fig. 57   Typical kitchen layout*

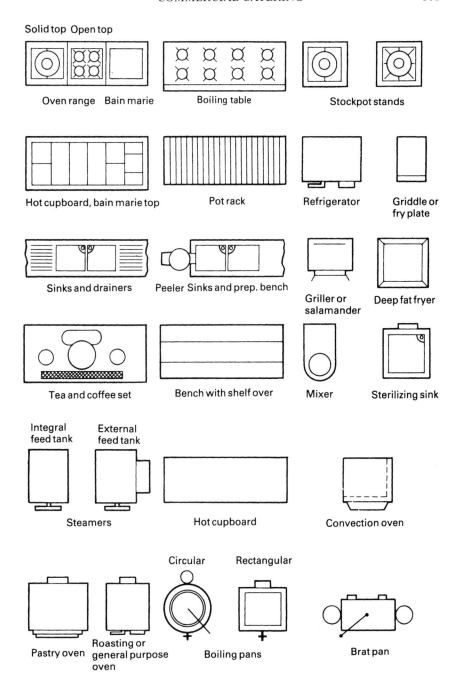

Fig. 58  Typical planning symbols

*Storage*

It is usual to keep the different kinds of food apart and in large establishments separate stores are required for dry goods, vegetables, meat and fish, frozen foods. Stores should be close to the reception areas.

Dry goods consist of tea, coffee, sugar, flour, tins, packets or jars. Approximate storage areas required are:

- 100 meals/day – 0.06 m²/person
- 1000 meals/day – 0.023 m²/person.

Vegetables are stored on racks or duck boards. Approximate storage required:

- 100 meals/day – 0.03 m²/person
- 1000 meals/day – 0.08 m²/person.

Refrigerated storage is required for meat, fish, poultry, dairy produce and cooked food. Approximate refrigeration capacity is from 0.14 to 0.28 m³/person.

Separate freezer capacity is required for ice cream and frozen food. The average weight which can be stored is 480 kg/m³.

*Preparation*

Preparation areas should be situated between the particular stores and the appropriate cooking appliances. Large kitchens should have separate areas for vegetables, meat and fish and pastry. Preparation areas require work benches, sinks, waste disposal and equipment for cutting, peeling, sawing, mincing and mixing.

*Cooking*

The conversion of the prepared new food is carried out in separate areas as close as possible to the servery.

*Serving*

Cooked food is temporarily held in hot cupboards and bains-marie. Depending on the type of service, the servery may also supply hot and cold drinks, cold dishes and ice cream. It may also dispense cutlery and take cash.

*Washing-up*

This should be situated away from the dining area to reduce noise and be equipped with sinks and machines to clean and sterilise the crockery and cutlery. It needs racks and waste disposal facilities. Clean tableware should be stored adjacent to the servery.

Pot washing is carried out separately and large sinks and racks are required. The clean pots should be stored close to the cooking area.

## Capacities of Equipment

*Average Portions*

These may vary considerably between establishments. Some approximate figures are as follows:

| | | |
|---|---|---|
| Soup or stew | 100 portions | = 18 to 28 litres |
| Custard or gravy | 100 portions | = 6 to 13 litres |
| Milk pudding | 100 portions | = 18 litres |
| Potatoes | 100 portions | = 15 to 18 kg |
| Green vegetables | 100 portions | = 15 kg |
| Root vegetables | 100 portions | = 7.5 to 10 kg |
| Fish | 100 portions | = 11 to 15 kg |
| Roast meat | 100 portions | = 9 to 11 kg |

Chicken or duck provides 4 to 6 portions
Turkey provides about 30 portions

*Beverages*

Boiling water required – hotels = 0.28 litres/person
                  – canteens = 0.42 litres/person
Tea      100 cups = 19 to 23 litres
Coffee   100 cups = 11 litres
(milk added extra)

*Equipment Sizes and Capacities*

1. *Ranges*
   Solid top about 0.84 m$^2$
   Oven capacity   0.17 m$^3$
   – shelf area     0.31 to 0.46 m$^2$
2. *Boiling Tables*
   150 litres/m$^2$

using vessels of up to 27 litres capacity.
Potatoes and root vegetables require 1 litre capacity for every 0.4 kg.
Green vegetables require 1 litre for every 0.15 kg.

3.   *Ovens*
  *General Purpose or Roasting*
    Poultry, 96 to 128 kg/m$^3$
    Roast meat, 128 to 160 kg/m$^3$
    Pies 19.5 kg/m$^2$ shelf area
    or    172 portions/m$^2$
  *Pastry*
    Small cakes, 250/m$^2$
    Meat pies, $108 \times 0.11$ kg pies/m$^2$
4.   *Grills*
  *Underfired*
    Steaks, 3 to 6 minutes, 65 to 85/m$^2$
  *Overfired*
    Steaks, 2 to 4 minutes with branding plate capacity as the underfired type.
    Toast, $1\frac{1}{2}$ to 2 minutes, 65 slices/m$^2$
5.   *Fryers*
  *Deep Fat*
    Fish or chips, 14.6 kg/m$^2$
  *Brat Pans*
    Dry fry, eggs,    600/hour
            bacon, 600 portions/hour
    Poach,  eggs,    800/hour
            fish,    200 portions/hour
    Sauté,  meat,    27 kg/hour
6.   *Steamers*
    Puddings,    individual, $172 \times 75$ mm/m$^2$ shelf area
                 roll tins, 258 portions/m$^2$ shelf area
    Root vegetables, 384 kg/m$^3$
    Fish,                72 kg/m$^3$
7.   *Boiling Pans*
    Green vegetables, 0.15 kg/litre
    Root vegetables, 0.4 kg/litre.

*Equipment Requirements*

One single oven range will cater for up to 50 persons. A double oven range is normally adequate for 100 persons although a separate overfired grill might be a useful addition. Above 100 persons more specialised equipment should be used, the type of appliance varying with the menu to be offered.

## Cook-Freeze and Cook-Chill Catering

These terms are used to describe processes in which food is prepared and stored in advance of consumption. Guidelines for both processes were updated by the Department of Health in 1989 and published under the title 'Chilled and Frozen – Guidelines on Cook-Chill and Cook-Freeze Catering Systems'. Users of these systems should be familiar with these Guidelines.

In both systems the sequence of operations is the same, i.e.

- receive
- store
- prepare
- cook
- portion
- pack
- freeze or chill
- store.

Food is then reheated and served, possibly at a later date or at another location. In cook-chill the storage life is limited to five days including those of production and consumption.

The cooking appliances used may be the conventional types and may include:

- large boiling pans (135 to 180 litres)
- brat pans
- steamers
- convection ovens
- convection steamers
- deep fat fryers
- baking ovens.

Large quantities of food are cooked at a time, possibly 500 portions. For very large quantities factory methods are applied and food travels on conveyors, through frying oil, boiling water and ovens as appropriate.

After cooking, the food is portioned into suitable containers which are labelled and coded before freezing or chilling. All precautions are taken to avoid hygiene hazards. Techniques such as Hazard Analysis and Critical Control Points (HACCP) can be used to ensure high standards of hygiene.

Freezing is usually carried out in a 'blast freezer'. This is an insulated tunnel in which refrigerated air is blown over the food at velocities of about 5 m/sec. Trolley loads of food have their temperature reduced to at least $-5°C$ within 90 minutes of entering the freezer and subsequently further reduced to $-18°C$ for storage. An alternative method used particularly for soft fruits, is to spray the food with liquid nitrogen at $-196°C$. This produces almost instantaneous freezing. Similar equipment is used for chilling, but in this case, the temperature is reduced to between $0°C$ and $+3°C$.

The 'end' or 'finishing' kitchens complete the cycle. They require stores maintained at the correct temperature for chilled or frozen foods and usually use forced convection ovens for reheating. It is most important that adequate temperatures are achieved during reheating, particularly for cook-chill products, to ensure that any harmful bacteria are destroyed. Some bacteria can survive chill temperatures but will be killed if the temperature of the food is raised to at least 70°C for 2 minutes.

Temperature probes should always be used to check food temperatures after reheating. End kitchens will also need ancillary equipment for cooking frozen vegetables, frozen chips and fried meals. This equipment can also be used to achieve a better quality when reheating some cook-chill dishes. In addition, end kitchens have to provide soup, sweets and beverages and will require a servery and facilities for washing up.

Cook-freeze and cook-chill systems may vary depending on the type and quantity of the foods produced and may be used separately or together. This determines the amount and type of equipment required by individual caterers.

## Mobile Catering Units

An increasing demand for good quality catering at outdoor events such as sporting tournaments, exhibitions and festivals has led to an increase in the use of mobile catering units fuelled by Liquefied Petroleum Gas (LPG). These units vary from the small roadside caravan serving simple meals, snacks, sandwiches and drinks, to the fast food outlets found at most public events. These units are usually either self-propelled or capable of being towed. Large events will often be served by full mobile kitchens capable of offering full restaurant facilities. These units are delivered on-site by low-loader and can consist of one self-contained unit or several independent specialist sections assembled together to satisfy the site requirements. All mobile units are covered by the Gas Safety (Installation and Use) Regulations 1998, the Gas Appliance (Safety) Regulations 1992, the Chemical (Hazard Information and Packaging) Regulations 1993 and the Health and Safety at Work Act (the characteristics of LPG and its use as a fuel are covered in Vol. 1, Chapter 3).

## Self-propelled Units and Trailers

All units in this category are subject to the Road Vehicles (Construction and Use) Regulations. Particular attention must be paid to the limitations on laden weight and, in the case of trailers, nose weight.

If more than two cylinders are carried on a catering or similar vehicle, the Road Traffic (Packaged Goods) Regulations require a Transport Emergency Card to be carried and adequate driver training to have been carried out.

## Cylinder Storage

When selecting a site for location and storage of cylinders or vehicle-mounted tanks, the following items must be taken into consideration.

1. The cylinders must be located in a position to minimise the risk of damage in an accident. The near side of the vehicle being preferred as offering better protection for the cylinders and also safer for cylinder changing. Preferably located in the open air or in a well-ventilated compartment mounted outside the vehicle or in a compartment sealed from the vehicle but recessed into it. The compartment must be constructed of materials to meet BS 476 Parts 20–22 to give thirty minutes fire resistance.
2. The compartment should be ventilated through the access door at high and low level, each vent being a minimum of 1/100 of the compartment floor area. There should be no ignition sources within an area extending 1 m horizontal radius outside the vehicle and from ground level up to 0.3 m above the high level vent.
3. Access to the cylinder compartment should be from outside the vehicle and must be secure from unauthorised interference. The cylinders should be secured in an upright position and be easily accessible for changing or removal.
4. The number of cylinders stored should not exceed those required to meet the demands of the appliances used. A warning notice should be attached to the outside of the compartment to indicate the presence of gas. This should consist of a red hazard diamond showing a flame with the wording 'EXTREMELY FLAMMABLE LPG'.
5. Cylinders may be temporarily stored outside the vehicle when parked provided that they are:

   - positioned close to the vehicle
   - on firm and level ground
   - not close to windows, doors or vents
   - safe from passing traffic
   - protected against unauthorised interference and not left unattended.

When parked for use there must be a minimum of 1 m between the vehicle and any flue terminal, vent or ignition source on another vehicle or building. Access must be provided for cylinder changing and there should be clear escape routes.

The high pressure stage gas flexible tubes between the cylinders and the vehicle must meet BS 3212, be as short as possible and have factory assembled connections. Multiple cylinder installations may be fed to a wall block or a manifold. If an automatic changeover device is used then an emergency control valve must be installed, and a warning notice displayed. Details on automatic changeover devices can be found in Chapter 7. Soundness testing of the high pressure stage can be found in Vol. 2, Chapter 4.

**Low Pressure Pipework Installation**

The installation of pipework for LPG systems generally follows the standards for natural gas – see Vol. 2, Chapter 3, BS 5482: Part 2, and the LPG Association Code of Practice 22 'LPG Piping System – Design and Installation'. The following points should be taken into consideration.

- Installation of pipes and appliances must be carried out by a suitably qualified and competent person when factory fitted. For installations carried out privately, or any subsequent alterations or servicing, the operative carrying out the work must be CORGI registered.
- Installation pipework should be as short as possible and of a size suitable to ensure that the maximum operating pressure drop between the regulator outlet and any appliance does not exceed 2.5 mbar on maximum load.
- The tubing used should be solid drawn copper with copper or copper alloy fittings, or stainless steel tube to BS 3605 with suitable soldered, compression or screwed fittings. Steel tube to BS 1387 with steel compression fittings or malleable cast iron fittings can be used but only on low pressure pipework.
- Any pipes installed below the vehicle should be protected as much as is practical from damage by stones etc. The pipework installation shall allow for vibration and flexing during transit.
- Connections between installation and appliances can be rigid or in the case of cookers and ovens by a flexible hose with factory fitted connections to BS 3212, BS 669: Part 1(strip wound hoses) or BS 669: Part 2 (corrugated metal hoses).
- All high pressure stage pipework must be outside the vehicle.

- All pipework installations must be tested for gas soundness upon completion (see Vol. 2, Chapter 3).

## Appliances

The type of gas appliances used on self-propelled or trailer type units will obviously be dictated by the use to which the mobile is being put, but in general, they are usually of the fast food type, typically fryers, griddles, grills, hotplates, ovens, refrigerators and water boilers.

All appliances must be approved by the manufacturer for use with LPG and conform to all relevant British Standards.

If manufactured after 1 January 1996 they shall have been verified by a 'notified body' as satisfying the Gas Appliance (Safety) Regulations 1992 and will carry a CE mark.

Appliances must be installed to manufacturer's instructions with particular attention being paid to the security of the appliances and their fixing to prevent movement during transport. Appliances must be turned off while the vehicle is in motion and the gas supply must be turned off at the cylinders.

### Space Heaters

These should be fixed heaters and preferably be of the room sealed type. Any flueless heater should comply with BS 5482: Part 2 and be fitted with an atmosphere-sensing device to shut down the heater if the carbon dioxide content of the atmosphere rises to between 0.8% and 1.5% volume in air.

### Refrigerators

LPG fuelled refrigerators shall have a flue and the burner should be protected by a flame failure device. A refrigerator with an alternative electrical heating source (vehicle battery) should be considered to enable the unit to remain 'on' during transit.

### Water Heaters

These should comply with BS 5258: Part 7 and BS 5386: Parts 1 and 2. All appliances should be flued, and preference should be given to room sealed appliances. All appliances must be fitted with a flame failure device.

## Catering Appliances

A description of these appliances and the regulations relating to catering equipment were covered earlier in this chapter. During installation, attention must be paid to the fixing of appliances to ensure security during transit. All appliances should be fitted with flame failure devices, with the exception being for the use of domestic cookers where the hotplate and grill would not normally require such a device.

## Non-mobile Units

These usually consist of either a self-contained unit containing a complete kitchen, or a number of specific purpose units (portable cabins, see Fig. 59) prefabricated together to form a much larger kitchen to serve a restaurant, canteen or a large marquee. These are often used as temporary kitchens while refurbishments are carried out to the original facilities. A typical prefabricated kitchen would have:

*Fig. 59   Typical prefabricated unit. Notice the ventilator outlets and the gas, water and electrical connections at the door*

- a storage unit containing refrigerators, freezers, storage shelving and food preparation benches and tables
- a baking unit for the cooking of bread and pastries
- a general purpose cooking area (equipment installed would depend on site requirements)

- a serving unit with hot cupboards and bains marie
- a wash-up and crockery storage unit with dishwashers and racks.

Most units are flexible in use and the contents can be varied to suit the needs of a particular site (see Figs. 60(a) and 60(b)).

The units are usually delivered on-site by a low loader vehicle and a crane is used to unload the units and position them on to prepared bases. Also available on-site, and planned in advance, are all the required services, electricity, water, drains and the fuel gas which can be LPG or natural gas (if a supply is available and the expected period of usage makes it worthwhile). Once the units are in position, short, purpose made, connecting passages are installed between the various buildings to complete the site, the services are connected and the kitchen is tested and commissioned.

## Ventilation

Wherever gas appliances are used, permanent, non-adjustable and properly fixed ventilation is essential at high and low levels, in addition to the normal ventilation provided by windows and doors. The ventilation can be either natural or mechanical depending on circumstances. In the case of catering units, in addition to the air requirements for the appliances, the ventilation must also take into

*Fig. 60(a)    Washing-up area using a balanced flue multipoint water heater*

*Fig. 60(b)    General cooking area with range, fryer, griddle, grill and ovens*

account the steam and vapours produced during the cooking process (covered earlier in this chapter). The ventilators should be designed to minimise draughts and screens should be considered to protect against vermin. The ventilators and screens should be accessible and easy to clean (see Figs. 61(a) and 61(b)).

The ventilation of catering establishments is an involved subject and reference should be made to the Heating and Ventilating Contractors Association 'Standard for Kitchen Ventilation Systems DW/171'.

As a general rule and guideline, the free area of ventilation required for the LPG burning appliances is a minimum of 6.5 cm$^2$ (1 inch$^2$) for each 0.3 kW (1,000 Btu/hr) input rating of all appliances or 97 cm$^2$ or (15 inch$^2$) whichever is the larger. This should be equally divided between high and low level.

Adequate flues should be provided for appliances that require them and reference should be made to the manufacturer's instructions. They should be constructed of suitable non-combustible materials and clear all combustible materials in the vehicle by a minimum of 25 mm. Flues should be firmly fixed and terminate in a position at high level away from any openings in the vehicle. Details of flueing regulations can be found in BS 5440: Part 1. Particular care must be taken in the removal of fumes from fryers and chip ranges. These

*Fig. 61(a) Typical high-level ventilation in mobile unit*

*Fig. 61(b) Fig. 61(a) with covers removed to show extraction fan*

should have a canopy or hood installed above the appliance extending at least 15 cm beyond the cooking area on all sides. The canopy should have a flue installed with a minimum cross sectional area of 27 cm$^2$ for every 1000 cm$^2$ of canopy base area. The canopy should be manufactured from non-flammable and non-corrosive materials, be furnished with the required filters and should be easy to clean.

## Safe Use of LPG

All mobile units using LPG or natural gas as a fuel must be checked frequently to ensure the safety of the people operating the kitchen.

## Daily Checks

Cylinders, pipework and appliances should be checked by a suitably trained person. This is usually a visual check of cylinders and connections, flexible hoses, visible joints and the purging, operation and correct combustion of all appliances. The vents should be checked for blockages and any mechanical extraction system tested.

## Inspections

The owner or operator must ensure that the gas installation, ventilation system and any flues are inspected and tested, and any necessary servicing work is carried out at least once per year by a competent gas installer. The owner or operator must keep a written record of the tests and maintenance carried out. For mobile units it is recommended that this inspection is done at six-monthly intervals.

The owner of the vehicle may be regarded as a 'landlord' if the vehicle is hired out, in which case the annual inspection is covered under the Gas Safety (Installation and Use) Regulations.

## Fire Precautions

Suitable fire extinguishers should be located in readily accessible positions close to exits. These should be to BS 5306: Part 3. Dry Powder extinguishers to BS 5423 are suitable for LPG and oil or fat fires. In addition, where frying takes place, there should be a fire blanket.

## Safety Instructions

The following instructions should be displayed in a prominent position on the vehicle.

### General Safety

(a) *LPG cylinders must always be used and stored in an upright position with the valves uppermost.*

(b) *When the vehicle is not in use, free-standing cylinders shall not be left unattended in any location which is not secure from public access or unauthorised interference. They shall not be kept inside the vehicle.*

(c) *When the vehicle is in motion the cylinder valves should be closed and all appliances turned off.*

### Changing cylinders (to be displayed on cylinder compartment door or as near to as possible)

(a) *Extinguish any ignition source.*

(b) *Where possible change cylinders in open air.*

(c) *Ensure that cylinder valves are closed before disconnecting or removing blanking caps or plugs.*

(d) *Make a firm connection using an appropriate spanner, or follow manufacturer's instructions. Use leak detection fluid to test for gas tightness. NEVER USE A NAKED FLAME.*

(e) *After reconnecting a single cylinder make sure that all appliances are turned off before opening the cylinder valve.*

### Action in the event of a gas leakage

*A gas leak may be detected by smell, or as a result of leak detection. Action must be taken. Do not ignore it.*

### If a leak is detected or suspected in any of the equipment:

(a) *All persons not required to take the following action leave the vehicle.*

(b) *Extinguish all naked flames and sources of ignition.*

(c) *Turn off gas supply at the emergency valve.*

(d) *Shut off cylinder valves.*

(e) *If the leak is inside the vehicle, ventilate by opening doors and windows to disperse the gas.*

(f) *Call a competent installer to repair the leak and test the system.*

### If a cylinder is found to be leaking, the following actions should be taken providing there is no risk to yourself:

(a) *Extinguish all nearby sources of ignition.*

(b) *Stop the leak by closing the cylinder valve or replacing the bung or cap.*

(c)  *If the leak cannot be stopped the cylinder should be carefully removed to a safe, well-ventilated area clear of drains, buildings, ignition sources and other cylinders. The cylinder should be clearly marked as faulty and a warning prohibiting smoking or other naked lights displayed. On no account should any attempt be made to dismantle or replace cylinder valves.*

(d)  *Advise the gas cylinder supplier.*

### Action in case of fire

*Anyone who discovers a fire in, or threatening, the vehicle should:*

(a)  *Raise the alarm.*

(b)  *Call the Fire Brigade and advise them that LPG cylinders are present.*

(c)  *If gas is ignited from a cylinder valve, and it is considered safe to do so, then attempt to turn off the cylinder valve or move any other cylinders away from the ignition source.*

(d)  *If in any doubt evacuate the area immediately and await the Fire Brigade.*

# Incinerators

Chapter 7 is based on an original draft by R.L. Brooks

## Introduction

Incinerators are special furnaces for burning waste material. Any process that uses combustion to convert any waste material (or gas mixture) to a less bulky, less toxic or less noxious material is called incineration. The residue may be reduced to between 7 and 12% of its original size and will be sterile. Almost every manufacturing process produces waste of one sort or another and the cost of the disposal of this waste adds to the total production costs. Where the waste is combustible, often incineration is a solution to the disposal problem and produces heat that can sometimes be used for other purposes. Gases from certain production processes containing solvents or hydrocarbons may also be passed through incinerators or thermal oxidisers to render them more environmentally friendly. Destruction of materials containing chlorine or fluorides need careful consideration.

The requirements of the Environmental Protection Act 1990 has had a dramatic effect on the incineration of waste and is encouraging the use of larger units and the centralised use of specialist incinerators especially in hospitals. See Chapter 13.

Historically the classification of domestic type waste in British Standard 3813 was 'rubbish, refuse and garbage'. Refuse represents household waste and contains equal parts of rubbish and garbage. Rubbish is harmless inoffensive waste which ignites easily and burns freely. Table 1 indicates details of that classification.

Garbage contains organic wastes which may be offensive, does not ignite easily and requires additional fuel for combustion. Table 1 indicates the general properties of such waste.

## Incinerator Design

This is governed by the quantities and different types of waste to be destroyed. The destruction time will vary according to the types of

**TABLE 1 Classification of Waste Materials for Incineration**

| Class | Type | Composition Approximate % by Weight | Constituents | Source | Moisture Content % | Calorific Value MJ/kg | Fuel Required to Incinerate kW/kg | Incombustible Solids % Volume |
|---|---|---|---|---|---|---|---|---|
| 1 | Rubbish | Rubbish up to 100% | Combustible waste, paper, cartons, wood scraps, rags, floor sweepings | Domestic, commercial or industrial | 25% | 15.15 | 0 | 10% |
| | | Garbage up to 20% | | | | | | |
| 2 | Refuse | Rubbish 50% | Rubbish and garbage | Residential | 50% | 10.02 | 0 | 7% |
| | | Garbage 50% | | | | | | |
| 3 | Garbage | Rubbish up to 35% | Animal and vegetable wastes | Restaurants, hotels, clubs, institutions, markets, commercial premises | 70% | 2.8 | 0.64 | 5% |
| | | Garbage up to 100% | | | | | | |

waste and to the temperature. The delivery of waste and the loading of it are important factors. Manual loading of some type of waste is acceptable but containment of waste and automatic loading has become more desirable and in some cases essential to meet the Health and Safety at Work Act.

Incinerators can be fed with waste by either top or front loading equipment, which in turn may be by bulk load or intermittent loading. Front loading is normally cheaper and is sometimes more convenient when the incinerator building is restricted in height. Top loading can be used more easily and cheaply if there is a two level or sloping site available.

The type of waste may dictate whether it requires pre-packaging before incineration. This may be essential if the waste is contaminated.

## Pyrolysis

Some incinerators employ a pyrolysis combustion process during incineration. This is where the waste material is burnt with insufficient oxygen. This is achieved by controlling the initial air and therefore oxygen at less than the stoichiometric air requirement in the initial combustion chamber. The waste gases at this stage are often called pyrogas. Oxygen deficiency prevents rapid cracking of free carbon and ensures an extremely clean combustion process. This is because the reduced volume of air leads to lower velocities of gas and minimises entrainment of solid matter so very low solid levels are present in the gas.

The pyrogas enters the secondary combustion chamber and is then mixed with excess air. This reacts rapidly to give a final clean inert state. The process is helped and the support fuel minimised by preheating the combustion air in an outer jacket prior to entry into the secondary combustion chamber.

## Waste Heat Recovery

Rising fuel costs make it desirable to take advantage of incinerator waste heat recovery. Using the energy in the waste material as a fuel is possible so long as no extra costs are involved in energy or waste disposal costs. Ideal situations are where high demands for steam or hot water exists and when the incinerator is near a boiler house, e.g. in a hospital or industrial premises. High efficiency waste heat boilers offer up to 70% recovery of gross heat input in many instances.

## Construction of Incinerators

Incinerators should be designed to comply with the requirements of the relevant British Standards and the Environmental Protection Act 1990.

- BS 3316 1987 Parts 1–4 'Large incinerators for the destruction of hospital waste'
- BS 3813 1964 'Incinerators for waste from trade and residential premises. Capacities from 22.7 to 453.6 kg/h'
- Environmental Protection Act 1990, Secretary of State Guidance:
  PG5/1 (91) Clinical waste incineration processes under 1 tonne an hour
  PG5/2 (91) Crematoria
  PG5/3 (91) Animal carcass incineration processes under 1 tonne an hour
  PG5/4 (91) General waste incineration processes under 1 tonne an hour
  PG5/5 (91) Sewage sludge incineration processes under 1 tonne an hour.

BS 3316 covers intensities of combustion less than 350 kW/m$^3$ but does not cover materials or methods of construction. No part of the incinerator that has to be touched should have a surface temperature of greater than 50°C. No other external part of the incinerator should be hotter than 80°C. Unburnt combustibles in the residue from incineration should not exceed 5% by mass.

BS 3813 covers surface temperature requirements and access and charge doors.

The Department of Environment Secretary of State Guidance Notes related to the Environmental Protection Act are based on BATNEEC – best available techniques not entailing excessive costs.

All emissions other than steam or water vapour should be free from visible smoke and the following emissions not exceeded.

| | |
|---|---|
| Chloride | 100 mg/m$^3$ (as hydrogen chloride) (250 mg/m$^3$ for general waste incinerators) |
| Total particulate matter | 100 mg/m$^3$ (200 mg/m$^3$ for general waste incinerators, 80 mg/m$^3$ for cremation) |
| Carbon monoxide | 100 mg/m$^3$ (averaged over/hour) |
| Sulphur dioxide | 300 mg/m$^3$ |

Organic compounds              20 mg/m$^3$ (as total carbon)
excluding particulate matter

Heavy metals                   5 mg/m$^3$

Emissions must be monitored with oxygen, carbon monoxide and particulate matter being continuously monitored at the outlet from the secondary combustion zone. Chloride emissions as hydrogen chloride must be tested once per week. Heavy metals should be tested at least once per year.

The temperature of the gases in the secondary combustion zone should be maintained at temperatures not less than 850°C (1,000°C for clinical waste in incinerators) and the minimum residence time for the gases should not be less than 2 seconds. The concentration of oxygen at the outlet of the secondary zone should not be less than 6% by volume.

In cases where the 2 seconds cannot be calculated or demonstrated, it is acceptable to demonstrate on commissioning that emissions of polychlorinated dibenzo dioxins and polychlorinated dibenzo furans do not exceed 1mg/m$^3$.

The height of the chimney must be more than 8 m above ground level and will be according to the 'Chimney Heights, 3rd Edition of the 1956 Clean Air Act Memorandum'. The minimum chimney height should be 3 m above the roof ridge height of any building within a distance of 5 times the uncorrected chimney height. The designer efflux velocity will be more than 15 m/sec in normal operation. The ductwork and chimney flues must be insulated to prevent condensation on internal surfaces.

## Types of Incinerators

Incinerators are designed for a variety of applications:

- Sanitary
- General
- Clinical and Pathological
- Special Industrial
- Cremation of Human Beings.

### Sanitary Incinerators

Due to the Environmental Protection Act (even though incinerators operating at less than 25 kg/h are exempt) and the development of macerators, there are now very few sanitary incinerators to be found on the district. The capacity varied from 0.003 to 0.009 m$^3$ for the destruction of sanitary towels and other small dressings. They were used in offices, factories and institutions, Fig. 1.

The general construction of a small sanitary incinerator is shown in Fig. 2. The primary chamber is of moulded or fabricated refractory brick above steel grate bars. The natural draught drilled ring burner is positioned below the grate bars.

Waste is fed through a front, drop-down loading door. This may be fitted with a hopper to limit the size of the load and keep the primary chamber closed while the door is open. The door is linked mechanically to the timer and to the grate bars. Opening and closing the doors starts the timer, so turning on the gas to the primary burner. This then fires for about 15 to 20 minutes. Movement of the loading door also shakes the grate bars to break up the ashes so that they fall into the ash pan below. The ash pan may be removable either through a front door or, as illustrated, from underneath.

Small sanitary incinerators generally do not have a secondary chamber or after burner. This is because the incineration of small dressings is mainly the evaporation of a little moisture with only a

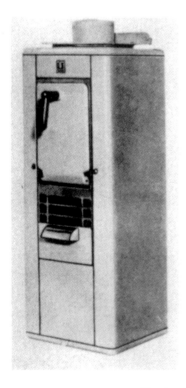

*Fig. 1   Sanitary incinerators*

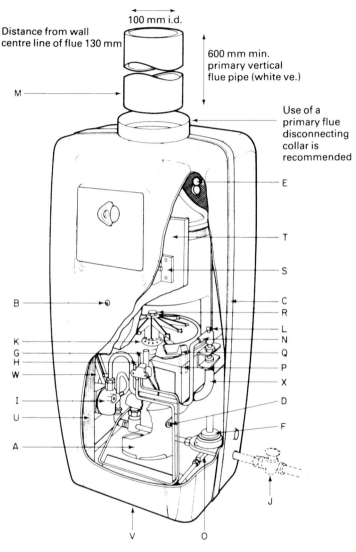

100 mm i.d.

Distance from wall centre line of flue 130 mm

600 mm min. primary vertical flue pipe (white ve.)

M

Use of a primary flue disconnecting collar is recommended

E

T

S

B

C
R
L
N
Q
P
X
D
F

K
G
H
W

I

U

A

V    O

J

| A | Ashpan | I | Clock | R | Location nut |
|---|--------|---|-------|---|--------------|
| B | Philips head securing screw | J | External gas cock | S | Door hinge |
| C | Bright strip | K | Aerated burner | T | Brick strap retaining screw |
| D | Screw retaining back | L | Pressure test point | U | Screws retaining timer |
| E | Studs | M | Primary flue pipe | V | Flame supervision device |
| F | Screw-in adjustable governor | N | Burner union | W | Actuating linkage arm |
| G | Thermocouple head | O | Pilot filter | X | Mixing tube |
| H | Pilot Jet | P | Brass nuts | | |
| | | Q | Combustion brick base plate | | |

*Fig. 2    Construction of small sanitary incinerator*

small weight of solid matter to be finally burnt. Combustion is normally complete within the primary chamber and very little smoke is produced.

The outer case is of steel sheet usually stove enamelled. The incinerator is wall fixing, suspended from slots in the rear section of the case.

The gas supply connection is $R_c\frac{1}{4}$ into a governor and a control cock must be fitted. The flue is 100 mm diameter. The heater is suitable for natural or mechanical draught from 0.57 to 2 m$^3$/min.

The controls are located in the lower section of the appliance and include:

- governor
- timer
- thermoelectric flame supervision device
- solenoid valve and air-flow switch, if flued by mechanical extraction.

The timers are clockwork and operate an electric switch which interrupts the thermocouple circuit.

Ignition is usually manual to a permanent pilot. With the pilot established and the flame protection valve operating, opening the loading door will riddle the grate, wind the timer spring and open the main gas supply to the burner. The loading door must be opened fully or the timer will not be properly set. This will result in a shorter firing period and probably incomplete incineration of the load.

Closing the door allows the clock spring to drive the timer escapement so that gas to the burner will be cut off after a pre-set time determined by the clock setting or the adjustment of the dashpot air valve.

*General Incinerators*

These incinerators are available in a wide range of sizes from about 25 kg/h up to 20,000 kg/h, capable of destroying a wide range of solid materials with the minimum of attendance. A typical size may be 400 kg/day. The units are usually loaded 2 or 3 times per day. One type is shown in Fig. 3. Closing the door starts the automatic burning cycle during which time the loading door will remain electro-mechanically locked until the cycle is completed and safe conditions for reloading exist. To meet the new regulations the minimum flue gas residence time must be 2 seconds at 850°C. This is required to meet the smoke-free emission standards for flue emissions mentioned earlier.

*Fig. 3   Incinerator showing heat recovery (Hoval)*

Waste heat is recovered from the secondary combustion chamber to produce hot water. Additional air is introduced into this combustion chamber to complete the drying, burning, degassing process started in the primary chamber. The secondary chamber with excess air and turbulent conditions causes oxidising of all the combustible gases.

Figure 4 shows another version with the hot gases passing through a waste heat boiler. This incinerator has mechanical charging and preheating of the combustion air. This incinerator has multi zones for pyrolytic incineration of the waste separating the different stages involved in the incineration process. The zones are for heating, drying and degassing of waste in the low temperature chamber; mixing, igniting and complete oxidation of the gases in the thermal reactor. Individual monitoring and control of each operation gives good incineration performance virtually independent of the waste composition. The waste is heated to the required ignition temperature. Only sufficient oxygen is admitted to give the necessary reaction heat to maintain the thermal decomposition and degassing of the waste. The gas released contains a large quantity of flammable energy-rich constituents. The gas passes to the mixing zone in the thermal reactor and secondary air is introduced to give an ignitable mixture which then combines with tertiary air and burns out in the ignition and reaction zones with a consequent rise in temperature.

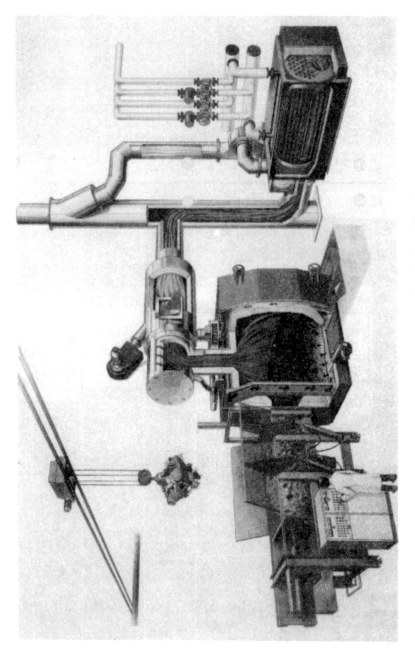

*Fig. 4   Incinerator with waste heat boiler (Hoval)*

Referring to Fig. 4, the air-cooled outer steel shell has an inner shell lined with thick insulation and refractory. The primary air passes between these shells providing cooling and at the same time preheating the incoming combustion air. This then passes into a plenum chamber before passing through small holes in the air-cooled cast iron bed plates. These holes are positioned for optimum distribution of the primary air so that sub stoichiometric conditions maintain correct degasification of the waste.

The operation of de-ashing is simplified by the flat surface of the air-cooled sectional bed plates. A large de-ashing door is fitted to one end of the chamber and the feed aperture is at the opposite end. A small waste ignition burner is situated adjacent to the waste entry point.

The thermal reactor is made up of a number of cylindrical steel sections with internal refractory lining. Basically there are three zones in the reactor:

- mixing zone with secondary air inlet
- ignition zone with modulating burner
- combustion zone with tertiary air inlet.

The temperature in the reactor depends on the quality of the gas from the low temperature chamber and up to 1,400°C can be reached. Hot gases can be drawn through a waste heat boiler to produce hot water or steam. If heat recovery is not required the gases pass directly up the chimney.

A centrifugal fan on top of the low temperature chamber supplies primary and secondary air. Primary air is controlled by the temperature in the chamber and the secondary air is preset. The tertiary air for combustion enters the thermal reactor so that good mixing and turbulence is ensured. The volume of air is controlled by the reactor temperature.

A wall-mounted or free-standing control panel houses all the components necessary for automatic operation.

The unit is made in a wide range of sizes to process between 300 and 20,000 kg of waste per day producing sterile ash, which can be removed automatically or manually or by vacuum cleaner.

*Controls and Operation*

The controls are panel-mounted for operation of the incinerator and the burners are packaged forced draught burners, Fig. 5. The flame safeguard units, safety shut-off valves are as described in Chapter 4. The de-ashing operation, if automatic will be operated from the control panel as will be the charging operation.

*Fig. 5    Packaged burner for incinerator (Weishaupt)*

### Clinical and Pathological Incinerators

Clinical waste is any waste containing any human or animal tissue, blood, body fluids, excretions, drugs, swabs, dressings, syringes, needles or pharmaceutical products. It also includes any other waste arising from medical, nursing, dental, veterinary, investigation, treatment care, teaching or research or the collection of blood for transfusion being waste which may cause infection to any person coming into contact with it.

Thus hospital waste for example can vary tremendously. Waste with a calorific value of more than 8.4 MJ/kg is generally self-supporting in combustion after initial ignition. Some incinerators can cope with plastic contents of between 50 and 100% and will adjust fuel requirements automatically. With a 50% plastic content one range of incinerators over an 8 hour charging period can deal with 40 kg to 2,800 kg using 8.4 m³/h and 22.5 m³/h of natural gas respectively. (This is equivalent to 87 kW and 232 kW).

Whilst manual loading may be acceptable for some waste products, containment of waste and automatic loading has become desirable if not essential in some instances because of the need to comply with the Health and Safety Act.

A large central incineration plant for 1 tonne per hour of clinical waste is shown in Fig. 6. In this plant waste from multiple sources is marshalled in 'one trip' boxes which are read from bar coded information. Many installations use containers which are returned after washing. The size of the containers is usually $2 - 5$ m$^3$. Relatively small quantities of waste are fed into a hydraulic loader and charged at a load cycle interval of $3 - 4$ minutes, so that the most uniform possible combustion conditions are achieved. The loader acts also as a ram pushing waste along the hearth, where it falls to two or three lower hearths, each one also provided with a hydraulic transfer ram.

Final burn-out occurs in the ash box which is also a burn-out chamber.

A fully enclosed ash container is mated-up to the ash hopper when ash discharge is necessary (about twice per 24 hours).

*Secondary Combustion*

In order to achieve the very high quality of combustion performance, the secondary combustion chambers provide 2 seconds residence time at 1,000°C – even before the start of waste burning. The chamber internal construction provides a tortuous path for gases, and design velocities of 15 m/s everywhere within ensure that turbulent flow and mixing take place.

*Energy Recovery – Gas Cleaning*

The opportunity exists for gas cooling by heat exchange (waste heat recovery) or by direct quenching with water. Thereafter absorption of acid gases and particulates removal (including heavy metals) can take place in one of the conventional scrubber systems.

A medium pressure hot water heat exchanger and impingement tray scrubber with pre-quench is used. Water circuit temperatures are $13 - 110$°C and the gas enters the quench at 275°C, becoming saturated at $65 - 70$°C for absorption on impingement tray scrubber elements.

Following the gas cleaning, warm air is added just before the induced draught fan to prevent liquid condensation and liquid corrosion within the fan and ducting. Plume discharge is prevented by the admission of a much larger quantity of air warmed to $115 - 120$°C. This technique is especially suitable when there are fluctuations in the gas volume produced by waste combustion.

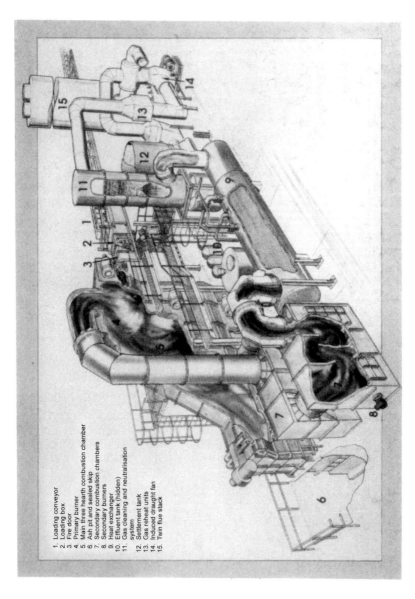

1. Loading conveyor
2. Loading box
3. Fire door
4. Primary burner
5. Main three hearth combustion chamber
6. Ash pit and sealed skip
7. Secondary combustion chambers
8. Secondary burners
9. Heat exchanger
10. Effluent tank (hidden)
11. Gas cleaning and neutralisation
    system
12. Settlement tank
13. Gas reheat units
14. Induced draught fan
15. Twin flue stack

*Fig. 6   Large clinical waste incinerator (Evans Universal)*

## Pathological Incinerators

Pathological incinerators will accept 100% pathological material such as animal waste and usually involve an increase in thermal input to the primary chamber and a means of retaining any fluids until they are incinerated.

The efficient incineration of pathological waste and animal carcasses demands a design of furnace different to that used for the disposal of refuse. Combustion chamber proportions, position and configuration of air ports, auxiliary heat input and its distribution across the hearth area have an important bearing on performance and running costs. The hot hearth principle is used where the hot products of combustion raise the hearth to an incandescent temperature, Fig. 7. This combusts the charge to a sterile ash without spillage of fats and without necessity to rake or manually disturb the firebed.

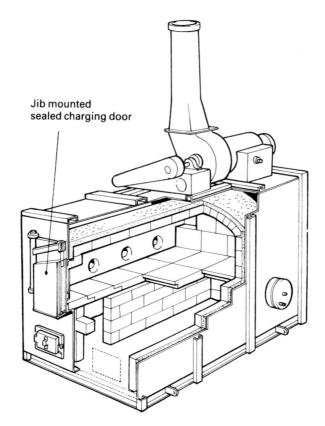

*Fig. 7    Pathological incinerator (Incol)*

The level of the L.E.L. to provide self-sustaining operation, influences whether the incinerator works on the regenerative or recuperative principle. Fig. 9 indicates the heat exchanger selection criteria. This shows that high heat recovery rates with less than about 8% L.E.L. must be regenerative and generally operate with high exhaust levels greater than about 1,400 standard cubic metres per minute.

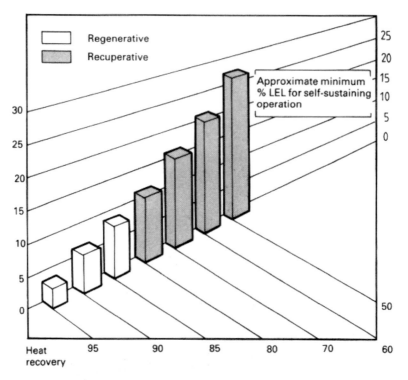

*Fig. 9    Heat exchanger selection criteria for thermal oxidisers*

The recuperative thermal deoxidiser described earlier is more common than the regenerative system. The waste heat recovered can be used solely to preheat the large volumes of cool gases containing the volatile organic compounds to reduce the amount of energy needed to oxidise them, or can be used to preheat air in addition to the polluted gases, Fig. 10.

Heat within the oxidiser is usually from induct burners as shown in Fig. 11. When fitted with a profile plate, the pollutant-borne gases must pass through the perforated holes within the burner and hence through the natural gas flame. In this way the oxidation temperature needed to destroy the volatile organic compounds is reached.

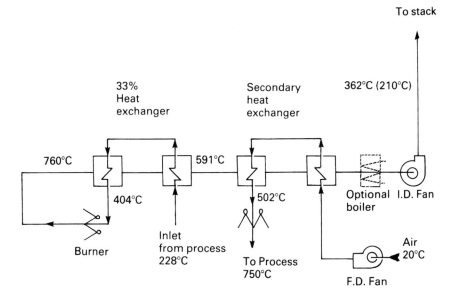

*Fig. 10    Line diagram showing heat recovery*

*Fig. 11    Induct burner*

## THE INCINERATION OF PROCESS PLANT WASTE STREAMS

### Introduction

The majority of process plants produce some liquid or gaseous waste which cannot be discharged directly from the plant. A large proportion of these waste streams contain combustibles and a surprisingly large number can be safely disposed of by thermal oxidation in a purpose designed incinerator.

Incineration has a justified claim to effectiveness and operating flexibility, both in terms of concentration and flow rate of the streams forwarded to the incinerator. In most instances thermal oxidation provides a complete answer to problems of pollution by combustible material. However, some waste streams containing inorganics or halogens will require downstream clean-up after incineration in the form of electrostatic precipitators, filters or scrubbers.

*Type of Waste*

The simplest organic compounds form the most easily incinerated wastes. Once mixed with the correct amount of air the combustion products can be discharged directly to atmosphere after completion of the oxidation reaction. As compounds become more complex the problems for the incinerator designer, grow accordingly. The complexity can be roughly associated with a grouping of the chemical elements contained in a waste:

Group 1 – C H O N
Group 2 – C H O N – S
Group 3 – C H O N – S – F Cl Br
Group 4 – As above with inorganics, Na K Ca P Si etc.

Group 2 compounds generally do not constitute a combustion problem but can require special material selection and compliance with the local chimney height regulations.

Group 3 compounds require special attention in the design of the thermal oxidation reaction as well as correct selection of materials. The resulting combustion products stream will require downstream clean-up but this can be reduced by correct reaction design.

Group 4 compounds add the problems of dust control and refractory slagging to those of Group 3.

*Support Fuel and Heat Recovery*

Many incineration reactions are self-sustaining in as much as the waste itself is readily combustible. In general a waste stream with a calorific value of 10,000 kJ/kg will burn without the addition of support fuels and raise the combustion products to sufficient temperature to complete the oxidation reaction.

When the waste stream calorific value is less than this, support fuel would be added to bring the reaction (ideally) to the minimum temperature required to complete oxidation. If use for the waste heat can be found, up to 80% of the otherwise waste heat can be recovered. With the Group 1 compounds above, almost all the heat could be recovered.

## Design Parameters

Thermal oxidation of a waste stream will be achieved if the waste stream is mixed with adequate oxygen and raised to sufficient temperature for the reaction to initiate and be maintained at temperature long enough for the reaction to complete. This is the traditional three 'T's of the combustion engineer:

Time
Temperature
Turbulence

*Time*: The majority of incinerators used on process plants attempt to mix the waste stream with the necessary air at the burner front. If this can be achieved residence times of half a second will ensure complete oxidation (at the correct temperature)!

*Temperature*: The lower limit for the incineration temperature is theoretically the ignition temperature of the combustibles in the waste. On this basis 600°C would generally be sufficient but most designers would consider this far too low since the speed of reaction will also increase at elevated temperatures. A typical minimum design temperature would be 700°C.

*Turbulence*: Achieving sufficient mixing of the waste and air streams is the heart of the incinerator designer's expertise. Techniques are obviously borrowed from the furnace designer but practices which would be acceptable in a furnace may not be so suitable in an incinerator. Consider for example an air heater: if only 50% of the air passed through the combustion zone, the resultant air, after downstream mixing, could still meet specification. The same design as an incinerator would only be 50% efficient.

*Stoichiometry*: With a known waste stream the quantity of oxygen and air to satisfy the oxidation reaction can be readily quantified. In many cases the combustibles may be very complex or unknown in an overall mixture. In such cases the approximate air requirements are 1 m$^3$ per 3,500 kJ of combustibles. This gives the stoichiometric air requirement for the reaction. To ensure mixing and reaction completion excess air is required. The excess air factor varies with both the design and waste itself – in all cases 100% excess air would be a comfortable figure.

### Design Types

A variety of incinerator designs to handle various waste streams are reviewed below:

### (1)  High calorific value gas or liquid streams

High calorific value (CV) waste streams are handled in the concept of a fuel and are mixed directly through the burner gun. Initiation is completed immediately and the reaction completed largely within the burner quarl. Support fuel may be required to heat the fabric of the incinerator on a black start and to compensate for variations in the flow rate of the waste streams.

### (2)  Gas streams with low calorific value

Gas streams with low CVs are usually solvent contaminant air streams. If the oxygen content of the stream is above 12% the stream is readily incinerated with an inline duct burner supplying the support fuel and mixing the stream as it passes across the burner section.

Low oxygen gas streams with low CV have to be mixed with the necessary combustion air in the heart of the incineration chamber. Tangential entries and low swirl systems are most suitable with the support burner located at the incinerator axis.

### (3)  Aqueous wastes with low combustibles

To incinerate low CV aqueous wastes it is necessary to atomise the liquid stream, mix it with air and introduce it to the heart of the flame of the support burner. Depending on the air requirement this is either achieved by spray systems located around the burner front or by injecting the liquid through a dummy burner on the incinerator axis with the support burner firing tangentially into the system.

### Halogenated Hydrocarbons

The incineration of halogenated hydrocarbons is a subject in its own right. It is included, in brief, to illustrate the conflicting balances and judgements required by the incinerator designer throughout the design process. Similar judgements have to be made in even the simplest incinerator units.

The design objective in a chlorinated hydrocarbon incinerator is to reduce the downstream scrubbing load by producing as little free chlorine as possible with the remainder converted to the easily scrubbed hydrogen chloride. Additionally free chlorine at high temperatures will react with silica and alumina and severely reduce the refractory life of the incinerator.

The formation of hydrogen chloride from chlorine is enhanced by:

elevated temperature which increases the equilibrium constant
increased steam
decreased oxygen
increased hydrogen:chlorine ratio

Options available to the designer are given below:

(a) Decreasing the excess air which will in two ways enhance the formation of hydrogen chloride but will put a much greater reliance on the incinerator mixing system. Excess air levels of 100% are not encountered on these units.

(b) Similarly, increasing the water vapour content will increase the formation of hydrogen chloride but will also reduce the operating temperature.

(c) Increasing the hydrogen:chlorine ratio by adding fuel will be doubly beneficial, since it will also increase the temperature in the unit, but will increase the running costs.

In general the complete destruction of the waste compounds is the overriding target in these units and high temperature operation is essential. With computerised design a reasonable optimum can usually be achieved.

## Crematoria

In England and Wales there are about 600 crematoria and most of them operate on natural gas. Fig. 12 shows an older design cremator with a twin chamber. The combustion chamber consists of one burner and a number of high pressure air jets which together facilitate rapid and complete combustion of the coffin and its contents. The burner situated in the upper back wall fires forwards and downwards so that the maximum intensity of the flame is concentrated where it is most effective. For appreciable periods the burner is unnecessary, air for combustion is supplied by jets in the roof. Separately controlled

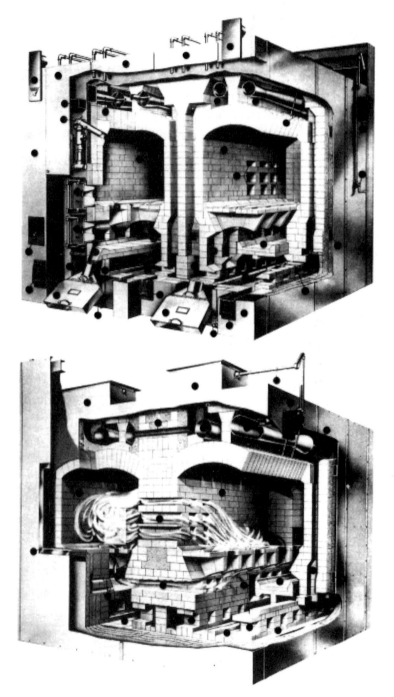

*Fig. 12   Cremator (Dowson and Mason)*

auxiliary or smoke consuming air also blows from one end of the secondary chamber under the hearth.

A very important feature is the venturi eductor system controlling suction at all times to govern the rate of combustion and prevent emission of fumes into the crematory area. Fig. 13 shows the venturi eductor system in more detail.

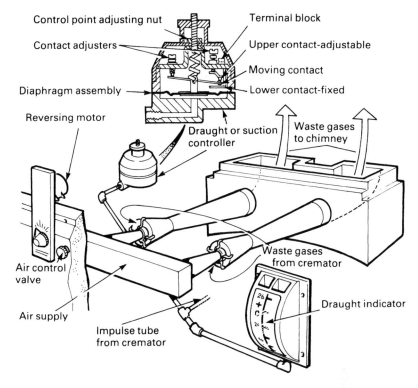

*Fig. 13    Venturi Eductor System (Dowson and Mason)*

At the rear are two large openings down which remains are raked at the end of each cremation. Air is added to complete combustion of gases which pass longitudinally through the middle chamber and out at one end through openings in each side. Direction is reversed in the third or bottom chamber, the gases being drawn up flues in the side wall and longitudinally into the venturi eductor system. The long and tortuous path through which the gases flow with varying velocity ensures complete combustion resulting in smokeless operation and high thermal efficiency.

A Reflux Twin Cremator consists of two single cremators in a common casing with interconnecting ports for heat transfer. The waste gas from one cremator preheats the second which is then charged, the flow reversal and the heat from the second is then transferred to the first. This results in low fuel consumption, higher and more uniform operating temperature, increased refractory life and assists in the control of odour.

*Controls*

The burner is usually a premix burner and controls consist of a push button operated electronic control box providing ignition and ultra-violet flame protection monitoring the main and pilot burner. The angle of the scanner tube makes it impossible for ultra-violet light to be picked up from any source of combustion other than the burner. A pre-purge is carried out before ignition. Dampers have limit switches fitted and a pressure switch protects against high gas pressure.

*Modern Cremator*

Fig. 14 shows a modern cremator. They are available as single or double chambers and may be single or double ended in operation and have long residence times and higher secondary combustion chambers of 1,000°C. The choice of manual or microprocessor control is available. The manual system incorporates a large number of fail safe and automatic control features in a fault monitoring system and leaves only the basic running of the cremator to the operator. The fully automatic system takes all control activities out of the operator's hands including the start up of auxiliaries such as air and hot gas fans.

Modern regulations of the Environmental Protection Act call for gas velocities of 15 m/sec at the flue exit and the pressure loss in duct systems frequently exceeds that which can be overcome by venturi eductors. Induced draught is the only solution to this problem and that of complex flue systems.

*Servicing*

It is important that the refractory brickwork is maintained in good condition and this is a specialist job as will be appreciated from Fig. 12. The air jets and burner can carbon up due to localised deficiency of air at points during the cremation. Conventional gas controls, safety shut-off valves, governors, pressure switches and manual valves are typical components described in Chapter 4.

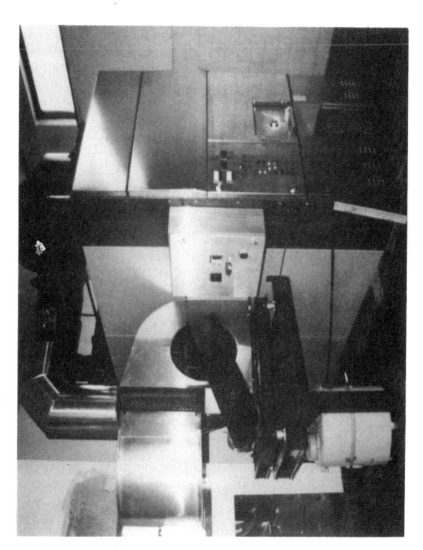

*Fig. 14   Modern cremator (Evans Universal)*

## INSTALLING INCINERATORS

### General

Note should be taken of the recommendations and statutory requirements of the following documents:

- the Building Regulations or the Building Standards (Scotland) Regulations
- the Gas Safety Regulations
- current BS codes of practice for flues and installation work
- the Clean Air Act, including Chimney Height Memorandum (3rd edition)
- the Environmental Protection Act and the relevant Secretary of State Guidance Notes
- National Air Quality Standards 1997 (see Chapter 13).

When incinerators are to be installed the approval of the local authority should be obtained. Generally if a proposed installation meets the requirements of the Building Regulations it is likely to meet those of the local authority.

### Location

Large general incinerators are usually housed in purpose-built locations. The requirements for their installation are similar to those for any gas-fired furnace. Because of the large amount of refractory brickwork involved, this type is often constructed on-site.

### Flues

Generally to meet the Environmental Protection Act flues should meet the Chimney Heights Memorandum 3rd edition and terminate 8 m from the ground. Some incinerators are fitted with a draught stabiliser, Fig. 15. This is a freely pivoted damper which is held closed by a weight. When the flue pull becomes excessive, as when a mainly combustible load is burnt, the damper opens to allow cool air into the flue. This has the effect of:

- reducing the flue gas temperature
- reducing excessive flue pull
- preventing over-fast combustion
- preventing overheating of the appliance.

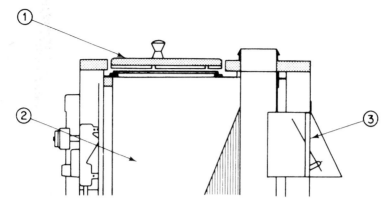

1. Loading door
2. Primary chamber
3. Draught stabiliser

*Fig. 15    Draught stabiliser*

The flue should be vertical, as far as possible. No part should be at an angle of less than 45° to the horizontal and only obtuse bends should be used. The amount of external flue should be kept to a minimum.

The Building Regulations require that provision is made for cleaning and inspecting incinerator flues. All bends should be of a type fitted with a cleaning door.

Flue pipe sockets should face the terminal. Existing brick flues may be used provided that:

- the brickwork is in good condition
- the flue does not communicate with another part of the building
- there are no obstructions
- all other apertures are closed.

Mechanical extraction may be used either for venting individual appliances or for a number of appliances sharing a common flue.

With multiple appliances sharing a common flue, the main flue must be sized, depending on the number of appliances connected.

A better performance may be obtained from a flue of constant diameter, rather than a stepped flue. The flow through each incinerator must be balanced and flue restrictors may need to be fitted to those appliances nearest to the fan. Any mechanical ventilation of the room should be taken into consideration when designing an extraction system.

Great care must be taken during the installation to ensure that the flue joints are sound. Leakage into the flue can seriously affect its performance.

When mechanical extraction is used, each appliance must be fitted with a flame protection device and an interlock system to cut off the gas supply to each appliance in the event of fan failure.

## Commissioning

The procedure for commissioning incinerators will vary with the type and the flue systems. Manufacturer's instructions should always be followed. On large installations the commissioning will usually be carried out by the manufacturer or the contractor responsible for building the equipment.

For smaller incinerators, the procedure includes the following operations:

- check the gas supply for soundness and purge
- light the appliance in accordance with the manufacturer's instructions; test any ignition device
- check that pilot flame is established, stable and correctly positioned relative to the flame monitoring sensor
- light up primary and secondary burners in turn, check burner pressure and gas rate at the meter if necessary (where a proportioning valve is fitted any adjustment must be made in the first two minutes after lighting)
- check controls, as fitted
  - flame protection device and flame sensors
  - timer
  - proportioning valve
  - flue thermostat controlling secondary burner
  - safety shut-off valves and flow switches on:
    - mechanical extraction
    - flue gas washer water supply if fitted
  - clock controls
- check operation of loading door, diversion flap and adjustment of linkages and interlocks.

When the appliance is operational it can be checked on its performance when incinerating an average load of appropriate waste. Check that the flue gases are extracted satisfactorily without leakage or spillage and that no smoke is emitted. Check the operation of any draught stabiliser. Check that the load is completely burnt in the recommended time.

Finally, instruct the operator, set any clock controls and leave the appliance in operation.

**Servicing**

The servicing of the smaller incinerators is mainly a cleaning operation, particularly where the pilot and main burner are below the grate bars.

The work on-site is reduced to cleaning out the combustion chamber, followed by an inspection of the refractory lining and flueways and attention to the door mechanism and linkages.

The general procedure for servicing general incinerators should include the following operations:

- question the operator on the performance of the appliance
- remove outer case or panels as required
- riddle the grate, check for freedom of movement, primary chamber clear of ash
- remove and empty ash drawer or pan
- remove and clean pilot burner
- remove and clean primary and secondary burners
- check that injectors are clear, burner tunnel clear and ports free from blockage
- examine condition of refractory lining to primary chamber
- clear any dust from secondary chamber
- check flueways clear
- clean and check door mechanism
- clean linkages and interlocks and adjust if necessary
- check gas control cocks, ease and grease if required
- replace gas burners
- check gas soundness
- light the appliance and check any ignition device
- check pilot stable and correctly positioned
- check burner pressures or gas rate at the meter if necessary
- check flue for pull and spillage
- check controls
  - flame protection device
  - timer set correctly
  - any other control devices
- check that the timer returns to zero, shutting down the appliance at the end of the combustion cycle.

*Fault Diagnosis and Remedy*

Most of the faults which can occur on incinerators are associated with the burners or the control devices including the mechanical devices associated with charging and de-ashing and have been dealt with in previous chapters or volumes.

Because of the amount of dust and debris associated with incineration, a number of faults may be due to accumulations of fly ash in burners and flame ports or deposits on jets or thermocouples.

Other common faults are given below:

### Timing Erratic

If the timer is operated by the loading door or by a remote knob, the linkage may be slack and require adjustment.

If the linkage is satisfactory, then the fault is in the timer which should be exchanged. It may be possible to repair an early dash pot type of timer but repairs to clockwork timers should not be attempted.

### Excessive Smoke from Terminal

This is usually due to the secondary burner failing to consume the smoke. It is commonly caused by overloading the combustion chamber, a blockage in the burner or an inadequate gas rate. The combustion chamber should be emptied, the burner cleaned and readjusted or the flueway and secondary air supply checked as appropriate.

The emission of smoke, dust and grit is subject to the requirements of the Environmental Protection Act 1990. A requirement of the Act is that dark smoke shall not be emitted from the chimney of any building, including dwellings.

It is an offence not to use all practicable means to reduce the emission of dust and grit from chimneys.

The Act also requires that furnaces burning solid fuel or waste at 50 kg/h or more shall be fitted with grit arresting plant, approved by the local authority.

The statutory requirements will be revised and the installation and operation of any gas-fired incinerator must be carried out in accordance with the current legislation.

Current updating of existing incinerators had to be completed by the mid-1990s.

# Steam Boilers

Chapter 8 is based on an original draft by A. J. Spackman

## Introduction

Steam is extensively used in industry and commerce chiefly for heating processes, power generation and space heating.

Its main characteristics are:

- it is produced from water which is readily available and cheap
- it is clean, odourless, tasteless and sterile
- it is easily distributed and controlled
- when condensed, it gives up heat at a constant temperature
- it has a high heat content
- it can be used to generate power and then to provide heating.

Steam may be produced in any one of the three following conditions:

- wet steam
- dry saturated steam
- superheated steam.

When steam is generated from water, it may be taken through several heating stages to produce steam in the condition required.

First, the water receives sensible heat to bring it to the temperature at which steam begins to form. This is 100°C at atmospheric pressure or a higher temperature if the pressure is increased.

Secondly, the water at the temperature and pressure of steam formation receives latent heat to evaporate the water into steam at the same temperature and pressure. The heat required is approximately 2,250 kJ/kg at atmospheric pressure. If the water is completely vaporised it is called 'dry saturated steam'. If, however, the steam contains a proportion of water in suspension, it is 'wet steam'. Wet steam has only received a fraction of the latent heat required to turn all its water content into dry saturated steam.

Finally, if required, the dry saturated steam may be given sensible heat to raise its temperature still further and produce 'superheated steam'.

The condition of wet steam is specified by its 'dryness fraction'. This is the ratio of actual steam to wet steam.

$$\text{Dryness fraction} = \frac{\text{weight of actual steam formed}}{\text{total weight of wet steam}}$$

For example, if 1 kg of wet steam has a dryness fraction of 0.8 then it contains:

    0.8 kg of actual steam
    0.2 kg of water
or   800 g of steam
    200 g of water

This 1 kg of wet steam has, therefore, only received 0.8 (or 80%) of the total amount of latent heat required for its vaporisation. That is:

    0.8 × 2250 = **1800 kJ**

## Types of Boiler

Boilers are generally sized by their steam output or 'evaporating capacity'. This is the quantity of water in kg/h which can be evaporated from and at 100°C. Working pressures vary with the type of boiler and the process requirements. They range from 350 mbar for small gas boilers, up to about 160 bar in modern power stations.

Boilers are classified as either 'fire tube' or 'water tube' types.

Fire tube boilers, or 'shell' boilers as they are called, are generally cylindrical in shape and have a combustion chamber with a minimum length to diameter of 3:1. They have often one, two or multi burners firing through tubes which pass through the water in the shell from one end to the other.

Fire tube boilers may be subdivided into either 'single-pass' or 'multi-pass' types.

### Single Pass Fire Tube Boilers

These boilers have one set of fire tubes with burners at one end and the flue at the other, Fig. 1. They may be mounted with the shell vertical or horizontal. The burners fire directly into each tube and are normally fanned draught on the horizontal boilers and natural draught on the vertical boilers. The boilers are designed for gas firing and have outputs of from 36 kg/h up to about 360 kg/h. A typical

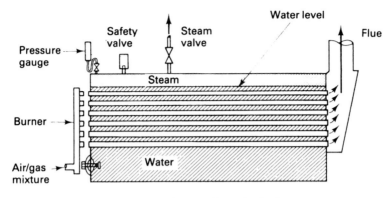

*Fig. 1   Single pass fire tube boiler*

vertical boiler is described later on in this chapter. Boilers of this type are commonly used by dry cleaners and clothing manufacturers to operate steam irons and Hoffman presses.

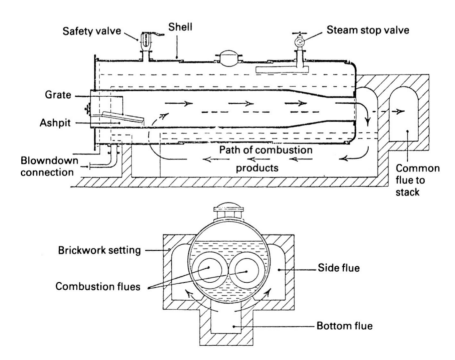

*Fig. 2   Lancashire boiler*

## Multi-pass Fire Tube Boilers

This boiler usually has a single main combustion tube with other sets of tubes passing the hot gases to the front of the shell and back again. One of the earlier designs was the Lancashire boiler shown in Fig. 2, these were originally designed for coal firing but were converted to natural gas firing. The range of diameter was from 1.85 m to 3.0 m with lengths of 5.50 m and 9.75 m respectively. These would give evaporative rates originally of 1,360 – 5,450 kg/h of steam. Steam pressures were of up to 17.7 bar.

Cornish boilers were similar to Lancashire boilers but only had one firetube and were smaller in diameter and length. They also operated at lower steam pressures of about 10.9 bar. The thermal efficiency of a Lancashire boiler is 64 – 78% with an economiser fitted. An economiser is a water/gas heat exchanger through which flue gases travel after leaving the boiler but before going up the flue. The Economic boiler developed from the Lancashire boiler and is shown in Fig. 3 in a dry back version.

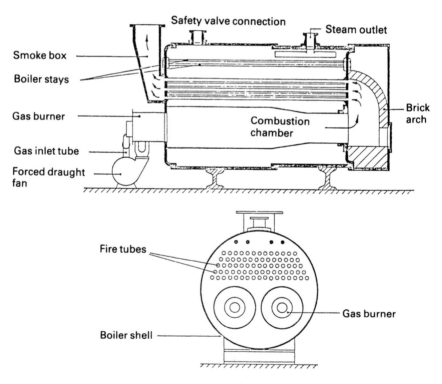

*Fig. 3   Economic boiler two pass dry back*

The hot combustion gases travel down the combustion chamber before being deflected by the brick arch into the bundle of tubes forming the secondary pass to the smokebox and flue. This type of boiler is a two pass dry back boiler.

The thermal efficiency of this type of boiler is generally about 73 – 77%. The modern packaged shell boiler is generally a three pass wet back boiler, Fig. 4, with no problems of air leakage and may be operated at thermal efficiencies of 78 – 83%. The packaged unit is supplied complete with burner(s) and controls.

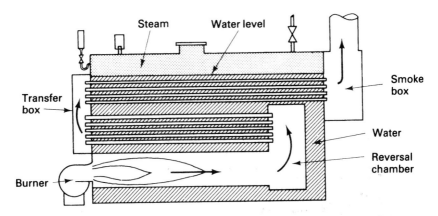

*Fig. 4    Three pass wet back boiler*

It may be oil, gas or dual-fuel fired. Outputs may be up to 31,800 kg/h at pressures up to 18 bar. Boilers over 16,820 kg/h output generally have two combustion tubes.

## Water Tube Boilers

These are the large, high pressure boilers used for industrial or power generation purposes. The hot gases from the burners pass around vertical banks of tubes containing the water. The boilers are roughly rectangular in shape and the tubes are connected to a water drum at the bottom and to a steam drum or manifold at the top, Fig. 5. There is usually a superheater above the main combustion chamber. Outputs are generally above 20,000 kg/h. Because of economic factors these boilers have generally been fired by pulverised coal or oil. Some have been converted to gas firing on an interruptible basis with dual-fuel burners.

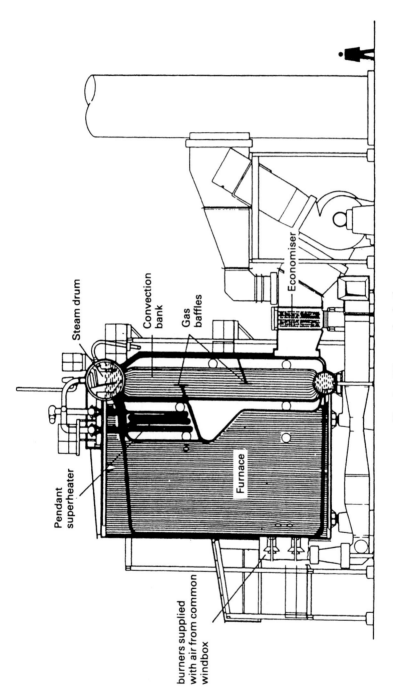

*Fig. 5   Water tube boiler*

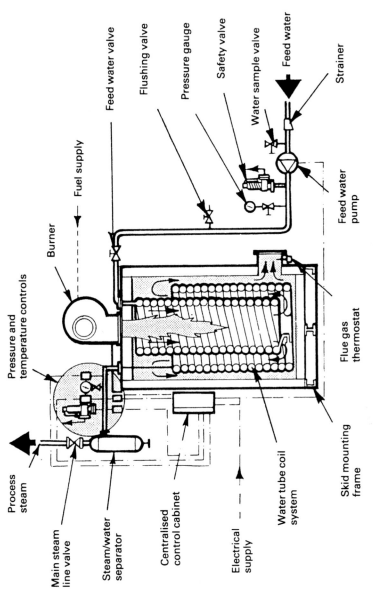

Process steam

Main steam line valve

Steam/water separator

Centralised control cabinet

Electrical supply

Water tube coil system

Skid mounting frame

Pressure and temperature controls

Burner

Fuel supply

Feed water valve

Flushing valve

Pressure gauge

Safety valve

Water sample valve

Feed water

Strainer

Feed water pump

Flue gas thermostat

*Fig. 6   Coil type boiler*

*Coil Type Boilers*

These are a form of water tube boiler with the water contained in sets of coils. The burner fires down into the centre of the inner coil and the products pass around the outer layers of coils, Fig. 6. These are sometimes referred to as steam generators or steam vaporisers. They are low water capacity boilers and produce small quantities of steam quickly in less than 5 minutes. Care must be taken with the water treatment. Standard sodium base exchange units combined with chemical additives are all that is normally necessary for feed water treatment. The outputs may range from 200 kg/h to about 9,090 kg/h at 40 bar. They are either gas or oil-fired by packaged burners.

*Other Boilers*

In addition to those described, there are small, gas-fired boilers which are used to provide wet steam for bakers' ovens, Fig. 7. This 'flash' steam is introduced into the oven for about 15 minutes during baking to give the loaves or rolls a rich brown, crusty surface.

The boilers are vertical and usually wall-mounted with outputs up to about 90 kg/h. They are used principally by small, specialist bakers. Working pressures are in the region of 2 bar.

## Safety Regulations

*Factory Acts*

All steam boilers and their installations must conform to the requirements of the Pressure Systems and Transportable Gas Container Regulations 1989. If the steam boiler is automatically controlled then it should conform with H.S.E. publication P.M.5 *Automatically controlled steam and hot water boilers 1989*. The overall intention of these Regulations is to prevent the risk of reasonably predictable danger from stored energy as a result of failure of a pressure system or part of it. The Regulations apply to all steam systems and systems in which gases exert a pressure in excess of 0.5 bar above atmospheric pressure.

Boilers must be fitted with various 'mountings' which include:

- a suitable steam safety valve
- a suitable stop valve connecting the boiler to the steam pipe
- a steam pressure gauge, showing the maximum permissible working pressure
- at least one water gauge of transparent material, fitted with a guard if the working pressure exceeds 2.75 bar

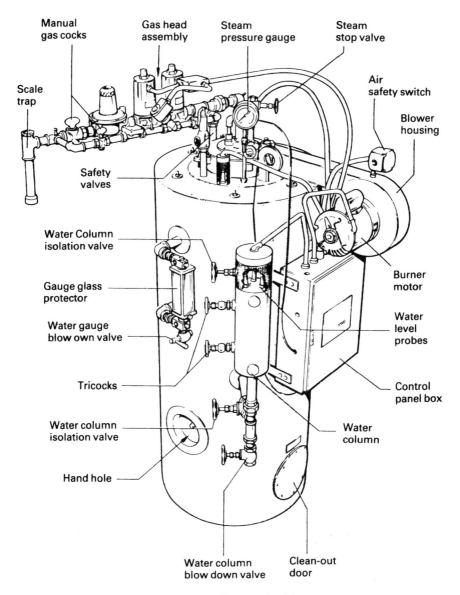

Manual gas cocks

Gas head assembly

Steam pressure gauge

Steam stop valve

Scale trap

Air safety switch

Blower housing

Safety valves

Water Column isolation valve

Gauge glass protector

Water gauge blow own valve

Tricocks

Water column isolation valve

Hand hole

Burner motor

Water level probes

Control panel box

Water column

Water column blow down valve

Clean-out door

*Fig. 7    Small steam boiler*

- where there are two or more boilers, each should bear a clearly visible plate showing a distinguishing number
- low water alarm device
- extra low water cut-off.

All boilers, together with their mountings, must be inspected, both internally and externally, at least once every twelve months, by a competent person. This person is usually the inspector employed by the insurance company.

*Insurance*

Steam boilers should be insured. Requirements vary, but premiums are usually related to boiler output. In addition to the annual inspection, most companies check the external mountings at six-monthly intervals.

The Insurance Companies' Associated Offices Technical Committee (AOTC) specifies the equipment which should be fitted to automatically controlled steam boilers as a condition of insurance. For boilers operated without supervision this includes:

- water feed control
- low water cut-off and audible alarm
- flame protection equipment.

Additionally, it is recommended that shell boilers and others having a perceptible water level which operate without regular supervision should be provided with an independent and separately operated device to cut-off the fuel and air supply to the burners and to operate an audible alarm when the water level reaches a predetermined low position. This is the extra low water cut-off. The device should require manual resetting to reinstate the flame.

## SMALL GAS-FIRED STEAM BOILERS

### Construction

This chapter is principally concerned with small, vertical, single pass fire tube boilers designed for gas firing. A cross-section of a typical boiler is shown in Fig. 8.

The boiler consists of a welded steel shell containing tubes of up to 50 mm diameter fitted between the end plates. The shell contains water up to $2/3$ of its height and steam in the top $1/3$. The water level is indicated by a gauge positioned between steam and water usually protected on three sides by glass plates. The shell has 'mud doors' which are removed periodically to clean out any deposits and inspection doors to allow the fire tubes to be examined.

The burner is fitted with jets similar to small cup burners shown in Chapter 4, which are positioned to fire up each of the tubes. Baffles or 'retarders' are suspended in the tubes to assist heat transmission to

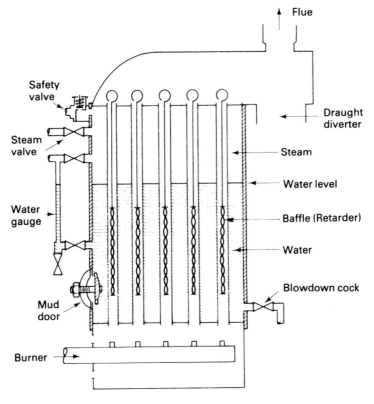

*Fig. 8   Cross-section of a vertical fire tube boiler*

the water. On some boilers the burner assembly is mounted on a swivel. When the burner is swung out, the gas rate is automatically reduced to the minimum necessary for manual ignition. On returning the burner the flames receive their full gas rate.

The top cover collects the products of combustion and directs them to the flue. A draught diverter is attached or incorporated.

## Controls

The gas controls fitted to the small gas boiler shown in Fig. 8 include:

- main gas control cock
- gas pressure governor
- pressurestat
- low water gas cut-off and alarm
- flame protection device
- low pressure gas cut-off

- interlocking pilot and main gas taps or burner swivel and cock for manual ignition.

The steam and water feed controls include:

- main steam valve
- safety valve
- pressure gauge
- water gauge
- injector
- automatic feed control
- feed pump
- blowdown cock.

A diagrammatic layout of the controls on a typical boiler is shown in Fig. 9. Details of the various controls are as follows:

*Main Gas Control Cock*

A cock should be fitted upstream of all the gas control devices to enable them to be isolated for servicing or repair.

*Gas Pressure Governor*

The gas supply should be governed to the required inlet pressure by a constant pressure governor.

*Pressurestat*

A direct-acting pressurestat was described in Chapter 6 and the small boilers used in bakeries normally have a similar device. The larger, fire tube boilers generally use an indirect or an electrical pressurestat. The former may be fitted on the weep pipe of an ordinary relay valve, Fig. 10, or be integral with the relay valve, Fig. 11. In the type illustrated, steam pressure from the boiler enters the device at A and is fed to the bellows B. As the pressure rises it compresses the bellows against the tension of the springs C. The gas valve meanwhile is held open by gas pressure under the diaphragm F.

When the required steam pressure is reached, the bellows pushes open the weep valve D. This allows gas to pass from below the diaphragm to the upper chamber through weep E. Because gas enters the chamber faster than it can escape through a small orifice at G, the pressures above and below the diaphragm are equalised and the diaphragm falls, closing the gas outlet. A small amount of gas is allowed to pass to the burner through the needle valve H, to maintain the flames.

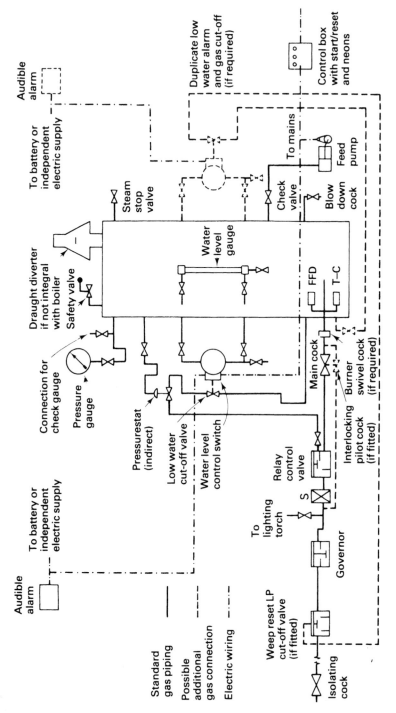

*Fig 9 Control layout*

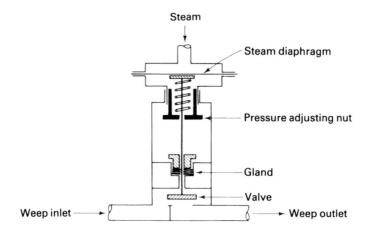

*Fig. 10    Indirect pressurestat*

When steam pressure falls, the bellows expands, the weep valve closes and the pressure above the diaphragm is dissipated to atmosphere through the weep at G. The inlet pressure below the diaphragm lifts the valve and restores the main gas flow.

The pressure at which the pressurestat will shut off the gas is set by adjusting the tension of the springs C.

### Low Water Cut-off and Alarm

The water level in a boiler is usually controlled automatically by means of a ball float. This can be made to operate mercury switches and a gas valve in order to:

- control an electrically driven water pump to feed water into the boiler
- activate an audible alarm
- open or close a small gas valve in the weep line to the pressurestat.

Water may also be fed to the boiler by a hand pump or a steam injector feed. One type of ball float control is shown in Fig. 12. The float is pivoted in a fulcrum plate at A and the float arm is sealed off by a flexible bellows B which acts as a gland. The outer end of the arm rises when the float falls and tilts the upper mercury switch C, to activate the pump motor. Further movement will operate the gas valve at E and shut off the gas through the pressurestat by equalising the pressures above and below the diaphragm. The lower mercury switch D is tilted and should make contact just before the float

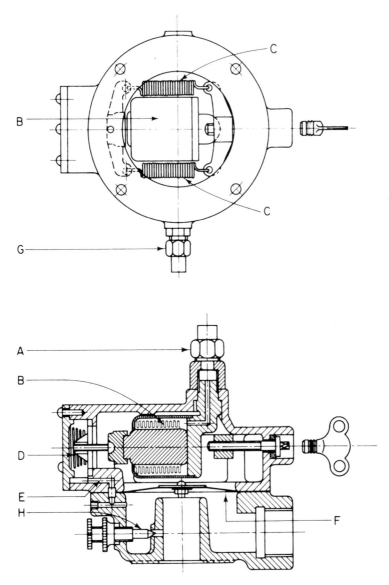

*Fig. 11    Indirect pressurestat with integral relay valve*

reaches its lowest level. This activates the alarm. The alarm may be either mains or battery operated. The pump switch should break contact just before the float reaches its highest level.

A duplicate low water alarm and gas cut-off is usually required as a condition of insurance. This could be as previously described, but

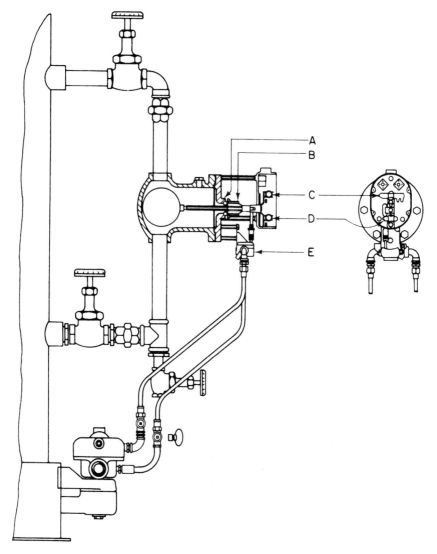

A
B
C
D
E

*Fig. 12    Ball float water feed control*

without the upper mercury switch. Alternatively a device may be used with one or two electrodes which complete a circuit when they are immersed in water. A single electrode control is shown in Fig. 13. If the water level falls below the bottom end of the electrode the sensing circuit is broken, the solenoid valve closes and the alarm is activated.

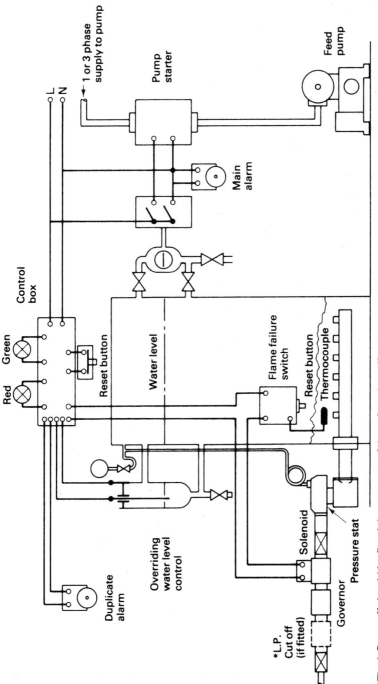

*The L.P. cut off should be fitted downstream of the safety shut off valve unless it has an internal weep reset.

*Fig. 13   Electrical controls, single electrode*

*Flame Protection Device*

The application of a thermoelectric flame failure switch is also shown in Fig. 13. It consists of a switch which can be closed manually by pressing and holding in a reset button. When the thermocouple is heated it energises a magnet which holds the switch in the closed position. If the flame fails, the magnet ceases to be energised and the switch opens. The device is wired in series with the gas solenoid valve.

Modern systems use flame rectification with a semi-automatic control box (Chapter 5).

*Low Pressure Cut-off Valve*

On boilers which are not electrically controlled it has been the practice to fit a weep reset low pressure cut-off valve immediately downstream of the main gas cock. The cut-off may be fitted with a weep line so that it can be operated by a low water level control.

*Interlocking Taps or Burner Swivel*

On boilers with manual ignition it is necessary to ensure that the main gas is not turned full on for lighting purposes. This can be done by having interlocking pilot and main gas taps so that the pilot must be turned on before the main gas tap can be turned. Alternatively a burner swivel with a reducing valve may be used, Fig. 14.

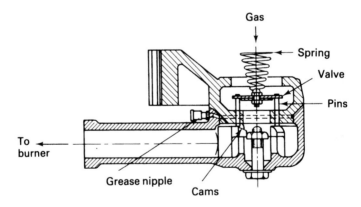

*Fig. 14   Burner swivel with reducing valve*

*Main Steam Valve*

This is a screw down stop valve fitted directly on to the boiler, near to the top. The main steam supply is connected directly into the valve.

*Safety Valve*

A totally enclosed, spring loaded safety valve, Fig. 15, is fitted towards the rear of the boiler near the top. The valve is protected from interference by a padlock.

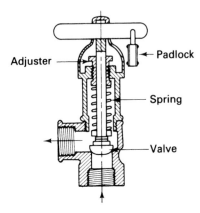

*Fig. 15   Spring-loaded safety valve*

*Pressure Gauge*

A bourden tube gauge complete with syphon and cock is mounted facing the front of the boiler.

*Water Gauge*

This consists of a stout glass tube held at top and bottom by packing glands, Fig. 16. It is fitted with steam and water inlet cocks and a drain cock. The tube is protected on three sides by heavy plate glass panels.

*Injector*

This is a device for feeding water into the boiler by the suction created when steam passes through a small jet, Fig. 17. It is brought into operation by first opening the feed check valve shut-off cock and the suction cock and then opening the injector steam valve fully and quickly. Once the injector has been brought into use it can normally be controlled by the steam valve only.

Although it may only be used as a stand-by for an electric feed pump it should be operated daily to maintain it in working order.

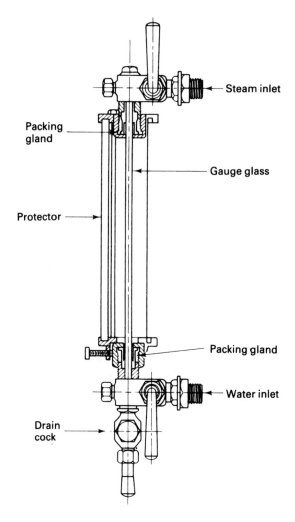

*Fig. 16    Water level gauge*

## Automatic Feed Control

This may be part of the low level control device as already described.

## Feed Pump

This may be either a simple hand-operated pump or an electrically driven feed pump as shown in the installation diagram in Fig. 19. Whether by automatic or manual means, water must be fed into the boiler frequently and regularly to maintain the level at the middle of the gauge glass.

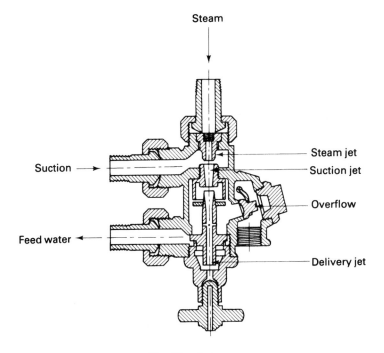

*Fig. 17 Injector*

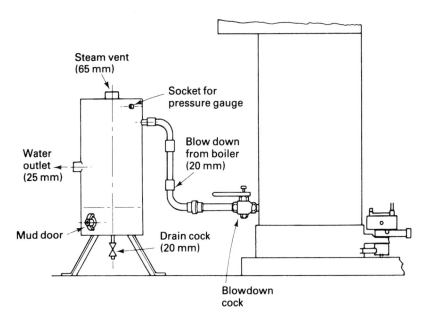

*Fig. 18 Blowdown tank and cock*

## Blowdown Cock

A special blowdown cock is mounted on the boiler and connected to the supply to a blowdown tank, Fig. 19. The cock is usually a fullway, lubricated plug type. It is fitted at low level at the rear of the boiler and its function is to prevent any accumulation of mud in the bottom of the boiler. This is done by regularly opening the cock when the boiler is under steam. The cock should normally be kept open until the water level in the gauge glass has fallen by at least 25 mm.

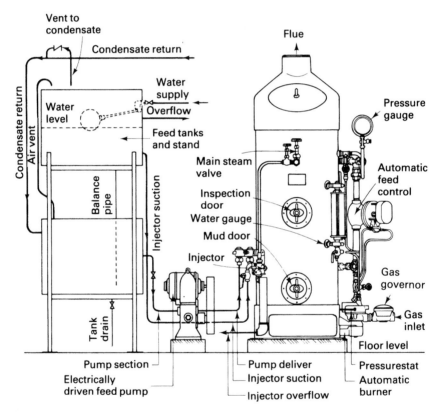

*Fig. 19   Boiler water supply installation*

## Installation

The manufacturer's installation instructions should be consulted before beginning any work.

The boiler should be mounted firmly and level on a concrete base about 100 mm thick. It should be sited to provide access to all ancillary equipment for servicing or repair. If the boiler has a swivel burner, space must be allowed for it to be swung out to its full extent.

Typical connection sizes are as follows:

- gas supply       25 to 50   mm
- flue pipe         75 to 200 mm
- steam supply   15 to 32   mm
- water supply
  - ball valve              15 mm
  - injector feed          15 mm
  - pump feed             20 mm.

The flue should be fitted with a draught diverter if this is not already integral with the boiler. It should be run in accordance with the recommendations given in Vol. 2, Chapter 5.

The floor must be of adequate strength to support the weight of the boiler which, when filled with water may be from 250 to 1,000 kg.

The cold water supply may be taken from a boiler feed tank fed by a cistern as in Fig. 19.

## Commissioning

After testing gas and water supplies for soundness and purging, fill the boiler with water to the normal working level. This may most easily be done by removing the safety valve. Replace the safety valve as soon as the boiler is full and before lighting the gas.

Light the boiler in accordance with the manufacturer's instructions. Check:

- burner pressure
- burner bypass rate
- flame picture, flames correctly positioned below fire tubes
- gas rate at the meter, if necessary
- flue for spillage at the draught diverter
- operation of flame failure device
- low pressure cut-off valve.

Vent the air from the top of the boiler shell through the water gauge. Close the gauge water cock and open the drain cock. Leave the drain open until steam is emitted. Close the drain and open the water cock. In the operating position all cock handles are normally vertical.

Turn on the main steam valve slowly and allow steam to pass to the plant. Check that the pressurestat shuts down the gas at the required working pressure. Check the automatic water feed and low-level cut-off and alarm. Switch off the power to the electric pump and allow the water level to fall so that the alarm sounds and the gas cut-off operates. Switch on the pump. The water level should return to normal, the gas supply be restored and the alarm switched off. The pump should stop at the normal working level. Check any duplicate cut-off and alarm system.

Finally check the blowdown cock and the safety valve and operate the injector system.

To shut down the boiler turn off the main gas cock and the steam valve. To prevent the alarm operating when the boiler cools down and the water level falls, it is necessary to fill the boiler to a high level when shutting down. This can be done either by using the injector or the hand pump. After filling, close the injector steam and suction valves. If an electric pump is fitted shut off the feed check valve cock and the pump inlet cock and switch off the power supply.

## Servicing

### Routine Checks

In addition to the normal periodic servicing, a number of regular daily checks and routine operations should be carried out as follows:

- in use, ensure that the boiler operates at the correct working pressure by observing the pressure gauge
- check that the water level is maintained at about the middle of the gauge glass.

### Water Gauge

Check regularly that the connections are clear.

Shut-off both the steam and water cocks and open the drain cock. Then open the other two cocks in turn to let steam and water blow through their connections and out of the drain. Finally close the drain cock and re-open the steam and water cocks.

### Automatic Water Feed Control

This should be flushed out daily to ensure that the connections and the float chamber are clear of deposits. The control is connected to the boiler in a similar manner to the water gauge and a somewhat similar procedure should be followed.

Close the steam and water valves and open the drain valve slowly. Open the water valve to flush out the connecting pipe and then close the drain valve. Close the water valve, open the steam valve and then open the drain valve slowly to clear the steam connecting pipe and the chamber.

Finally, close the drain valve and open the water valve to restore the valves to their normal operating position.

*Injector*

This should be operated at least once a day to maintain it in good working order. It operates by opening, in turn, the following cocks or valves:

- feed check valve cock
- suction cock
- injector steam valve.

If water issues from the overflow, the suction cock should be gradually closed until this stops.

*Blowdown*

The blowdown cock should be opened regularly and frequently to prevent deposits building up around the bottom of the fire tubes. If the cock has a lubricated plug it should be re-packed regularly with lubricant and the plug operated to distribute the lubrication. This should be done when the cock is hot.

*Safety Valve*

The safety valve should be tested frequently to ensure that the valve is free to move and not choked with scale.

**Major Servicing**

A full service and clean out is usually timed to coincide with the visit from the boiler inspector.

The operations to be carried out should include:

- question the operator and examine the boiler for visible faults
- shut down the boiler and isolate any electrical supplies
- remove and clean the burners, replace jets as required
- remove the top cover and tube baffles, brush out the fire tubes, clean and replace the baffles and the cover
- drain the boiler, remove the mud doors and clean out the interior
- replace the doors, refill the boiler and check for leakage
- check gas taps and ease if necessary
- check gas soundness
- check any non return valve on pumped water supply
- grease swivel joint
- check and clean spiral retarders if necessary

- light and reposition the burner
- check burner pressure and flame picture
- check burner bypass rate
- check flue for spillage at the draught diverter
- check all controls and adjust as necessary
  - flame failure device
  - low pressure cut-off
  - pressurestat
  - automatic feed and low water cut-off and alarm
  - any duplicate alarm and cut-off system
  - water gauge
  - injector
  - blowdown cock
  - safety valve
- check all steam and water cocks operate satisfactorily
- carry out flue analysis and adjust as required.

**Fault Diagnosis and Remedy**

Faults on governors, low pressure cut-off valves and flame failure devices were dealt with in Vol. 1.

Two common faults, failure of the electrical or gas supplies, would have the following consequences.

*Electric Supply Failure*

This would close any solenoid valve and the feed pump would cease to operate. The boiler would shut down. In the case of a boiler with a feed pump but no solenoid valve, the water level would fall until the gas cut-off operated to shut down the boiler. Any independently supplied low water alarm would be activated.

*Gas Supply Failure*

In this event the flame failure device and the low pressure cut-off valve would close and remain closed until manually reset. The boiler would shut down.

*Other Common Faults*

Any major leakage would obviously require the boiler to be shut down immediately. However, small leakages from glands, joints or any other source should also receive immediate attention to prevent dangerous situations developing.

Faults on specific devices are as follows:

*Safety Valve*

Safety valves often develop slight leakages and for this reason it is tempting to leave them alone when all seems well. Testing the valve to ensure that it is free may start it leaking but the test is essential to ensure that the valve will operate in an emergency. Any slight leakage should be stopped immediately. Shut down the boiler, dismantle and clean the valve, regrinding the valve and seating. Reassemble, test and adjust to the correct pressure.

*Injector*

The injector failing to operate may be due to:

- feed check valve not seating properly
- leaks in suction or delivery pipes
- mud or scale in the injector
- injector overheated due to leaking steam valve or several unsuccessful attempts to operate
- feed water too hot.

The injector and the feed check valve may be isolated for cleaning by closing the injector steam valve and the feed check valve shut-off cock while the boiler is still under pressure.

*Automatic Feed Control*

The operation of the float can be checked by manually operating the switch levers between the float chamber and the switch box. The switches are adjustable by altering their angle on the switch carrier. The gas cut-off valve is adjusted by rotating the knurled sleeve on the push rod after first slackening the lock nut. The float may be cleaned or renewed while the boiler is under pressure by closing the steam and water valves and opening the drain valve. Then isolate the electrical supply and close the pressurestat weep cocks. The switch gear and the float chamber can then be dismantled. In this situation the water level in the boiler must be maintained by means of the injector or the hand pump.

If the bellows gland develops a leak, shut the steam and water valves to stop steam or water escaping and open the drain valve. The water pump can be controlled by the main switch and the pressurestat kept operative by closing the relay weep cocks. There is then no protection against low water level and great care must be taken to ensure that the boiler operates safely. The gland must be replaced before the boiler can be left unsupervised.

If, for any reason, the water level should become too low and not be visible in the glass, turn off the gas supply immediately. Allow the boiler and its ancillary equipment to cool down naturally and completely before admitting any cold water. If cold water is introduced into an empty, pressurised boiler it could cause an implosion or leaks around the fire tubes due to rapid contraction. In either case the boiler would be damaged beyond repair.

*Pressurestat*

On any type of pressurestat a leaking steam bellows will result in excessive steam pressure, since the main gas valve will not shut down. This may result in the safety valve blowing off if the higher pressure is not noticed quickly.

With indirect pressurestats, faults on the weep line, the indirect controls or the diaphragm may have different effects on the various types of relay valve or main control valve.

On an ordinary relay valve controlled by an indirect pressurestat, the gas valve will stay open when:

- the pressurestat bellows is leaking
- the pressurestat valve does not close or close completely
- the weep line is broken or leaking
- the relay valve diaphragm is leaking
- the relay valve weep jet is choked.

The gas valve will stay closed when:

- the pressurestat valve does not open
- the weep line is blocked.

On the pressurestat shown in Fig. 11, the gas valve will stay open when:

- the pressurestat bellows is leaking
- the weep valve is closed or sticking.

The gas valve will stay closed when:

- the weep valve does not close or close completely
- the main diaphragm is leaking.

In all cases when dealing with faults on any device it is advisable to consult the manufacturer's instructions before carrying out any work.

CHAPTER 9

# Overhead Heating

---

Chapter 9 is based on an original draft by B. Gosling

---

## Introduction

Industrial and commercial buildings may be heated by either radiant or convected systems. In industrial premises in particular, floor space is usually at a premium and fairly high, open roofs offer a location for heaters, emitters and air ducts.

So industrial space heating is often by overhead heating.

In commercial premises suitable heaters or ducting can be accommodated in suspended ceilings.

The types of heater in use include:

- radiant heaters — high temperature, luminous radiant panels
  - radiant tube heaters
  - air heated radiant tubes
  - steam or water heated radiant systems
- convection heaters — unit air heaters
  - direct fired air heaters
  - indirect fired air heaters
  - make-up air heaters.

Radiant heaters are normally mounted at high level on the walls or suspended from the roof trusses.

Of the convection heaters, unit air heaters are similarly mounted and make-up heaters are roof-mounted. The direct fired air heaters may be fitted overhead, in the form of horizontal or inverted heaters, or stand on the floor. All convection heaters may be fitted with or without ducting.

## RADIANT HEATING

### Radiation

Radiation was dealt with in Vol. 1, Chapter 10.

It was established that:

$$Q = \sigma EAK^4$$

where $Q$ = heat energy
    $\sigma$ = Stephan-Boltzman constant ($5.67 \times 10^{-8}$W/m$^2$K$^4$)
    $E$ = emissivity (black body = 1)
    $A$ = surface area
    K = absolute temperature, Kelvin.

Radiation also obeys the inverse square law, that is, the intensity of radiation is reduced in inverse proportion to the square of the distance between the emitter and the receiving surface, the total energy remaining constant, Fig. 1.

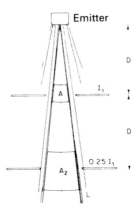

*Fig. 1   Inverse square law*

The following additional points have a bearing on the design of overhead heating schemes.

When rays from two or more sources are projected on to a surface, the total energy received is the sum of the output from each source, Fig. 2.

Infra-red rays may be projected in any direction without affecting the amount of energy received on the absorbing surface. Each of the surfaces at (a), (b) and (c) in Fig. 3 will receive an identical amount of heat.

The maximum intensity of radiation from a flat panel is received at a point on a perpendicular line from the centre of the panel that is, point $A$ in Fig. 4. This line at right angles to the panel is called the 'normal'. When heat is received at points which are at angles of less than 90° to the panel's surface, the intensity of radiation is reduced. When the angle with the surface becomes zero, the amount of radiation is zero. The intensity of radiation at any point $B$ is actually

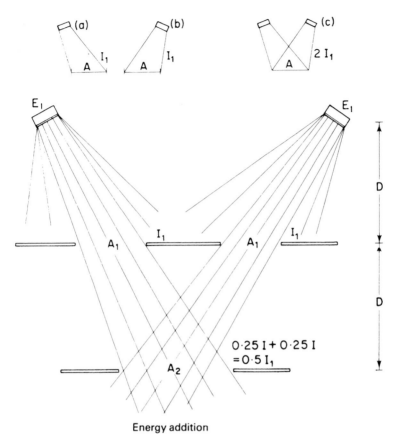

Fig. 2 Energy addition

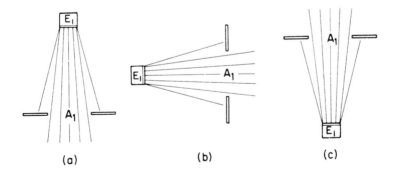

Fig. 3 Direction of emission: (a) downwards, vertically; (b) horizontal; (c) upwards, vertically

the intensity of $A \times$ cosine $\propto$, where $\propto$ is the angle between $EB$ and the normal $EA$, Fig. 4. This means that a single heater will not heat an area evenly and several heaters may be necessary to give an acceptable heat distribution, Fig. 5.

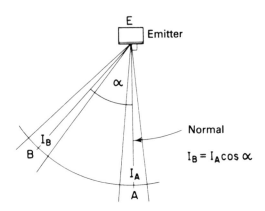

*Fig. 4   Reduction in radiant intensity with reduction in angle of emission*

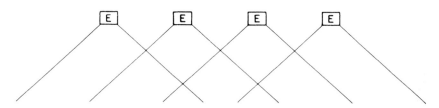

*Fig. 5   Arrangement of heaters to give even heat distribution*

Gas fired heaters radiate on nearly all the infra-red wavelengths, from 0.8 to 500 micrometres. However, for any particular heater, the emission is at a maximum for a particular wavelength. The maximum emission is related to the temperature of the emitter in accordance with Wien's Law, discovered by a German physicist. Actually, the wavelength for maximum emission,

$$\lambda_{max} = 2884 \times 1/K \text{ micrometres}$$

where K is the absolute temperature of the emitter.

Luminous heaters emit rays of a relatively short wavelength, $\lambda_{max} =$ 2.5 micrometres. They operate at temperatures of from 800°C to above 1,000°C.

Non-luminous heaters emit rays of longer wavelengths, $\lambda_{max} = 4$ micrometres. Their surface temperatures are usually between 300 and 600°C. These conditions produce weaker rays and more gentle heating.

## Application of Radiant Heating

Radiant heating has advantages when used in the following situations:

- enclosed areas with a height of more than 6 m
- a partially enclosed area, for example, a loading bay or sports stadium
- selected areas of a large enclosure, for example, a work bench or a particular machine in a large production area
- areas requiring heat for short periods.

Figures 6 and 7 show a workshop. In Fig. 6 it is heated by two air heaters and in Fig. 7 it is heated by overhead radiant heaters.

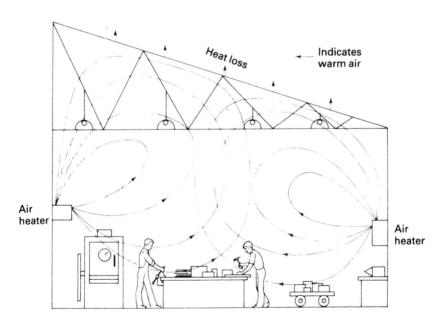

*Fig. 6   Workshop heated by air heaters*

As the air is warmed, in Fig. 6, it becomes less dense and rises above the colder air. So there is comparatively cooler air at ground level and warmer air under the roof. The temperature gradient

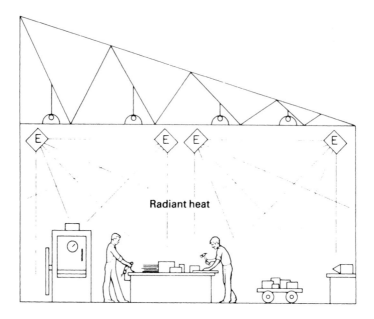

*Fig. 7    Workshop heated by radiant panels*

between floor and roof may be up to 11°C and it varies in proportion to the height of the space and the mixing achieved by the heater. This can result in a high heat loss from the roof. Large air changes will also increase running costs.

The air is circulated through the heaters by fans. High air velocity can cause draughts near return air grilles and may stir up dust in some environments.

In Fig. 7, the radiant panels emit infra-red rays which are immediately absorbed by the people and equipment at ground level. Static air in contact with the roof is generally at a slightly lower temperature than that in the working area, which gains a little heat from the contents of the workshop. The air plays no real part in the transmission of energy and the air temperature may be quite cool without detracting from comfort. Temperatures may be 6°C below those required with convection heating.

Luminous radiant heaters rapidly reach their operating temperature. So people in the path of the radiation are quickly made comfortable although the air and the building structure are still cool.

If a radiant heating system is used for 8 to 10 hours per day, it is generally only necessary to turn it on at the time that heating is required. After a weekend break an initial warm-up period is required, particularly in cold weather.

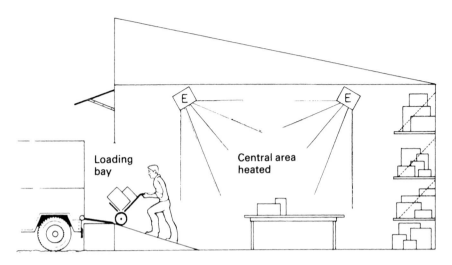

*Fig. 8 Loading bay heated by radiant panels*

The directional nature of radiant heaters allows them to be used in spaces which are not totally enclosed. Fig. 8 shows a typical loading bay. The partially enclosed area should be screened from high winds.

**Luminous Radiant Heaters**

These heaters utilise the radiant burners described in Vol. 1, Chapter 4, usually with natural draught injectors. To give flame retention on natural gas a stainless steel gauze may be fitted about 5 mm below the outer surface of the ceramic plaque. A typical heater is shown in Fig. 9.

One model uses a ring burner firing on to a tapered ceramic fibre gauze, surrounded by a stainless steel mesh, Fig. 10. A circular reflector of polished aluminium alloy concentrates the heat over a relatively small area.

Luminous heaters are not flued and the products of combustion are discharged directly into the premises. This is acceptable where the roof is high and there is a fairly high rate of air change. A certain amount of heat is carried away by the products and there is a possibility of condensation occurring on cold walls or steel work.

Heat outputs of about 15 kW may be obtained from a typical heater, although heaters with greater outputs are available if required. A single plaque burner unit has a heat input of about 4 kW.

*Fig. 9   Luminous radiant panel*

*Location*

The number of panels required is determined by calculating the heat losses from the building as detailed in the C.I.B.S.E. (Chartered Institute of Building Services Engineers) guides A1–A5 and in Vol. 1 Chapter 10. Additional heaters may be required to cover loading bays and frequently opened doors.

The information required when planning a heating scheme includes:

- use of the premises
- environmental temperature required
- type and area of walls
- window sizes and positions
- type and size of doors, including loading bays
- roof area and insulation
- floor construction and area
- height to ceiling, if fitted
- air changes per hour
- location of occupants, passageways, light fittings and equipment

*Fig. 10    Circular luminous radiant heater*

- position of roof trusses and girders suitable for supporting heaters
- presence of air-borne contaminants, dust, paint spray, any solvents, or corrosive substances
- type of heater suitable and minimum fixing height specified by manufacturer.

When planning the installation, check:

- position and size of gas meter and gas supply
- location voltage and phase of electricity supply, if required.

The closer that a radiant heater is to the ground the smaller the area heated but the greater the radiant intensity. An intensity of 79 $W/m^2$ is normal for general heating at floor level. At working levels 158 $W/m^2$ is comfortable but 236 $W/m^2$ at head height should be regarded as a maximum value and is only required for exposed locations. Table 1 gives typical intensities and mounting heights.

**TABLE 1 Radiant Intensities and Mounting Heights**

| Radiant Intensity W/m² | Mounting Height (m) |
|---|---|
| 230 | 3.7 |
| 183 | 4.3 |
| 104 | 4.9 |
| 91 | 5.5 |
| 85 | 6.1 |

Intensity for a given heater is at its maximum when the radiant surface is horizontal. Heaters may be fitted at up to 60° to the horizontal, Fig. 11. When the heater is angled the products of combustion escape more readily, allowing cooling air to pass over the heated surface.

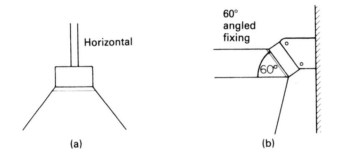

*Fig. 11   Angle of mounting: (a) horizontal; (b) 60° to horizontal*

When the number of heaters has been determined they should be arranged as evenly as possible over the area to be heated. Generally each point should be heated from at least two and preferably four directions to avoid shadowing, Fig. 12.

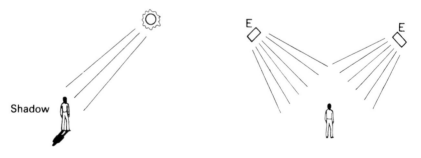

*Fig. 12   Avoidance of shadowing*

Examples of heater locations are given in Figs. 13 and 14.

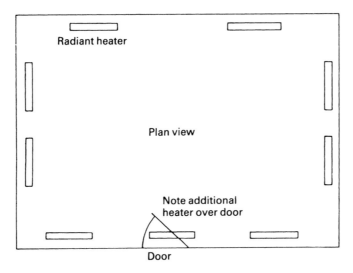

*Fig. 13    Typical heater locations*

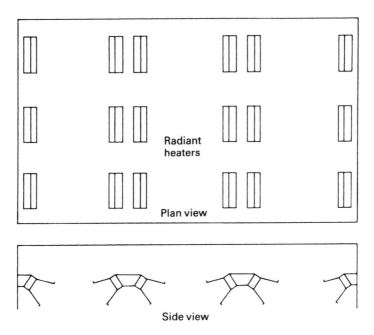

*Fig. 14    Heater locations to cover large area*

*Installation*

A wide range of ignition and control systems may be used with luminous panels. These range from a simple lever cock with chains and a permanent pilot to semi-automatic and automatic systems which may incorporate thermostats and timers. In recent years there has been a tendency with new installations to control these heaters with energy management systems.

The installation of gas-fired overhead radiant heaters for industrial and commercial heating is specified in BS 6896: 1991.

Heaters may be suspended from:

- the gas supply
- brackets
- chains.

Only lightweight heaters should be suspended from the gas supply. If the connecting pipe is more than 600 mm long a cup and ball joint should be fitted close to the main gas supply, Fig. 15.

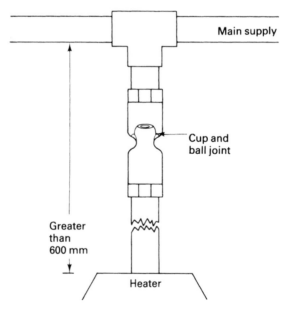

*Fig. 15    Method of suspension*

Brackets are usually provided by the manufacturer. Some types are adjustable for either horizontal or angled mounting.

Chains may be supplied by the makers. They support the heater and prevent strain on the gas supply. The chains are secured to joists or girders by lugs, clips or screws, Fig. 16.

*Servicing*

Radiant heaters should be serviced annually prior to the heating season. Servicing should include cleaning injectors, burners and pilots and checking taps and ignition and control devices. In addition the reflectors should be cleaned and polished.

Dust from the atmosphere builds up on the rear face of the ceramic plaques and can cause local overheating and light-back. It may be removed by a fine jet of compressed air at a pressure of about 5.5 bar. This should be directed first into the holes in the plaque and then into the venturi and finally into the holes again. Care must be taken not to damage the plaque or dislodge the sealing material holding it to the burner body. Some manufacturers recommend washing the plaques using a detergent.

It may be necessary to replace damaged or badly linted plaques. This is best done on a bench. The new plaque must be sealed by the material appropriate to the particular model which may be metal strips, fireproof cement or insulating material.

**Non-luminous Radiant Heaters**

*Radiant Tube Heaters*

These are heavier, longer and more robust than the luminous heaters. The principle of operation is shown in Fig. 17 and a burner at Fig. 18.

The burner fires into one end of a medium or heavy duty mild steel tube. Natural gas is combusted at the mouth of the steel tube which is 75 – 100 mm in diameter and 6 – 11 m long. Air for combustion may be provided by a fan blowing air around the burner, as in Fig. 18, or it may be drawn in via an induced draught fan at the far end of the tube away from the burner, as in Fig. 19.

This develops a temperature of about 450°C on the outside of the tube within about 2 m of the burner which will fall to about 150°C at the tube outlet with a 'U'-shaped configuration. This heat is emitted downwards with the aid of a stainless steel or polished aluminium reflector. The thermal input ranges from 10 – 38 kW.

These heaters are suspended from the roof trusses or girders or at angles as indicated in Fig. 11 (b).

One manufacturer's heaters may be connected together in series with one large fan serving up to about 16 heaters, Fig. 20. Most

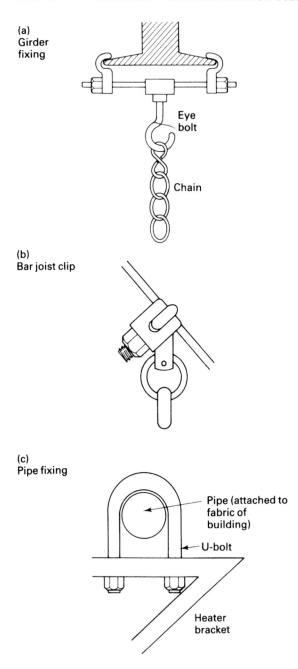

**(a) Girder fixing**

Eye bolt

Chain

**(b) Bar joist clip**

**(c) Pipe fixing**

Pipe (attached to fabric of building)

U-bolt

Heater bracket

*Fig. 16   Fixing devices: (a) girder fixing; (b) bar joist clip; (c) 'U' bolt*

Radiated heat

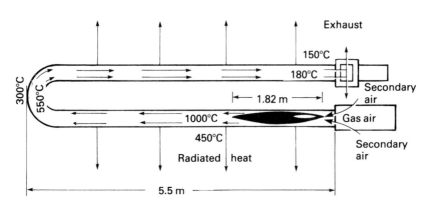

Fig. 17   Radiant tube heater

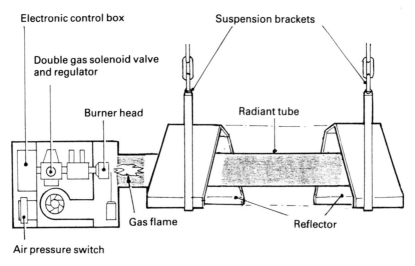

Fig. 18   Radiant tube with fan-assisted burner

models have one length of tube in a 'U' shape, Fig. 21. The heaters are generally fitted with automatic controls. These normally incorporate a purge period, automatic spark ignition, flame monitoring and an air-flow switch. The gas supply is controlled by a governor or a zero governor so that a constant gas/air ratio is maintained irrespective of fluctuations in fan suction. Thermostats and timers may be connected to the system. Energy management systems may also be used to control the whole heating functions.

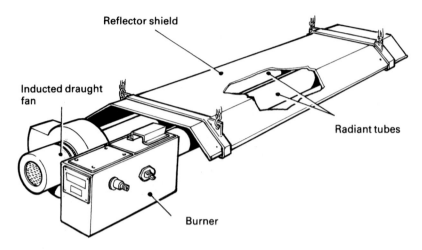

*Fig. 19   Radiant tube burner with induced fan at the end of tube*

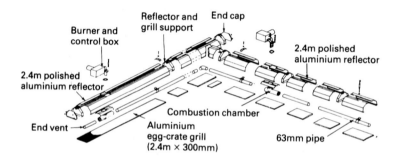

*Fig. 20   Multiple tube installation*

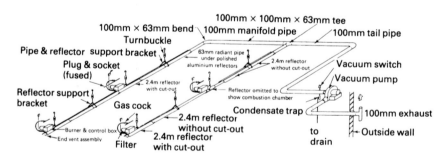

BS 7186: 1989. Specification for non-domestic gas-fired overhead radiant tube heaters, covers the design requirement of these heaters.

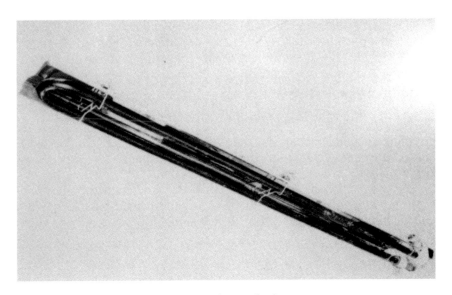

*Fig. 21   Radiant tube heater*

Heat outputs of about 18 kW are obtained from a 7 m tube. The final gas connection to the heater should be by armoured flexible tubing.

## Air Heated Radiant Tubes

This system is shown in Fig. 22. It consists of a direct-fired air heater connected to a bank of circular metal ducts forming a closed circuit. The air heater and fan may be fitted in the roof space or mounted on the floor. A flue is taken to outside air.

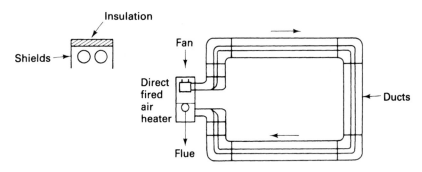

*Fig. 22   Air heated radiant tubes*

The metal ducts are each about 600 mm diameter and may be in banks of up to three tubes. The tubes are fitted with an insulated top and with side shields to protect them from draughts and to concentrate the radiation downwards. The ducts and shields are suspended from the roof trusses or girders.

*Steam and Hot Water Radiant Systems*

Medium or high temperature hot water at 120 to 180°C or steam is piped from a central boiler through overhead panels or strips. These are usually mounted horizontally near to the roof.

The strips, which have superseded the unit panels, consist of continuous lengths of panel made of a plate heated by contact with one or more pipes. Heat output from a single-pipe strip may be about 525 W/m.

## CONVECTION HEATING

### Unit Air Heaters

Although this term has been used to cover all types of flued, forced convection air heater, it was first given only to those heaters which were fitted overhead. It is in this context that it is used here.

Some unit air heaters were indirectly heated by steam or hot water piped from a central boiler. Air is blown by a fan through a finned water-to-air heat exchanger and directed on to the area to be heated by a bank of louvres at the front of the heater. This type is now largely being superseded by indirectly heated gas-fired models operating in a similar manner. The products of combustion from the burners pass up through a tubular steel or clamshell heat exchanger to the flue. Air is blown by a centrifugal, or a propeller-type fan around the heat exchanger and out through the louvres at the front of the heater, Fig. 23.

Guidance is given in BS 5991: 1989. Specification for indirect gas-fired forced convection air heaters with rated heat inputs up to 2 MW for industrial and commercial space heating.

Unit air heaters are suitable for installation in stores, halls and workshops where floor space is at a premium and quick heating up is required to eliminate cold spots. They are available for free air discharge or for use with ducting systems in the following variations.

*1. Duct Model*

A basic heat exchanger without fan or motor but with inlet and outlet spigots for installation in duct work with a separate forced convection system, Fig. 24 (a).

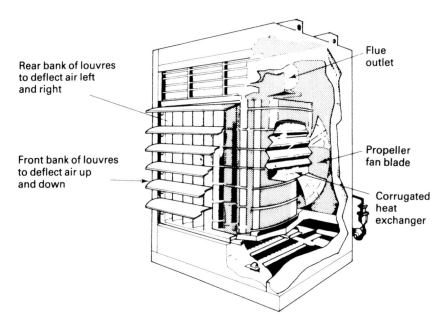

Rear bank of louvres
to deflect air left
and right

Front bank of louvres
to deflect air up
and down

Flue
outlet

Propeller
fan blade

Corrugated
heat
exchanger

*Fig. 23    Unit air heater*

## 2. Propeller Fan Model

Self-contained unit with propeller-type or axial fan and adjustable
louvres to direct the flow of warm air. A filter cannot be used, Fig. 24
(b).

## 3. Centrifugal Fan Model

Similar to propeller type but with a centrifugal fan or blower, giving
an increased air throw. Fitted with louvres for free discharge or a
ducting spigot, Fig. 24 (c).

## 4. Fan Compartment Model

With an enclosed centrifugal fan enabling a filter and return air
ducting to be fitted, Fig. 24 (d).

The heaters are fitted with controls similar to those used on
domestic warm air heaters. These may include:

- ignition, often by permanent pilot
- flame protection, thermoelectric, usually in conjunction with a
  multifunctional control
- electric limit thermostat and fan control

The gas burner is automatically ignited and controlled on and off by a thermostat located in the room or air ducting to the appliance. An electronic programming control unit operates on command and provides a sequential programme each time the heater is required to operate. This programme consists of a prepurge of the combustion chamber, spark ignition, flame proving and main gas ignition. In the event of flame failure for any reason, a lock out will occur.

Overheating of the heater is protected by inbuilt thermal switches which cause the burner to shut down in the event of overheating. The burner heats a sealed heat exchanger from which products of combustion are exhausted by a fan to the outdoor atmosphere.

These heaters operate in the condensing mode because of the heat extracted by a second alloy heat exchanger. Approximately 1 litre of condensate per hour per 30 kW of gross thermal input is produced. Therefore proper provision must be made to evacuate the condensate and to avoid nuisance problems. This condensate, mainly water, is acidic with a pH value of 3.5 to 5 and should be run on the inside of the building to prevent the pipe becoming blocked during freezing conditions.

*Flueing*

The flue gases are drawn through a heat exchanger by a heavy duty centrifugal fan built into the appliance. This fan also serves to discharge the exhaust gases which are at low temperature (30°C) and below the dew point, to the outside of the building via a 100 mm diameter vent pipe. Air for combustion is drawn into the appliance by the same fan and therefore this may also be connected outside making the air heater ideally suited for use in dirty, dusty environments.

*Installation*

Guidance on installation is given in BS 6230: 1991. Installation of gas-fired forced convection air heaters for commercial and industrial space heating.

Heaters are usually suspended from roof trusses or girders although some models may be fitted on walls. Mounting height is about 2.5 m to the bottom of the heater with at least 1 m clearance between the heater and a ceiling. Manufacturer's instructions should be followed in all cases. Care must be taken to ensure that the suspension is reliable and unaffected by vibration.

The louvres should be set so that warm air is directed on the area or personnel to be heated. Where several heaters are mounted in a room they should be positioned so that they all contribute to a general circulation of warm air around the enclosure. Fig. 26 shows typical locations for heaters to provide even heat distribution.

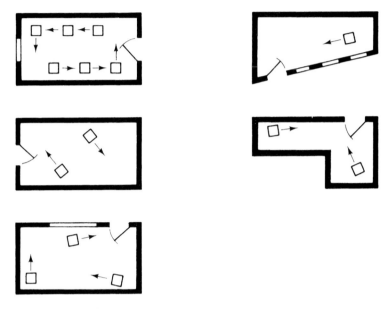

*Fig. 26    Typical locations for unit heaters*

Installation of unit heaters in garages or car showrooms requires extra care. They must be mounted at a minimum height from the floor of 1.8 m and all incoming air to the heater must be fresh air.

## Horizontal and Inverted Heaters

These large output heaters are modifications of floor standing models, as described in Vol. 2, for installation overhead. They are normally supplied mounted in a steel framework, complete with a purpose-made servicing platform and guard rails, Fig. 27. Because of their weight and size care must be taken to ensure that they are securely supported, usually by the structural brickwork. Outputs range from about 44 kW to 1,025 kW.

Both types of heater may be fitted with several discharge nozzles which may be swivelled through 360° to supply heated air to all parts of the working area. Alternatively they may be connected to nozzle extensions or ductwork to convey warm air to cold spots or air curtains.

Most heaters are flued but some may be direct-fired and allow the products of combustion to mix with the discharging warm air. In this

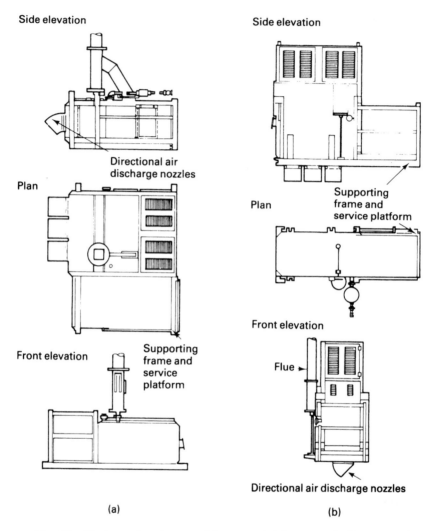

*Fig. 27 Horizontal and inverted heaters: (a) horizontal; (b) inverted*

case the heater must conform to BS 5990 and BS EN 525. Specification for direct-fired heaters with rated heat inputs up to 2 MW for industrial and commercial space heating.

## Direct-fired Air Heaters

The British Standards 5990 and BS EN 525 referred to above give guidelines on the maximum levels of constituents from products of combustion allowed into the atmosphere. These levels are shown in Table 2.

**TABLE 2 Maximum levels of constituents from products of combustion allowed into the atmosphere**

| Component | Limiting Concentration |
|-----------|------------------------|
|           | ppm V/V |
| Carbon monoxide | 10 |
| Carbon dioxide | 2,800 |
| Nitric oxide | 5 |
| Nitrogen dioxide | 1 |
| Adehydes | 0.4 |

The heaters are often floor mounted but they may be suspended from the roof girders. Some models are available with weatherproof housings for mounting externally on a flat roof.

Figure 28 shows a smaller, cylindrical type. This is fully automatic with spark ignition and electrical controls. These include:

- air flow switch
- low gas rate start
- safety shut-off valves
- limit thermostat
- summer/winter switch
- flame rectification or UV flame protection
- full sequence automatic control.

*Fig. 28    Direct-fired air heater*

The axial fan is mounted at the rear of the body and may be operated for ventilation only.

Large rectangular models may also be mounted at high level and are used for normal heating or make-up air heating.

## Make-up Air Heaters

Make-up air heaters are defined as a type of direct-fired forced convection air heater that is provided with an air inlet connection so that fresh air and the products of combustion of the fuel gas are discharged from the heater outlets at a temperature close to that of the heated space.

Air extraction systems are fitted in a number of commercial and industrial premises. For example, these are found in:

- restaurants and canteens – to ventilate the dining rooms and to remove cooking odours, steam and products of combustion from the kitchen
- garages – to remove flammable vapours
- factories – to remove unpleasant, toxic or flammable fumes.

As the polluted air is extracted, fresh air must be taken from outside the building to replace it and this replacement air is called 'make-up air'. Where rates of air extraction are high, special provision must be made for the entry of make-up air and for heating or cooling it according to its temperature. In this country, cooling is less common and it may only be necessary to blow in fresh air to obtain a cooling effect. However, cooling coils may be fitted to some types of heater.

The principle of the make-up air heater is shown in Fig. 29. Air is drawn in from outside the building by the suction caused by the circulating fan. The air enters through an inlet protected by angled louvres and a wire mesh grille to prevent the entry of birds, rain or leaves and is usually filtered before passing to the burner. Some of the air is used for combustion and the remaining air is heated. The products of combustion mix with the heated air and are propelled by the fan, at a slight pressure, to the outlet grilles.

One type of burner, Fig. 30, has perforated baffle plates attached to it which create a low pressure area in front of the burner ports into which combustion air is drawn. The profile plate, Fig. 29, controls the volume and velocity of the make-up air in conjunction with the fan.

The heaters are generally designed to provide a slightly greater volume of air than is being extracted. This allows a slight pressure to build up which distributes the warm air through the building and also prevents the ingress of cold draughts. The systems may either give general heating throughout the whole building, Fig. 31, or be ducted

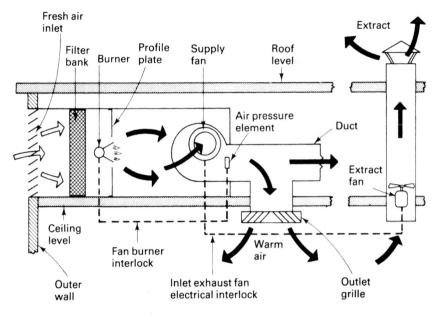

Fig. 29 *Principle of make-up heater*

Fig. 30 *Make-up air heater burner*

to a particular enclosure, like a paint spray booth or kitchen, to compensate for the air being extracted from that environment, Fig. 32.

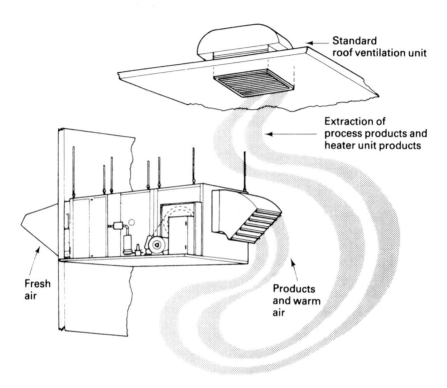

*Fig. 31    Make-up air unit for general area heating and ventilation*

Heaters are available in a variety of sizes, typically from about 118 kW to 880 kW. They are usually fitted in a roof space and have lifting lugs for easy handling and support, Fig. 33. Some models may be supplied with a weatherproof casing for external installation.

The controls supplied vary from one manufacturer to another. The heaters are generally fully automatic and a full sequence control is incorporated. This gives a purge period and has spark ignition usually to an interrupted pilot burner which shuts off when the main gas is established. Flame protection is by ultra-violet detection or flame rectification.

The fan on the heater is often interlinked with the extractor fan and will only run, on 'winter' setting, when the extractor fan is already running. When the heater fan is running, an air flow switch operates a

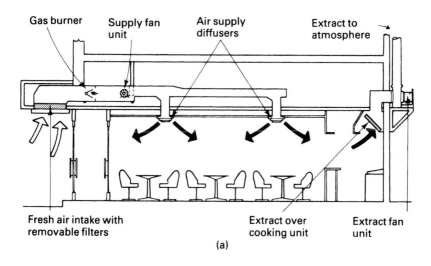

Gas burner    Supply fan    Air supply    Extract to
              unit          diffusers     atmosphere

Fresh air intake with                    Extract over     Extract fan
removable filters                        cooking unit     unit

(a)

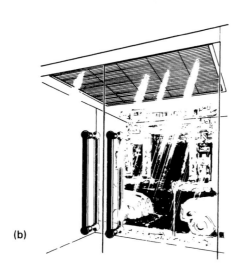

(b)

*Fig. 32    Compensating system replacing air extracted from canopy*

second link with the burner controls. If either link fails, the sequence
shuts down and must be manually reset.

Temperature control is usually by means of a thermostat in the exit
duct operating on a modulating valve in the main gas supply. Turn
down rates of 35:1 are possible.

Tests have shown that the concentrations of $CO_2$ and $CO$ in the
heated air can meet the requirements of 5,000 ppm and 50 ppm

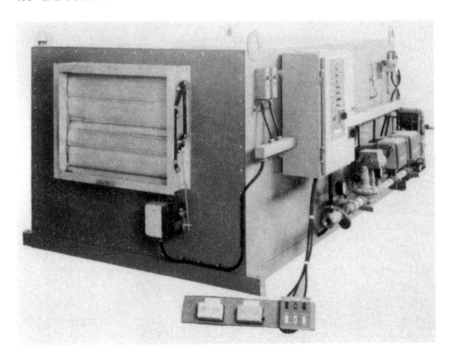

*Fig. 33   Make-up heater*

respectively as listed in the HSE occupational exposure limit of time-weighted averages (threshold limit valves).

The high dilution of the products of combustion by large volumes of make-up air results in acceptable concentration of products of combustion as can be seen from comparing the figures in Table 2 with the above Health and Safety exposure limits.

CHAPTER 10

# Combined Heat and Power

Chapter 10 is written by R. Proffitt

## Introduction

Combined heat and power is often abbreviated to CHP. It is not a new concept and has been known for many years. In the late 1960s, the term 'total energy' was used to describe the situation where only one energy form was supplied to provide all the energy needs of a building. This is achieved by supplying gas to generate electricity from gas engines or gas turbines and use the waste heat to heat the building or to use for process purposes. This technique allows about 80 – 85% of the energy supplied to be used with a loss of only 15 – 20% compared to the electricity generation process at the power station being 34% efficient in producing electricity. This is shown diagrammatically in Figs. 1 and 2.

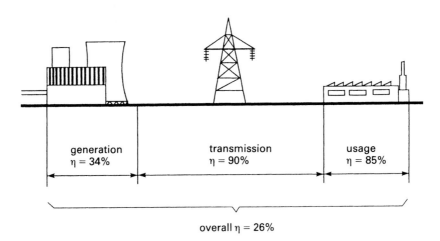

*Fig. 1   Electric power supply efficiency*

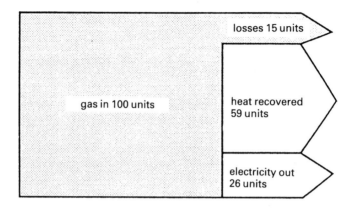

*Fig. 2    Sankey diagram for electricity production with waste heat recovery*

Fig. 1 is based upon information supplied by the Electricity Council and shows that the overall efficiency of electricity supply and usage from the national grid is only about 26%. The ratio in price between electricity and gas is 4:1 and this reflects the loss of over 70% of the energy supplied to the power station.

In CHP schemes much of the waste heat is recovered. In Fig. 2 the output of electricity is the same but in addition there are 59 units of useful heat. Fig. 3 shows this principle applied to a practical installation at a leisure centre containing a swimming pool. The increase in the effectiveness of energy use is at the cost of increased engineering complexity on the customer's site and as a result, capital and maintenance costs have to be set against savings in electricity and heating bills.

The economic case for CHP only becomes viable when the waste heat is recovered and the user's requirement for electricity and heat match those being generated by the CHP system.

**Components of a CHP System**

The essential requirements are:

- an engine or turbine
- an electricity generator
- a heat recovery system
- a control system.

These are shown in Fig. 4 in a schematic form and in Fig. 5 as a more practical working diagram.

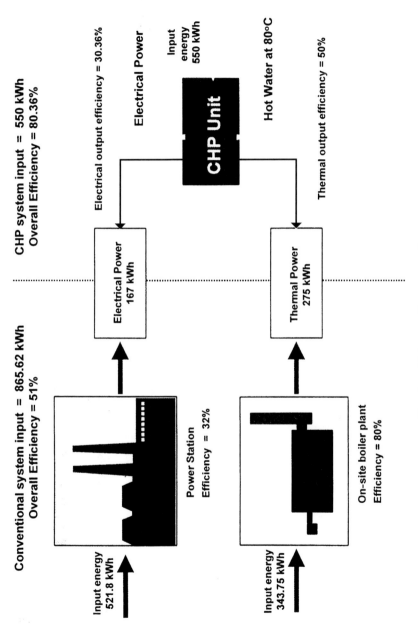

*Fig. 3   CHP installation at a leisure centre*

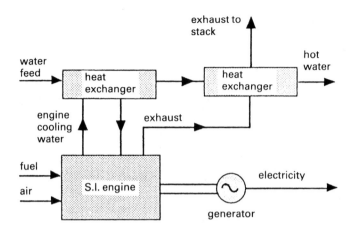

*Fig. 4   Small scale spark ignition CHP plant*

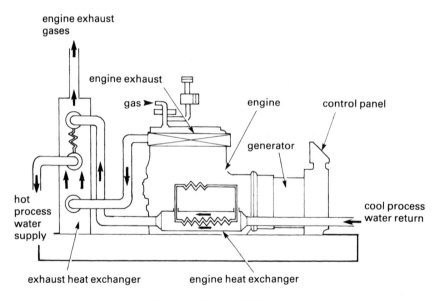

*Fig. 5   Typical CHP set*

## Engines

These are generally used when the amount of electricity to be generated is less than 500 kW. There are three types used – industrial gas engines, automotive derived gas engines and diesels. The smallest size of engines start at 18 kW but are made in excess of 4 MW of electricity. Gas engines are usually of the spark ignition type and can be an automotive derivation or an industrial type. Fig. 6 shows an actual CHP unit rated at 167 kW of electricity with 275 kW of heat supplied simultaneously.

*Fig. 6    167 kW CHP Nedalo unit*

The industrial type engines are heavy rugged stationary types that have been developed for the purpose of providing reliable power with very low maintenance costs. Like ship engines they are built with large bearing surface areas for low wear and are constructed for ease of maintenance. There are examples of such engines running 50,000 hours.

Automotive derived engines are based on de-rated and modified lorry engines to run on natural gas. This is usually achieved by altering the pistons, cylinder heads and valve gear to cope with the different requirements of the spark ignition gas fuelled engine. Because the running speed is much slower and constant at about 1,500 rpm under steady conditions, the engine life is extended to 10,000 – 30,000 hours.

Diesel engines may also be used for CHP although they are generally used for standby electricity to protect vital electricity supplies and will be larger than for optimum operation for CHP.

## Generators

There are two types of generators used in CHP systems known as asynchronous and synchronous. A synchronous generator always rotates at a fixed integer multiple of the mains frequency e.g. 1,500 revs/min for 50 Hz mains. With an asynchronous generator the rotor speed is slightly faster and will vary with power output.

Mains excited asynchronous generators are the commonest, simplest and cheapest form used in CHP. They are identical to a large induction motor and are thus often used to start the engine, becoming a generator when the motor has run up to speed and the engine is able to provide the necessary power. They are more efficient in smaller sizes than their synchronous counterpart but care is required to limit starting currents when used to crank the engine. Because they are normally connected to the mains to provide the exitation current which supplies the operating magnetic field they cannot readily be used as a standby generator without modification.

Synchronous units have the advantage of being able to fully double as standby generators and do not require power factor correction. They are more complex than asynchronous as it must maintain its own frequency standards and requires equipment to enable synchronisation with the mains.

Synchronous generators with outputs below 100 kW tend to be more expensive than their asynchronous counterparts. The additional equipment needed to start the engine, control the generator and interface it with the mains adds to the cost of the smaller CHP units. This may not be justified unless there are real benefits from the provision of a steady supply. In general above 100 kW output costs advantages of asynchronous over synchronous types disappear.

## Heat Recovery Systems

From Fig. 2 it can be seen that it is possible to save up to about 59% of the fuel input in the form of waste heat. This heat has to be recovered by using heat exchangers and several may be used. About 33% of recoverable heat can be obtained from the engine jacket at 120°C, whilst about 22% can be recovered from the exhaust at 650°C. The heat from the engine jacket can be used to produce hot water at 70 – 85°C using a water to water heat exchanger. The most common type is a shell and tube heat exchanger.

Gas to water heat exchangers can also be used with the exhaust gases and can be arranged to connect directly with the water heating system of the building. Another advantage is that it is possible to by-pass the heat exchanger with the exhaust gases for example in the summer when less heat is required. Also if the CHP set is being used simply as an electricity standby set, the waste heat may not be required. The temperature of the products of combustion after the heat exchanger is likely to be about 120°C.

If additional heat exchange area is introduced into the system, it is possible to take even more heat out of the products of combustion and to cool them to less than about 55°C. This allows latent heat from the moisture in the products to be extracted in addition to the extra sensible heat. Connection to the direct source of cold water supply will allow this saving to be achieved. It should be noted that this can be done with natural gas because of the lack of sulphur in the products of combustion. Diesel contains sulphur which would give high levels of acid sulphur dioxide and would cause corrosion of the secondary heat exchanger area.

**Control Systems**

The start up and shutdown sequence and the mechanical and electrical conditions need to be controlled. These functions are usually achieved with relay logic, solid state systems or full micro-processor control sometimes with remote communications facilities. Some units provide pre-circulation of oil and battery starting with full system proving before on-load operation. On shutdown, the mains and fuel supplies must be cut off but the water pump may continue to run to cool down the engine block.

The requirements of the safety system are:

- interlocks with heating system pumps
- flow switches in pipework
- control and limit thermostats
- low engine oil pressure
- emergency over-temperature thermostats
- low gas pressure
- high gas pressure
- overspeed protection
- low speed protection
- electrical power overload.

On some systems air/gas ratio control is used to achieve lean burn conditions to minimise oxides of nitrogen.

The products of combustion which will contain carbon monoxide must be safely passed to the outside via a flue.

In order to control noise from the engine, they are sometimes installed in an acoustic enclosure although this is usually only when particularly quiet surroundings are required.

## Types of CHP System

It is convenient to discuss CHP systems in terms of large and small scale units and although it is difficult to define these exactly, a figure of 500 kW (0.5 MW) of electricity generation is used as the reference point.

## Small Scale CHP (less than 0.5 MW)

Two types of prime movers are available and are the spark ignition engine and the gas turbine. It can be seen from Table 1 that the gas turbine, at present, has low electrical generating efficiency due to the shaft efficiency being low and is unlikely to be used since in addition the installed costs are 1.5 – 3 times more expensive than the reciprocating engine.

TABLE 1   Small scale CHP under 0.5 MW

| Prime Mover | Size Range | Electrical Generating Efficiency | Heat/Power Ratio |
|---|---|---|---|
| Spark Ignition Engine | under 0.5 MW | 25 – 33% | 1.5 – 2.5:1 |
| Gas Turbine | under 0.5 MW | 10 – 15% | 3 – 10:1 |

The heat/power ratio refers to the amount of heat that will be generated as waste heat for each unit of electricity produced, so that a spark ignition engine generating 0.5 MW of electricity will produce 0.75 – 1.25 MW of heat. A similar gas turbine rated at 0.5 MW of electricity will produce 1.5 – 5 MW of heat.

The applications for small scale CHP based on the spark ignition reciprocating engine are in hotels, hospitals, swimming pools and leisure centres.

## Large Scale CHP

For electrical loads greater than 0.5 MW electrical output the selection of equipment is more complex as shown in Table 2.

**TABLE 2    Large scale CHP over 0.5 MW**

| Prime Mover | Size Range | Electrical Generating Efficiency | Heat/Power Ratio |
|---|---|---|---|
| Gas Turbine | 0.5 MW to over 100 MW | under 2 MW 15 – 23%<br>over 2 MW 20 – 40% | 2.5:1<br>1.5 – 4:1 |
| Gas Turbine + Steam Turbine (Combined Cycle) | 1 MW to over 100 MW | 35 – 50% | 1 – 2:1 |
| Spark Ignition Engine | 0.5 MW to 2 MW | 30 – 38% | 1.5 – 2:1 |
| Dual Fuel Engine | 1 MW to 6 MW | 33 – 38% | 1 – 3:1 |

Table 2 shows the wide size ranges that are available and the widely differing electrical generating efficiencies and wide heat to power ratios that are available. The important factor is that the system matches the site electricity and heat requirement profile as closely as possible. A 24 hour profile is shown in Fig. 7 but the system more importantly also has to meet the month by month profile.

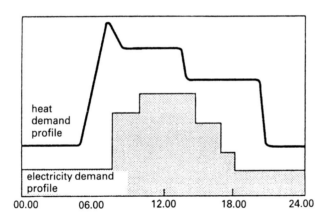

*Fig. 7    Example heat and electricity demand profile*

## Gas Turbine CHP

The gas turbine system shown in Fig. 8 is more complex but offers more flexibility in its output than a small scale system. Here the exhaust gases from the turbine can be used directly to generate steam (which is the common case) or can be used for drying or some other process application.

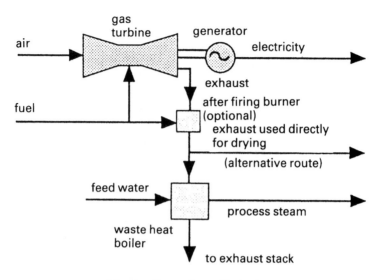

*Fig. 8   Gas turbine CHP plant*

Supplementary heat is often added to products of combustion from the turbine since combustion can be supported because the oxygen level is in excess of 17%. The actual heat added can raise the heat to power ratio effectively from that shown in Table 2 to about 9:1. This additional heat can be produced and controlled very easily to match the site requirement. A schematic of a gas turbine is shown in Fig. 9. The burner used for supplementary heat is often known as an 'after burner'.

Gas turbines systems are generally larger and the number of potential sites is lower than is the case for smaller CHP sites. An example of a 3.6 MW turbine is shown in Fig. 10.

### Microturbines

Microturbines are small gas turbines capable of generating 30 kW to 100 kW electrical power and are beginning to emerge after years of development. Fig. 11 shows a microturbine with the associated heat transfer unit in the form of a packaged unit.

A 45 kW electrical unit will produce between 104 and 209 kW of heat depending upon recuperator setting with outlet water temperatures of 90°C. The recuperator gas inlet temperature is about 700°C.

The natural gas input to such a machine is 214 kW giving rise to a heat to power ratio of between 2.3 and 4.6. The overall efficiencies are 69 – 85%. They have numerous advantages:

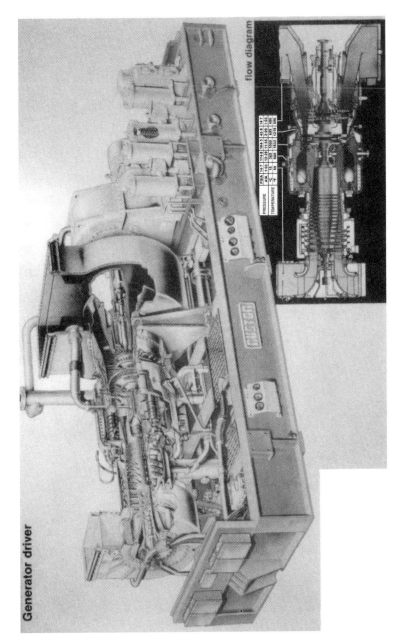

*Fig. 9   Gas turbine (European Gas Turbines): (a) Generator driver; (b) Flow diagram*

*Fig. 10   3.6 MW gas turbine (Centrax)*

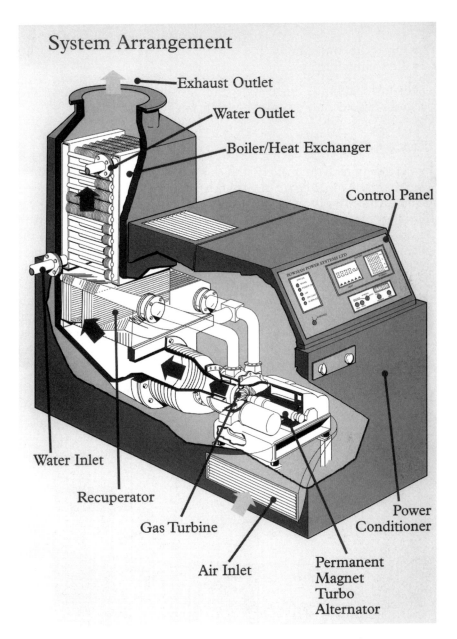

## System Arrangement

Exhaust Outlet

Water Outlet

Boiler/Heat Exchanger

Control Panel

Water Inlet

Recuperator

Gas Turbine

Air Inlet

Power Conditioner

Permanent Magnet Turbo Alternator

*Fig. 11   Microturbine (Bowman Power Systems)*

- Single stream of high temperature heat
- Competitive operational costs
- Low maintenance costs (although engine speeds may be 116,000 rpm)
- Operational flexibility
- Low emissions.

The installation of this new type of gas turbine – the microturbine – should become more common in the next decade.

### Combined Cycle CHP

In the combined cycle CHP the gas turbine produces waste heat which is used to generate steam in a waste heat boiler for use in a steam turbine to produce even more electricity. The heat from the gas turbine is often increased by the use of supplementary gas burners to produce the amounts of steam required. The exhaust gases from the steam turbine are still available for process use. From Table 2 it can be seen that up to 50% of thermal energy input can be converted to electrical energy. The size ranges from 1 MW – 250 MW. Fig. 12 shows a schematic of the combined cycle CHP plant. This technique is therefore used where the objective is primarily electrical generation – for example in gas fuelled power stations. A recently commissioned (1991) power station at Roosecote in Cumbria producing 220 MW of electricity used this technique.

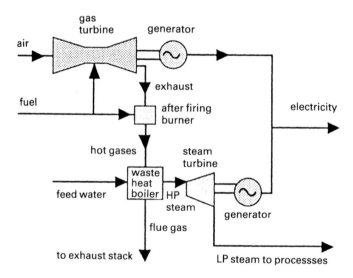

*Fig. 12 Combined cycle CHP plant*

It is also possible to increase power output from a gas turbine by injecting steam into the turbine, however this practice is only done in one application in this country at the present time. This technique also reduces oxides of nitrogen content in the exhaust gases.

## Spark Ignition Engine CHP (over 0.5 MW)

Fig. 13 is a schematic of a large spark ignition engine CHP system. It is very similar to that for the small scale CHP system except that the exhaust heat can potentially be used to raise steam due to the higher quantity of exhaust gases. An example of a 750 kW engine is shown in Fig. 14.

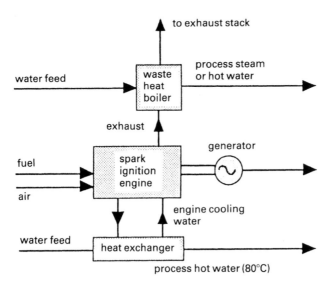

*Fig. 13    Large-scale spark ignition CHP plant*

## Dual Fuel Engine CHP

A dual fuel engine based CHP system is shown schematically in Fig. 15. In this type of engine between 5 – 10% of the total fuel input comes from a gas oil pilot injection via a diesel type injection system. The remainder of the fuel input comes from a lean mixture of natural gas and air introduced by either a conventional carburettor system mixing the gas and air prior to a turbocharger, or by direct injection of the gas into the cylinder at high pressure. The compression of this gas will consume about 2% of the engine shaft power. Waste heat from the jacket cooling water and the lubricating oil appears as hot water at around 80°C whilst heat recovered from the exhaust gases may be used

*Fig. 14    750 kW spark ignition engine (Brons M·A·N)*

to produce further hot water or, if required, process steam. An engine producing 3.5 MW of electricity and a low heat:power ratio of 1.5 is shown in Fig. 16. In addition the exhaust gases raise 3,200 kg/h of steam which can be increased to 13,500 kg by supplementary firing into the exhaust gas prior to entering the steam boiler.

**Maintenance of CHP Systems**

The maintenance of CHP systems will include the prime mover, the electrical equipment and the boiler or heat exchangers. However in this section only the prime movers will be considered – that is gas turbines or gas or dual fuel engines. These items will generally be maintained by the manufacturers, their agents or sometimes specialist companies such as energy consultants. A large number of reciprocating CHP units have various parameters monitored, which indicate whether a fault has developed or maintenance is needed. These parameters are sent automatically via modems and telephone lines to the manufacturer's headquarters, who will then initiate a maintenance visit.

The typical spark ignition engine under 0.5 MW will have the maintenance schedule shown in Table 3.

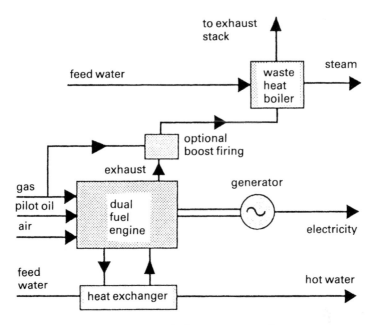

*Fig. 15   Dual fuel engine CHP plant*

*Fig. 16   3.5 MW engine (NEI Crossley-Pielstick)*

**TABLE 3  Maintenance schedule for spark ignition engine under 0.5 MW**

| Frequency | Work Done | Duration |
|---|---|---|
| Daily | Check for leaks, fluid levels, temperature. | 10 minutes |
| 4 – 6 weeks | Adjust valve clearances. Change oil and oil filter. Check and clean spark plugs, replace if necessary. Clean air filter, replace if necessary. Inspect distributor cap and rotor, replace if necessary. | 2 – 3 hours |
| 18 – 36 months | Top end overhaul<br>– replace piston rings<br>– valves<br>– valve seats<br>– and other parts as required. | 1 – 2 days |
| 54 – 72 months | Major overhaul Replace or recondition all major internal components as required. | 4 – 5 days |

Whilst the engine is being maintained, it is obviously not possible to generate electricity so electricity from the grid has to be used.

The gas turbine maintenance is shown in Table 4.

Tables 3 and 4 both assume that the prime movers are running continuously for 8,700 hours per year. The major overhaul of the gas turbine may involve temporary replacement of the turbine to maintain production.

The costs of maintenance are related to the size of the prime mover and are higher for spark ignition engines and dual fuel engines than for gas turbines. In fact the turbines will be about ten times cheaper to maintain than the engines.

**TABLE 4 Maintenance schedule for gas turbine**

| Frequency | Work Done | Duration |
|---|---|---|
| Daily | Check for leaks, fluid levels, temperature | 10 minutes |
| 1 – 6 weeks | 'Hot wash' of compressor blades | 5 – 10 minutes |
| 3 – 4 months | Borescope inspection of internal components | 6 hours |
| Annually | Detailed internal visual inspection | 2 – 5 days |
| 12 – 24 months | Replace oil and air/filters | – |
| Approx 2 years | Partial overhaul of 'hot end' components | 2 – 5 days |
| 3 – 4 years | Major overhaul Replacement of components as necessary | 6 – 8 days |

## Gas Compressors

Gas turbine systems require fuel supply pressures of between 10 and 25 bar, which cannot generally be supplied directly from the distribution system. A gas compressor is therefore required and this may absorb up to 5 – 7% of the turbine shaft output power.

Compressors may be of the screw or reciprocating type and their characteristics are shown in Table 5.

**TABLE 5 Choice of gas compressor**

| Screw Type | Reciprocating |
|---|---|
| Limited inlet pressure range<br>Good flow modulation<br>control (15 – 100%) | Wider inlet pressure range<br>Flow modulation varies from poor<br>(0 or 100%) to very good (10 – 100%)<br>depending on complexity of the<br>control system. |
| Does not induce pulsations<br>in the gas supply | Induces pulsations in the gas<br>supply possibly affecting<br>metering. |
| Vibration free | Comparatively high vibration<br>level. |
| Relatively expensive | Relatively cheap. |

The screw compressor is more costly but has good flow modulation control whilst the reciprocating compressor needs a costly complex control system to equal the performance of the screw compressor. The reciprocating compressor induces pulsations into the gas supply that can cause pulsations upstream and affect the accuracy of the gas metering. Sudden changes such as crash stops, changeover of fuel on line, shutdown etc in the gas supply to the turbine create metering inaccuracies.

These are taken into account when specifying the metering installation which may incorporate a positive displacement meter or a turbine meter. Because such loads are large, care must be taken with flow compensation for temperature and pressure conditions – see Chapter 2. Advice on installations is given in British Gas plc IM24 'Guidance Notes on the Installation of Industrial Gas Turbines, Associated Gas Compressors and Supplementary Firing Burners'. The IGE are to redraft and update this advice as Report UP/9.

## Government Influence

The 1983 Energy Act has been very important in stimulating the high level of interest in CHP and in achieving a high number of installations. The important aspects of this Act are:

1. the electricity companies are required to publish a tariff for the purchase of privately generated electricity.
2. the electricity companies are required to supply private electricity generators and advise the potential operator about equipment suppliers.
3. the electricity companies are required to permit connection to and the use of the national grid.
4. removal of restrictions on the sale of electricity from private generation as a main business.
5. for electricity companies to adopt and support schemes for CHP.

The Act therefore removed all operational restrictions to the use of CHP and enabled connections to and the use of the grid for load balancing purposes.

However in 1989, the electricity industry was subjected to privitisation and the Act of Parliament enacting this legislation repealed the 1983 Act.

Electricity was generated in England and Wales by the Central Electricity Board, which was replaced by National Power, Powergen and the nuclear stations. In addition, there are a number of independent generators operating combined cycle gas turbines. If they are all above 10 MW capacity they must be licensed and supply their electricity to the pool.

This power is transferred from the generators through the transmission network owned by the National Grid Company to the local distribution systems owned by the 'local' electrical companies and then to customers' electric meters. These supply companies buy their electricity from the pool or non-pooled generators and have to pay transmission and use of system charges to deliver the power to their customers.

In Scotland there is no pool. Scottish Power and Scottish Hydro-Electric own power stations and distribution systems and have supply businesses. In addition, they buy power from Scottish Nuclear and sell power to the pool in England and Wales.

In Northern Ireland, there are three generating companies that sell their output to Viridian (Northern Ireland Electricity). Viridian own and operate the transmission and distribution system and is responsible for the electricity supply.

CHP units that generate surplus power can export it to other customers by using the electrical distribution wires, for which there is a charge. In addition, special half-hour electrical meters are required at the CHP unit and the receiving end to identify the time that the electricity is exported to enable an accurate bill to be generated. If the CHP unit was exporting more than 10 MW, a generation licence would be required.

The Government encourages CHP and has set a target of 10,000 – 17,000 GW power generation by the year 2010. This is over and above the earlier target of 5,000 GW by the year 2000. In July 1997, the level of CHP capacity installed in the UK was 3,600 GW, with the majority installed in the industrial sector.

The electrical installation and controls must comply with the Electricity Council Engineering Recommendations G59. This sets out the conditions to be met by a private generator, which briefly are that the CHP system must be isolated from the grid:

(a) in the event of the failure of any one phase of the distribution grid
(b) if the difference between the declared supply voltage and that of the generator exceeds plus or minus 10%
(c) in the event of the loss of the electricity companies supply
(d) if the frequency of the generator departs from 50 Hz by more than +1% or – 4%.

## Site Assessment for CHP

For a site to be suitable to install a CHP system it must meet a number of requirements:

(a) it must operate at least 4,500 hours per annum
(b) it must have suitable natural gas supply
(c) it must have a need to use both the electricity and the waste heat produced at the same time.

Provided the above criteria have been met a more detailed examination of the site can be made to obtain the following information:

(i) heat and power profiles for typical days throughout the year.
(ii) annual fuel and power consumptions giving total amounts used for the year.
(iii) detailed operating modes of the electrical and gas plants used on-site.
(iv) electricity price structure and tariffs for purchase of electricity and export to grid prices.

It is obviously a difficult task to fully assimilate the above information into an accurate forecast to enable the size of the CHP system to be optimised. Some of the problems are:

(a) matching the heat demand and importing or exporting electricity as appropriate.
(b) matching the electricity requirement and if required using independent boilers to supply additional heat.
(c) providing heat and electricity for the base loads and importing electricity from the grid and heat from boilers when required.
(d) using single or multiple CHP units.
(e) adding an absorption chiller to provide cooling in the summer when conventional heating is not required but use waste heat conventionally in the winter.

Each of the above factors can dramatically affect the cost of the CHP installation, the running costs, maintenance costs and consequently the savings.

# Air Conditioning

Chapter 11 is based on an original draft by A.J. Spackman

## Introduction

The purpose of air conditioning is to provide an environment in which the temperature, moisture content and movement of the air are maintained at a level required to ensure the comfort conditions desired by the occupants. In some instances it is used to produce the special conditions necessary for a manufacturing process or a particularly sensitive piece of equipment, such as a computer.

In a fully controlled air conditioning system, air-borne dust from outside is eliminated by filtering. So dust is limited to that generated within the conditioned environment.

Where the environment must be closely controlled, it is essential that make-up air is only allowed to enter through the air conditioning system. So windows should not be opened and all external doors should be fitted with air locks. The environment must be maintained at a slight pressure so that any air leakage can only be outwards.

To maintain the desired conditions it is necessary to remove and replace some of the air in the controlled environment continually. The temperature and moisture content of the replacement air will be different from that of the air withdrawn and from that entering the unit from outside. The unit must continually process the make-up and return air to give the desired characteristics to the recirculated air.

## Basic Principles

A number of the following principles were introduced in Vol. 1, Chapter 4.

### Moisture Content

Because part of the air conditioning process is concerned with controlling the moisture content of the air it is necessary to determine the amount of moisture contained.

429

This is usually measured in grams of water vapour (moisture) associated with 1 kg of dry air. It may also be in kg/kg. There is a limit to the quantity of moisture that air can contain and this limit varies with the temperature of the air. This limit is low when the air temperature is low and becomes greater as the temperature rises. When air contains the maximum quantity of moisture that it can hold at a particular temperature it is said to be 'saturated'.

*Relative Humidity*

'Saturation vapour pressure' is the pressure exerted by the maximum quantity of water vapour (moisture) that a given volume of air can contain under particular conditions of temperature and pressure, and is expressed in millibars.

Normally air is not completely saturated, so it has an actual vapour pressure which is lower than the saturated vapour pressure. It is useful to compare the actual moisture content to the maximum possible for that temperature and this ratio is the 'Relative Humidity' (RH).

$$\text{Relative humidity} = \frac{\text{actual vapour pressure}}{\substack{\text{saturation vapour pressure} \\ \text{(at the same temperature)}}} \times 100\%$$

The saturation factor for a given temperature is expressed in millibars and is obtainable from published hygrometric tables.

*Dew Point*

If air which is saturated with water vapour is cooled, some of the moisture will be precipitated in the form of mist or dew. Dew point is the temperature at which precipitation begins to appear. Expressed in another way, dew point is the temperature at which the actual vapour pressure becomes equal to the saturated vapour pressure. That is, the temperature when the relative humidity becomes 100%.

*Measurement of Moisture Content*

The normal method of determining the moisture content of the air is to compare the temperature readings of wet and dry bulb thermometers. The dry bulb temperature is measured by an ordinary mercury in glass thermometer. This is affected only by sensible heat and indicates the air temperature.

The wet bulb temperature is obtained by a similar mercury in glass thermometer, but the bulb is covered by a muslin sleeve which is kept wet by a small reservoir of water. Air is made to pass over the bulb and, if the relative humidity is below 100%, water will be evaporated from the muslin. The evaporation takes heat from the air and from the bulb of the thermometer which consequently gives a lower reading than the dry bulb thermometer. The difference in the readings is called the 'wet bulb depression'.

The rate at which water can be evaporated from the muslin around the wet bulb depends on:

- the relative humidity of the air
- air temperature
- velocity of air past the thermometer.

Wet and dry bulb temperatures can be accurately obtained by using a 'Sling psychrometer', Fig. 1. This consists of a flat frame in which are set similar wet and dry bulb thermometers. The frame is pivoted on the handle so that it can be rotated to move the wet bulb through the air. A small water reservoir is situated at the other end of the frame. The rate at which the instrument is rotated and the minimum time of rotation required for an accurate reading are specified by the manufacturer. Wet and dry bulb psychrometers are used at meteorological stations but there the thermometers are static and enclosed within a 'Stevenson Screen'. This allows air to circulate around the instrument but gives a shaded, draught free area. There are considerable differences in the wet bulb readings obtained from sling and screen methods and for air conditioning calculations sling temperatures are used.

Hand-held electronic instruments are obtainable which give direct digital read-outs.

*Psychrometric Charts*

Various authorities produce sets of tables and graphs which combine together the information required for calculations relating to the condition of the air. One of the charts in common use is that produced by the CIBSE. This psychrometric chart brings together the following information:

- dry bulb temperature
- wet bulb temperature
- moisture content
- relative humidity
- total heat
- specific volume.

There are various types of psychrometric chart and a typical psychrometric chart is shown in Fig. 2. This has dry bulb temperature plotted on the horizontal ($x$) axis and moisture content on the vertical

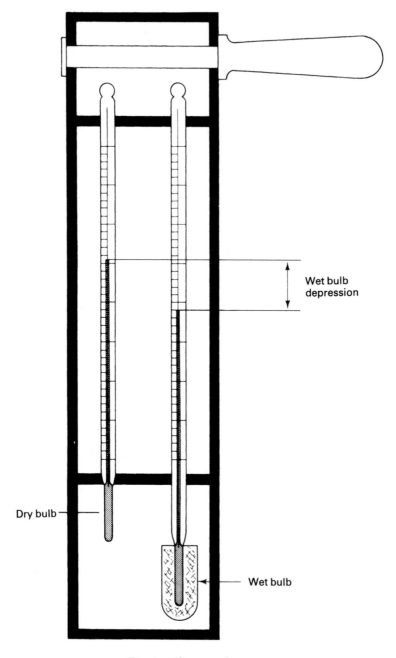

*Fig. 1   Sling psychrometer*

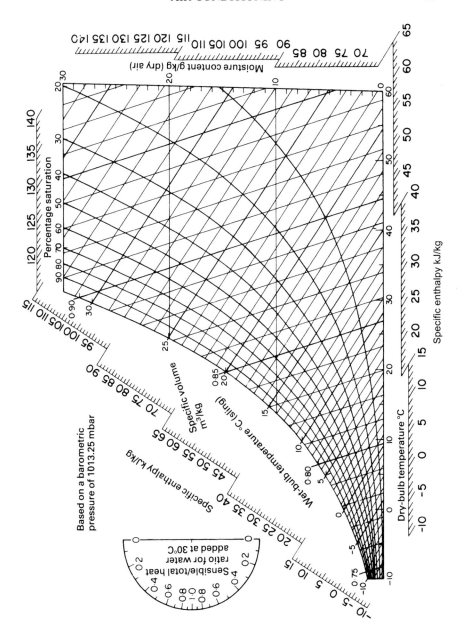

Fig. 2  Psychrometric chart

($y$) axis. On some charts a series of curves are plotted showing moisture content against dry bulb temperatures for a range of relative humidities from 10 to 100%. Wet bulb temperatures obtained by the sling psychrometer are plotted along the 100% RH curve, where they are identical to the dry bulb temperature. The wet bulb temperatures appear as diagonal lines across the chart, sloping downwards to the right.

Total heat lines may also be plotted, they run diagonally from top left to bottom right. They are not drawn on the main part of the chart because they are almost, but not quite, parallel to the wet bulb lines. The values may be read by placing a ruler or straight edge across the chart.

### Total Heat (Enthalpy)

Total heat is the sum of the sensible and the latent heat contained by a volume of moist air. Sensible heat is indicated by the dry bulb temperature. If more sensible heat was added to the air, its temperature would rise in proportion. Latent heat is the heat required to change the state of a substance. For example, 2,250 kJ are required to change 1 kg of water into steam at 100°C. So, if 1 kg of water is evaporated into a volume of air, heat must have been added to the water.

If the heat was drawn from the air, then the air temperature will have been reduced. In effect, some of the sensible heat will have turned into latent heat and the total heat will still be the same as before.

If the heat to vaporise the water is supplied from an outside source, the air temperature will remain constant. But the total heat will have been increased by 2,250 kJ, or the latent heat of vaporisation of the 1 kg of water.

So,

total heat = sensible heat + latent heat. The total heat energy content of moist air is called 'enthalpy' (symbol H) measured in kJ per kg of dry air, above a datum temperature.

Using this concept it is possible to calculate the total heat required to raise the temperature of a given volume of air and to change its RH to the required value by evaporating a quantity of water. In this way the output of the air conditioning plant required may be determined.

The output of a cooling plant is sometimes expressed in 'tons of refrigeration' (TR). This is based on the amount of heat absorbed by melting a 'ton'. This is the 'short' or 'net' ton, 2,000 lb.

A ton of refrigeration is based on melting 2,000 lb of ice at 0°C to water at 0°C in 24 hours, which is equivalent to 3.52 kW.

Air conditioning plant operates to provide filtered air at the desired temperature and relative humidity. From the previous section it can be seen that cooling air to a temperature below its dew point will cause water vapour to condense out. This is normally achieved by passing the air over a chilled water coil at 6 or 7°C. The air is then reheated to the required temperature for entering the room.

Two basic types of chiller are available. These are:

- absorption chillers which use heat as the primary driving force
- vapour compression chillers which use mechanical energy.

An alternative way of achieving dehumidification is by the use of a desiccant drying system. This absorbs moisture from the air stream and then rejects the moisture outside the building through a separate air or hot gas stream.

## Absorption Units

Absorption chillers use heat to produce chilled water via an evaporating, condensing and absorbing cycle. The heat can be supplied directly, by a gas burner, or indirectly by hot water or steam from a boiler or combined heat and power (CHP) system.

The most commonly used refrigerant pairs are lithium bromide/water and water/ammonia. Lithium bromide is a salt which changes the boiling point of water (the refrigerant) depending upon the concentration and surrounding pressure. In a commercial chiller using lithium bromide, the pressure is always below atmospheric and ranges from about 10 mbar up to 800 mbar absolute pressure at different points in the system. The lowest possible chilled water temperature, that can be produced, is about 5°C. The ammonia systems operate at pressures above atmospheric and some of them can achieve even lower cooling temperatures because the ammonia is the refrigerant, not water.

The chillers usually reject heat to a separate cooling circuit through a wet cooling tower or large radiator, although some chillers have direct air cooling. Cooling water temperatures range from 20 to 35°C and the quantity of heat to be removed is the sum of the heat needed to run the chiller and the chilling capacity.

The operation of an absorption chiller is shown in Fig. 3. Heat from the boiler, CHP or a gas burner, generally operates the generator (or concentrator) at a temperature of between 80 to 160°C depending upon the design. The generator boils refrigerant (water) out of the solution and this vapour passes to a lower pressure zone in the condenser where heat is removed through a cooling coil. The refrigerant vapour condenses and passes to an evaporator where it is sprayed over the chilled water coil. The 'warm' chilled water entering the chiller at around 12°C gives up some heat through the coil and this evaporates the refrigerant at very low pressures. The chilled water leaves at around 6 or 7°C to supply the building or other chilling load.

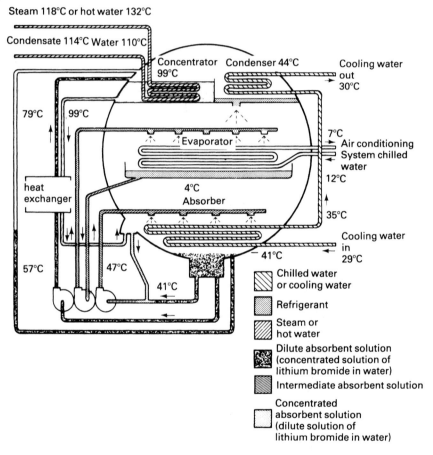

Steam 118°C or hot water 132°C

Condensate 114°C Water 110°C

*Fig. 3   Operation of an absorption unit*

The refrigerant vapour is absorbed into concentrated solution returning from the generator as this is sprayed into the absorber. Some heat is taken out of this solution by a heat exchanger before the solution enters the absorber and by a cooling water coil inside the absorber. After absorbing refrigerant vapour, the weak solution from the absorber is driven to the higher pressure generator by a small solution pump to complete the continuous cycle.

Absorption chillers are available as single effect or double effect depending upon the design and temperature of the heat source. Double effect chillers need a high temperature (160°C) to operate them. They have two generator stages and use heat recovered from the refrigerant vapour leaving the first generator stage to operate the second stage. This makes them more efficient and a working coefficient of performance (COP) of around 0.9 can be expected. Single effect chillers operate from a lower temperature source, 80 to 120°C, and have a COP of 0.6 to 0.7.

$COP_c$ – Coefficient of Performance (Chilling). This is the ratio of the amount of chilling produced from a given system to the amount of input mechanical energy (vapour compression system) or thermal energy (absorption chiller).

$COP_h$ – Coefficient of Performance (Heating). This is similar to $COP_c$, but is a ratio of the heat output from the system to the energy input. $COP_h$ can be, and often is, higher than $COP_c$ because the heat output from a system includes the energy from a low grade heat source which is raised to a higher grade through work being put into it. When chilling, the low grade heat (which becomes the system load) is no longer useful and must be discarded.

### Direct-fired Absorption Chillers

The smallest commercially available absorption chiller range is rated at 3, 4 and 5 TR, see Table 1. These are single effect units using an ammonia/water absorption cycle. They have a COP of between 0.4 to 0.5 and are air cooled so that they can be mounted outside on a rooftop or other convenient location. The units are available with a heating option to cope with Summer cooling and Winter heating requirements using one unit.

The units are self-contained with an integral water circulation pump and air cooler operating from a common fractional horsepower single phase electric motor. Installation is very simple and the compactness, low maintenance costs and low electrical power consumption are attractive features. Several hundreds of these units are in use in the UK.

A range of medium sized modular chillers is available in 10, 20, 30 and 40 TR sizes with multiple units up to 120 TR on a common frame. These units use a lithium bromide/water absorption cycle.

The 10 TR size of chiller is available as single effect unit only, whilst the 20 TR chiller is available in both single and double effect designs. The larger 30 and 40 TR units are available in double effect design only and have a COP of around 0.9.

**TABLE 1 Absorption chiller units**

| Make | Size Range TR | Single/Double Effect | Heat Source |
|------|---------------|----------------------|-------------|
| YORK/MITSUBISHI | 120–1377 | Single | LPS/HW |
| | 100–1500 | Double | HPS |
| | 100–1400 | Double | DGF |
| CARRIER/EBARA | 90–1500 | Single | LPS/HW |
| | 90–1100 | Double | HPS |
| TRANE | 100–1660 | Single | LPS/HW |
| | 385–1060 | Double | HPS |
| SANYO | 10–20 | Single | DGF |
| | 20–950 | Double | DGF |
| | 10–250 | Single | HW |
| HITACHI | 200–1500 | Double | EGH |
| SERVEL | 3–5 | Single | DGF |

HW    Hot water at 85°C
DGF   Direct gas-fired
EGH   Exhaust gas heated at 500°C
LPS   Low pressure steam at 120°C
HPS   High pressure steam at 160°C – 180°C

Each chiller in this range requires a liquid cooling system to reject heat using a wet evaporative cooling tower or dry radiator cooler.

All of these units are available with both dual chilling or heating functions if required. Low electrical power consumption, efficient performance and low maintenance costs are key advantages.

A range of large direct-fired units is available with outputs from 100 to 1,500 TR. These are double effect chillers using lithium bromide/water. These chillers require an external cooling water supply like the medium size modular range.

## Indirectly-fired Absorption Chillers

A wide choice of steam/hot water driven absorption chillers exists with equipment being available from at least four manufacturers. The equipment is available in nominal sizes from 100 TR up to 1,600 TR. Both single and double effect units are produced with COPs of around 0.6 and 0.9 respectively.

These absorption chillers can be operated from steam or hot water provided by a central boiler system or alternatively this may be provided by a gas engine or turbine CHP system using a heat recovery boiler fitted in the exhaust gas stream.

The growing popularity of CHP schemes means that a relatively cheap source of heat is becoming available to many users and absorption chillers can provide a good use for this heat.

With engine driven CHP, the waste heat is available in two forms. Hot water is produced by the jacket and hot gases are available from the exhaust. In many cases, the jacket heat will be available at 90°C. The exhaust heat is available at a relatively high temperature (550°C). This can be combined with the jacket heat in a water/exhaust gas heat exchanger but the temperature will be reduced to around 90°C to satisfy the jacket temperature limitation. The total amount of heat available can then be used to operate the chiller.

Engine driven CHP will normally link with a single effect chiller due to the relatively low temperature available.

For gas turbine CHP systems, the exhaust gases from the turbine will be at a relatively high temperature and probably with 300% excess air. A waste heat boiler will be required to produce steam at 160°C and further fuel gas may be combusted in the exhaust stream to supplement the heat if required. With high pressure steam available, a double effect chiller will be the most appropriate.

Double effect chillers are about 20 to 30% more expensive to purchase than single effect chillers for the same output, but they use less heat per unit of chilling output and have a 20% smaller cooling water heat rejection requirement. The resulting savings in capital cost of the coolers can largely offset the extra capital cost of the double effect chiller.

## Desiccant Dehumidifiers

The use of desiccant dehumidifiers is particularly attractive in situations where there is a need to dry the air rather than to chill it. One typical situation is in supermarkets where a low relative humidity can avoid the problem of condensation on freezer cabinets.

The desiccant drying system uses a honeycomb shaped wheel, lined with moisture absorbing salt, to take up water vapour as air is passed through the honeycomb, Fig. 4. The wheel is slowly rotated through separate chambers containing the cold incoming air and a stream of heated air or combustion gases. Moisture is absorbed from the cold air stream into the desiccant but it is then given up to the hot stream to reactivate the wheel as it rotates.

Units capable of handling from 85 m³/h up to 37,500 m³/h of air flow are available. Reactivation is possible by electric heating, or by steam or direct gas firing on some of the larger units.

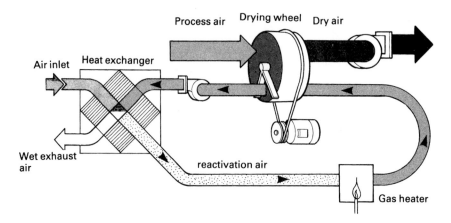

*Fig. 4    Desiccant dehumidifier (Munters)*

## Vapour Compression Units

Vapour compression chillers use mechanical energy to produce chilling via an evaporating and condensing cycle. This mechanical energy is usually provided by an electric motor but it can also come from a gas engine, or other source of shaft power such as a steam or gas turbine.

Chlorofluorocarbons (CFCs), such as R11 or R12, have been used widely as refrigerants for vapour compression chillers. These are excellent refrigerants because they are non-flammable and non-toxic, but they are now considered to have a long term damaging effect on the earth's ozone layer. Replacement refrigerants, such as halofluorocarbons (HFCs), are being used in new equipment to reduce ozone depletion and other longer term refrigerants are under development.

In the working cycle, Fig. 5 refrigerant vapour is compressed in a reciprocating, centrifugal or rotary compressor whereupon the temperature rises. The hot vapour passes through a condensing heat exchanger and turns to liquid at high pressure. The chiller rejects this heat to atmosphere either indirectly through a water cooling circuit or directly by passing refrigerant through an air cooled radiator.

Liquid refrigerant is expanded through a throttling valve and cools as it turns to vapour. This cold vapour passes through an evaporating heat exchanger where it takes heat from the building or other chilling load. The evaporator can operate through a chilled water circuit or it can be in the form of several heat exchangers in each room or in the building air supply ducts.

The refrigerant vapour then returns to the compressor to complete the continuous cycle. As well as refrigerant, oil is circulated around

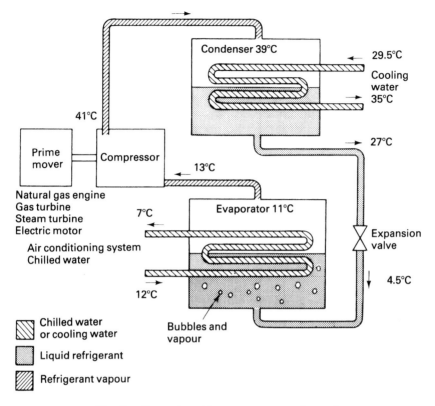

*Fig. 5   Operation of vapour compression unit*

the compressor to lubricate it and this oil is removed from the refrigerant by an oil separator situated after the compressor.

Reciprocating piston compressors are used for small sizes and up to about 600 kW (170 TR) operating at typical electric motor and engine speeds. Centrifugal compressors generally operate at high speed with outputs from around 350 to 3,500 kW (100 to 1,000 TR). Rotary screw and sliding vane compressors are becoming available in smaller sizes and have good volumetric efficiencies.

Vapour compression chillers often have a COP of between 2.5 to 3.5 based on the ratio of chilling output to power input although higher performances are possible depending upon the design temperatures. They appear to be much more efficient than absorption chillers but, of course, the shaft energy to drive the compressor, whether it be electric motor or engine, ultimately comes from a heat engine and the efficiency of this needs to be included in any comparison.

Chillers which can be operated in reverse cycle to act as a heat pump and provide heating in cooler periods are becoming more popular. As the standard of building insulation continues to improve and there are more internal heat gains from electrically operated office equipment, so the balance of Winter heating to Summer cooling loads changes and the need for large boiler plant decreases. Heat pumps are particularly attractive in these circumstances because one piece of plant can fulfill two functions.

Gas engine driven chillers produce heat from the engine jacket and exhaust which can be recovered and used to provide hot water simultaneously with chilling. If operated as a heat pump, the heat recovered from the engine can be added to heat from the refrigeration circuit to provide a substantial heating capacity from one unit.

## Gas Engines

The previous chapter covered the use of gas engines to provide the prime mover in the case of combined heat and power. The waste heat from the process is often the source of heat for the air conditioning unit. The two main types of engine are the dual fuel (compression ignition) and the spark ignition engine.

### Dual Fuel Engines

Similar to diesel engines, these can be run either on oil or on a mixture of gas and 'pilot' oil. The air/gas mixture is compressed in the cylinders and is ignited by the injection of a small amount of oil. The injected 'pilot' oil is usually between 5 and 10% of the full fuel requirement of the engine.

Engines are available in sizes from 0.5 MW to 6 MW. They are heavy for their output and require frequent and highly skilled maintenance. In consequence they are used for large installations and are not generally marketed as packaged units with compressors.

### Spark Ignition Engines

In these engines the air/gas mixture is ignited by a high tension electric spark as in a petrol engine. They are lighter and easier to maintain than dual fuel engines and their initial and installation cost is less. The types available may be either industrial engines or converted automotive engines. Industrial engines are available in sizes from 37 to 1,566 kW with running speeds ranging from 1,200 to 2,400 rpm and compression ratios of up to 10:1. Converted automotive engines are simply ordinary car or lorry engines modified to run on natural gas instead of petrol. A range of converted engines is available and these can be obtained linked to chilling compressors of 70 to 5,275 kW (20 to 1,500 TR) output. An example is shown in Fig. 6.

*Fig. 6    Reciprocating compressor, dual service set*

## Installation of Gas Engines

*Mounting*

A heavy dual fuel engine must be mounted on a separate base which is isolated from the building structure so that vibration is not transferred. Spark ignition engines are generally light enough to be coupled to the compressor and the whole unit mounted on a steel frame or 'skid'. The skid is then set on anti-vibration mountings as in Fig. 6. Large spark ignition engines are mounted on mass concrete bases which are considerably cheaper than an equivalent steel base. There is generally less trouble with vibration from spark ignition engines because of their lighter weight and higher speed.

*Noise*

The compressor of a chilling unit is sufficiently noisy to require careful siting and the substitution of a gas engine for an electric motor makes it essential that steps are taken to reduce the noise level.

The additional engine noise comes from:

- mechanical noise from the engine
- the cooling equipment
- the exhaust system.

Acoustic cladding of the plant room walls and possibly hoods over the engine and compressor can reduce the effect of mechanical noise. Fan and radiator cooling systems tend to be noisier than water to water heat exchangers. Exhaust noise is reduced by suitable silencers and by siting the discharge away from any sensitive areas.

The unit should be connected to its outlet ducting by flexible joints to avoid the transmission of noise from the plant.

*Air Requirements*

Air is required for combustion and ventilation. Combustion air is required for the engine and any other fuel burning equipment. For the engine this is about 10 times the gas flow rate.

For ventilation the air requirement depends on:

- the permitted air temperature rise in the room
- the method of cooling the engine water jacket
- the degree of insulation of radiating surfaces, in particular the exhaust manifold and system.

In each case the plant manufacturers should be consulted and any other equipment in the room also taken into consideration.

A major engine manufacturer recommends the following formula:

$$\text{m}^3/\text{s} = \frac{0.1406 \times \text{kW}}{\text{t}^\circ\text{C}}$$

where:

    $\text{m}^3/\text{s}$ = cubic metres per second (air required for ventilation)
    kW  = rated power in kilowatts of the engine
    $\text{t}^\circ\text{C}$  = permitted rise in air temperature.

*Engine Cooling*

Engines are water-cooled. Water circulates through passages in the cylinder block and a difference in temperature of 10°C should be maintained between the inlet water at 70 to 80°C and the outlet at 80 to 90°C. Actual temperatures vary with different makes of engine. Thermostats are used to control the flow of cooling water and a high temperature cut-out is often included.

The hot water leaving the engine is cooled by means of:

- water to air heat exchanger (radiator)
- water to water heat exchanger.

Radiators with fans are simple but noisy and are generally used for smaller engines.

Water cooling of the engine water requires a cooling tower but a tower is already required for the refrigeration condenser and one cooling tower can serve both units. The temperature of the cooling water leaving the condenser is low enough to be used for the inlet water to the heat exchanger and only one circuit is required.

*Exhaust*

The temperature of the exhaust gases depends on:

- type of engine
- fuel used
- engine loading.

Turbo-charged engine exhaust temperature is typically 420 to 460°C. Other engines are generally 550°C or below and a few may reach 650°C. The hot gas must be conveyed safely to discharge outside the building in accordance with the recommendations of the manufacturer.

The following points must be considered:

- the position at which the exhaust terminates
  - clear of air inlets to the building
  - unlikely to cause noise nuisance
  - clear of adjacent buildings
  - in accordance with planning requirements
- exhaust pipe diameter should be at least as large as the manifold connection
- total backpressure of piping and silencers to conform to makers specification
- exhaust piping to be capable of withstanding maximum exhaust temperatures
- a high-temperature flexible connection to be used between the engine exhaust manifold and the exhaust pipe
- the piping must be protected by insulation or ventilated sleeves so that it does not present a fire hazard inside the building
- precautions should be taken to avoid condensation; where moisture is likely to be trapped, drainage points must be provided
- provision must be made to accommodate the thermal expansion of the exhaust piping

- adequate support is necessary to avoid damage to the flexible connection.

## Engine Starting

Large dual fuel engines are usually run up to starting speed by injecting compressed air. Smaller engines and spark ignition engines are usually started by means of electric starter motors powered by lead/acid batteries. The batteries are trickle charged from the mains supply. They require only a weekly check on acid level and an occasional check on specific gravity.

Some spark ignition engines have a compressed air motor mounted in place of the electric starter.

## Gas Supply

The gas meter and the gas supply to the engine must be capable of meeting the maximum consumption demanded by the engine on full load, without reducing the pressure to any other appliances on the same supply.

Industrial engines must be fitted with a low pressure cut off switch in the supply, as close to the engine as possible but upstream of the final governor.

Engines with natural induction require gas at pressures below 200 mbar.

Turbo-charged engines require gas at pressures of about 2 bar. Usually the pressure is supplied by a compressor. It is important to check that the gas meter, pipework system, pressure reservoirs on the inlet and outlet of the compressor are correctly sized to avoid pressure transients and hunting. This can be done using a computer based programme to indicate pressure changes on start up, control and shutdown of the compressor.

Warning notices should be fitted at the engine starter and at the meter inlet, Chapter 2.

## Dual Service Sets

Supermarkets and some other commercial undertakings require emergency lighting at about 40% of the normal load in case of electric supply failure.

When a gas engine is used to drive a compressor, it can also be coupled to an alternator. Figure 6 shows a dual service unit giving up to 60 kVA in which the alternator is driven by three Vee belts and the coupling to the compressor is through an electric clutch.

When the electric supply fails, the electric clutch disengages the compressor, the engine speed is automatically adjusted to 1,500 rpm and current is available in the stand-by circuits at 50 Hz a.c. A reasonable level of emergency lighting is provided at the expense of the chilled water. However, the cooling load is reduced because of the reduction in lighting. If the engine is stationary at the time when the supply fails, it is automatically started, declutched and run up to generating speed.

## Controls

Controls are fitted to the sets to prevent overheating, ensure safe operation and obviate any false starts. These may include:

- oil pressure switch, to shut down the engine if the oil pressure falls too low
- time delay switch, to prevent too rapid cycling by stopping the compressor for about 5 minutes after each shut-down
- low pressure switch, to stop the compressor if suction becomes so low that it would produce freezing in the chilled water
- high pressure switch, to shut down if pressure rises excessively for any reason.

Most plants have safety switches fitted in the chilled water circulation to stop cooling if the water temperature approaches freezing point.

## Plant Servicing

Much of the maintenance work on air conditioning plant is carried out by specialists in this field. Organisations which have large installations often have a resident plant engineer. The plant engineer carries out daily and other periodic checks and operations. Annual servicing is normally done by a specialist firm of refrigeration engineers.

### *Water Chillers*

Servicing of gas-fired chillers and chiller/heaters should include:

- cleaning combustion chamber and flueways
- checking and adjusting gas rate
- checking operation of gas and electrical controls
  - overheat switches
  - thermostats
  - safety cut-offs.

Some machines require non-condensable gases to be purged and on air-cooled chillers the air ways must be cleaned and the air flow checked. Belt drives should be examined and belt tension checked.

### Gas Engines

Major services are carried out on the basis of the number of hours of running. Most engines are fitted with an 'hour-run' meter. It is quite common now for engines to be remotely monitored and results obtained via a modem. Deterioration of sensed parameters are quickly and easily spotted. This allows remedial work to be carried out before performance drops too much or before premature failure occurs.

Oil and water levels should be checked daily and battery acid levels every week. Spark plugs, points, oil and filters should be changed periodically as specified by the manufacturer. The requirements for servicing are less than those for a comparable petrol engine.

Servicing schedules are provided by most manufacturers and some offer maintenance contracts.

*Fig. 7    Air curtain, door heater*

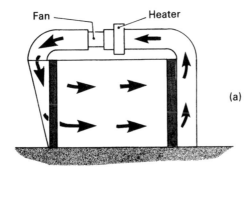

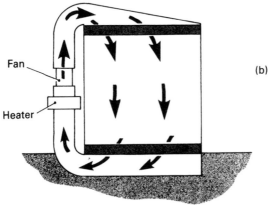

*Fig. 8    (a) Horizontal recirculation by collection duct and (b) Vertical recirculation by collection duct*

## Absorption Chiller Maintenance

The typical maintenance requirements for an indirectly heated 100 to 150 TR size absorption chiller involve four site visits per year as follows:

Annually – Major inspection of heat exchangers for scaling and fouling, and to analyse the refrigerant solution.

Every 6 months – Check the refrigerant charge level.

Every 3 months – Check the low temperature cut out and other safety interlocks.

Daily (customer) – Log the machine purge, either manually or automatically via a BEMS. This is a motorless purge which removes hydrogen, air and non-condensables from the system.

## Air Curtains

Many air conditioned buildings have two sets of outside doors forming an air lock to prevent any sudden in-rush of un-conditioned air. Where only single doors are used, the incoming air may be dissipated by a stream of air directed downwards or across the opening. Some air curtains use only cold air but more frequently, in this country, warm air curtains are employed. The warm air is produced by a directly or indirectly fired unit or by a steam or water to air heat exchanger. Larger doorways may be heated as shown in Fig. 8a and 8b.

A typical gas-fired heater for large doors is shown in Fig. 7. This is a direct-fired air heater, with the burner situated in the path of the air flow through the heater. Twin centrifugal fans expel the hot air downwards through the outlet louvres. The heat input can be varied from 102 kW upwards to suit a particular doorway. Mounting heights are generally between 3.5 and 7.5 m and the air inlet is located outside the building, unless the operation of the heater is interlinked with the door opening and closing.

The design, installation, commissioning, operation and maintenance of mechanical ventilation and air-conditioning systems is covered in BS 5720 (1979) Code of Practice for Mechanical ventilation and air-conditioning in buildings.

CHAPTER 12

# Large-scale Heating and Hot Water Systems

---

Chapter 12 is based on an original draft by B. J. Whitehead

---

## Introduction

Whilst hot water systems are heated by various designs of gas-fired boilers, heating systems can utilise either boilers or warm air heaters. The choice of system depends on a variety of factors including:

- whether building is existing or projected
- type and construction of building
- size and purpose of building
- frequency of occupation
- environmental requirements of activities or processes carried out within each room
- limitations imposed on projected capital and running costs.

In many cases buildings or parts of buildings may be used for very different activities, each requiring its own level of heating. For example, in hotels and some public buildings a room used for a conference during the day may be used for a dance in the evening and the system must be flexible enough to provide for both. In buildings where not all of the rooms are in continuous use a zoned system or individual independently controlled emitters may be required. Libraries, museums and art galleries must maintain constant levels of temperature and humidity and electrostatic filters and humidifiers may be used. Many of the supermarkets and departmental stores have systems which need to exhaust vitiated air and introduce a quantity of fresh air to maintain a pleasant environment for their customers, however 'total loss' systems are now outdated, in view of the energy situation.

## Warm Air Heating

Warm air heating units are available in four main forms:

451

- unit heaters – for overhead suspension with or without ducting systems
- horizontal and inverted heaters – for overhead installation
- direct-fired air heaters (e.g. make up air heaters)
- floor standing heaters – with or without ducting systems.

All types are generally suitable for installation in existing and projected buildings and have the added advantage of providing fresh air circulation or air recirculation at ambient temperature during warm weather. Units may also be fitted with filters to clean fresh or recirculated air. This is desirable in shops, restaurants and public houses.

The first three types of heaters were introduced in Chapter 9.


## Floor Standing Heaters

*Construction*

The units are designed to draw in cool air at floor level and to discharge heated air through one or more outlets at the top of the heater, Fig. 1. The outlets may take the form of fixed or swivel directional nozzles or a connection to a duct system.

Most heaters are indirectly fired and the heat exchanger is usually tubular, made of stainless steel and may have baffles in the tubes to improve heat transfer. The air is forced through the heater by one or two large capacity centrifugal fans situated in the lower section. The fan motor is often 380/440 V a.c. 3 phase and may range from 1.0 kW up to about 11.0 kW.

Automatic forced draught burners are used in conjunction with full sequence control and flame rectification or ultraviolet flame monitoring. Other controls may include:

- air pressure switches; activated by combustion air or flue fan pressures
- limit stat
- fan delay unit
- thermostats; appliance, day and night or night set back types
- clock.

Heater outputs range from about 44 kW up to 1,000 kW. Gas connections are usually Rc1 to Rc1½ and flues from 200 to 250 mm diameter. Heaters are open flued which may have fan assisted or fan diluted systems added.

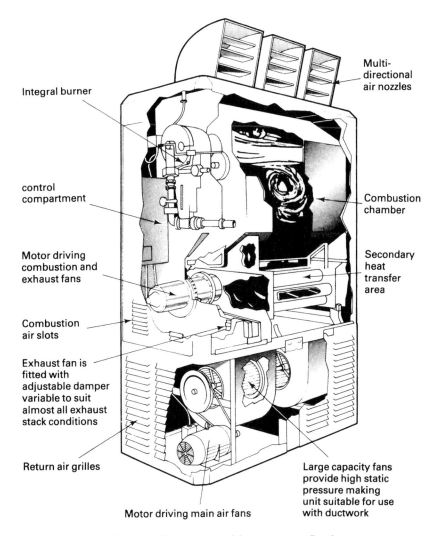

Multi-directional air nozzles

Integral burner

control compartment

Combustion chamber

Motor driving combustion and exhaust fans

Secondary heat transfer area

Combustion air slots

Exhaust fan is fitted with adjustable damper variable to suit almost all exhaust stack conditions

Return air grilles

Large capacity fans provide high static pressure making unit suitable for use with ductwork

Motor driving main air fans

*Fig. 1    Floor mounted heater, open flued*

The British Standards covering detailed requirements including safety and performance and methods of test for forced convection air heaters up to 2 MW are:

BS 5990: 1990 Direct gas-fired forced convection air heaters with rated heat inputs up to 2 MW for industrial and commercial space heating: safety and performance requirements (excluding electrical requirements) (2nd family gases).

BS 5991: 1989   Indirect gas-fired forced convection air heaters with rated heat inputs up to 2 MW for industrial and commercial space heating: safety and performance requirements (excluding electrical requirements) (2nd family gases).

BS 6230: 1991   Installation of gas-fired forced convection air heaters for commercial and industrial space heating. (2nd family gases).

## Warm Air Systems

### Free Discharge Systems

When designing a heating system based on direct discharge from a floor mounted heater, the position of the unit is of paramount importance. The following points should be noted.

Warm air must be free to discharge evenly over the area to be heated without obstruction. The manufacturer's data normally gives information on the effective throw of the heated air output. This may be from 15 to 45 m.

A clear space must be left around the base of the heater to allow the return of cool air for recirculation. When return air is taken in on all sides a distance of 450 mm should be allowed between the heater and any obstruction. Recirculation intakes must not be restricted or blanked off without prior agreement with the heater manufacturer.

The movement of return air to a heater may be noticeable within a radius of about 2 m depending on heater size. The heater should be sited so that no permanent places of work are within this area.

Typical locations for effective warm air distribution are shown in Fig. 2.

### Ducted Systems

Where consideration of building layout, furniture and fittings, appearance or noise factors make it impracticable to fit the heater in the working area a duct system can be used. Systems may be designed to distribute heat evenly throughout a large area or to convey varying quantities of warm air into separate rooms, wherever practicable diffusers should be sited or designed so that the air flow does not come into contact with the occupants of the room. Duct-work is more easily installed in buildings during their construction. It may, however, be fitted in existing buildings particularly where it can be hung openly under ceilings, or concealed in false ceilings or under suspended floors. In certain situations fire dampers or smoke detec-

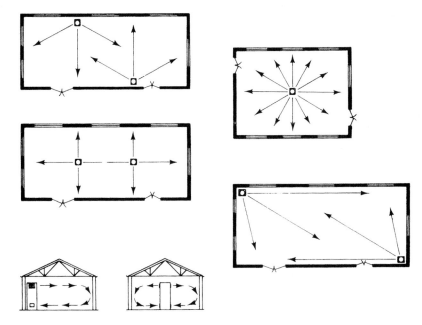

Fig. 2    Location of free-standing floor heaters

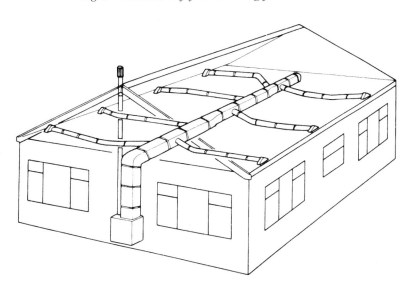

Fig. 3    Simple overhead duct system

tion equipment may be necessary under Local Authority by-laws or fire regulations. Examples of various duct systems are given in Figs. 3 to 7.

Fig. 3 shows a simple overhead duct system from a floor mounted heater. Warm air is delivered by ceiling diffusers supplied by flexible circular ducting from a stepped duct system. Air returns via the grilles on the heater unit. All ducting in the roof space is insulated with 50 mm of glass fibre or similar material. It would be possible to save floor space by using a horizontal heater mounted overhead.

Another application of an overhead duct system to a single storey building is shown in Fig. 4. This is a shop where the heater is fitted unobtrusively in a store room at the back. Ducts are run under the

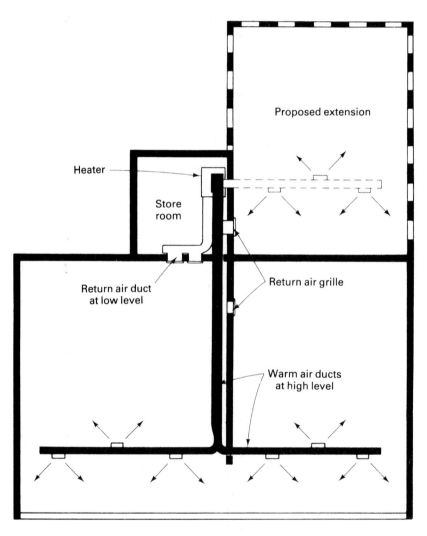

*Fig. 4   Duct system with high level warm air discharge*

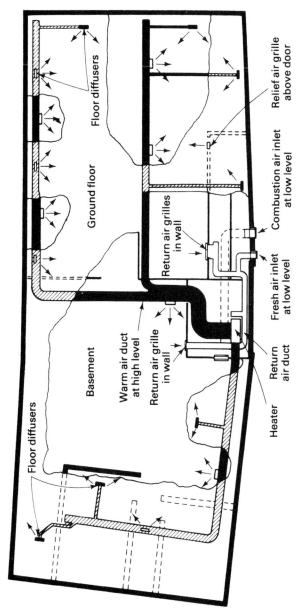

Fig. 5  Duct system in two-storey premises

ceiling with high level outlets. As in the previous example, this requires air velocities of about 3.5 to 5 m/s to take the air down to floor level. The high velocity creates noise which, in some cases, may be unacceptable but would probably be unnoticed against the background noise in a town centre. It is possible at the design stage to calculate the noise emission factor.

Figure 5 shows a single level duct system serving two storeys in a building. The heater is fitted in the basement and the ducting run on the basement ceiling. Ceiling level outlets serve the basement and the ground floor is heated by floor diffusers.

Larger buildings with several floors may be heated by the type of system in Fig. 6. The ducting is concealed above suspended ceilings

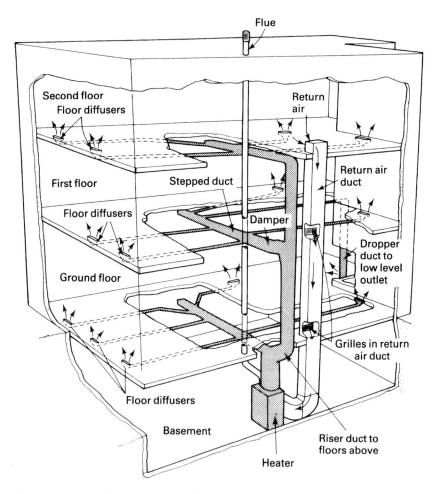

*Fig. 6    Three-storey building with ducts in suspended ceilings*

using floor diffusers to supply the floor above. The return air is ducted from each floor by grilles at low level in the return air duct.

Where a building is composed of various sized rooms with widely different heat requirements, the use of two or more heaters should be considered. This allows simple duct systems to be used and gives flexible operation of the whole system. Figure 7 shows three heaters installed in a church and operating independently according to building use. The heater serving the sanctuary draws its combustion

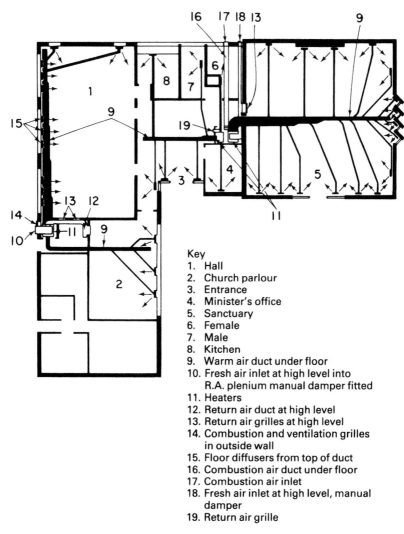

Key
1.  Hall
2.  Church parlour
3.  Entrance
4.  Minister's office
5.  Sanctuary
6.  Female
7.  Male
8.  Kitchen
9.  Warm air duct under floor
10. Fresh air inlet at high level into
    R.A. plenium manual damper fitted
11. Heaters
12. Return air duct at high level
13. Return air grilles at high level
14. Combustion and ventilation grilles
    in outside wall
15. Floor diffusers from top of duct
16. Combustion air duct under floor
17. Combustion air inlet
18. Fresh air inlet at high level, manual
    damper
19. Return air grille

*Fig. 7   Church with three independent heaters*

air from outside via a duct below the floor. A fresh air inlet is also provided to feed into the high level return air duct. 10 to 15% of the air returned is fresh air.

## Installation of Floor Mounted Units

*Location*

In addition to any space required for return air, access must be provided for servicing or replacement of the equipment. The manufacturer's instructions should be consulted. Where the heater is in contact with combustible materials, suitable heat insulating shields should be provided in accordance with the maker's specification.

*Air Inlets and Discharge Outlets*

The heater depends for efficient operation on the free flow of air through the heat exchanger. Any restriction of inlets or outlets can adversely affect the operation. Precautions must be taken to prevent obstruction occurring and guard rails may need to be provided.

In addition, where there is a risk of combustible material being placed near to the warm air outlets, suitable barrier rails should be fitted to maintain a distance of 1 m between the material and the outlet.

The minimum distance between the heater inlet and any obstruction should be 450 mm for 117 kW rising in proportion to the heater output.

British Standard 6230: 1991, gives guidance on the installation of gas-fired forced convection air heaters for commercial and industrial space heating. Heating of industrial and commercial premises is often from air heaters in plant rooms or heater compartments and in this case ventilation should ensure that the temperature does not exceed 32°C.

In special circumstances where air may be polluted by dust or shavings, the air inlets should be positioned to avoid contamination and screens may be necessary to trap the pollution.

When return air inlets are close to extraction systems carrying flammable vapours, precautions must be taken to prevent any vapour being entrained with the return air. This may entail ducting the return air from a safe distance. In such cases system design is even more critical and compliance with the additional safety requirements essential.

The installation of heaters in buildings where there is a fire hazard from flammable vapours should conform to the requirements of the local petroleum or fire regulations. As a general rule, combustion air should be ducted to the heater from outside the building. The duct must be of adequate size and its inlet should be positioned clear of any low pressure zones, areas of wind turbulence, and high enough above ground level to prevent accidental blockage from snow, dead leaves, etc. All ducting must be independently supported of the air heater. Joints and seams of supply ducts and fittings must be securely fastened and made airtight.

## Prevention of Noise Transmission

Where noise level is important, the heater should be insulated from the structure of the building by standing it on felt pads or anti-vibration mountings. When this is done it is essential that all fuel, ducting, flue and electrical connections are made flexible. Otherwise vibrations may still be transmitted and the connections may be damaged.

Noise transmission into ducting can be reduced by using a non-flammable, sound insulating connection between the heater and the duct run. Where a heater is installed in a separate compartment, it may be necessary to fit sound 'attenuators' or reducers in the duct run before it enters a sensitive area. Manufacturers can advise on this aspect.

## Heater Compartments

To comply with architects' requirements or local regulations or to cater for those installations where fire hazard, noise or risk of tampering must be minimised, it may be expedient to site the heater in a separate compartment. Requirements for compartments are broadly similar to those described in Vol. 2. The main points to be considered are as follows:

The compartment must be constructed of stable, fire resistant materials. It may be acoustically lined to minimise noise.

There must be adequate access available only to authorised persons for servicing, inspecting and renewing the equipment.

Provision must be made for:

- a supply of air for combustion
- circulation of cooling air to remove heat emitted from the flue and any duct work in the compartment
- entry of return air.

Any air inlets to a compartment should be clear of sources of smells or fumes. If noise may be emitted from an air inlet, the inlet should be positioned to cause the minimum of nuisance.

*Flues*

Flues should be supported independently of the heater unless otherwise specified by the manufacturer.

Precautions must be taken to prevent rain water or condensation from the flue from entering the heater. A condensation trap should be provided at the base of the flue.

## Servicing

Servicing should be carried out annually or after every 1,000 hours operation whichever is the shorter.

The servicing of large warm air heaters follows a similar procedure to that specified in Vol. 2, Chapter 11 for domestic models. The major differences are that the larger heaters are normally fitted with:

- 415 V a.c. 3 phase fan motors
- automatic forced draught burners
- flame rectification or ultraviolet flame sensors.

Methods of checking and servicing these components have been dealt with in previous chapters.

The remaining components which require special attention are the heat exchanger and the filter, if fitted.

The heat exchanger and the combustion chamber should be examined for corrosion or blockage, using an inspection lamp. Follow maker's instructions on methods of gaining access. Clean out the heat exchanger tubes using a scraper or a brush and remove deposits with an industrial vacuum cleaner.

The normal atmosphere carries a considerable quantity of dust and filters can quickly become choked if not cleaned regularly. A regular cleaning schedule should be established.

## Central Heating by Hot Water Boilers

The introduction of natural gas has been followed by new developments in commercial boiler design and in boiler systems. The big advance has of course been condensing boilers. Many of the older designs of boilers are still to be found on the district and the principal types to be met are:

- cast iron sectional boilers
- steel shell boilers
- modular boilers.

## Cast Iron Sectional Boilers

These are built up of a number of cast iron sections which contain the waterways and form the combustion chamber, Fig. 8. The sections are joined by nipples to form a flow header at the top, and left and right return headers at the bottom.

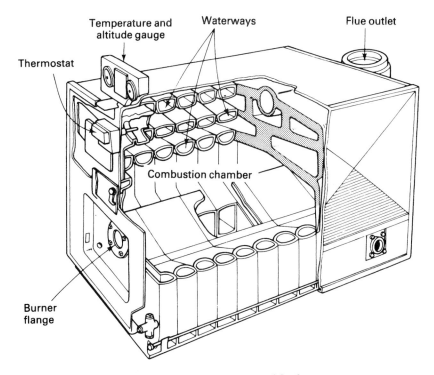

*Fig. 8    Cast iron sectional boiler*

Boilers are made up of the number of sections required to give the desired output capacity. They may be fitted with jetted bar natural draught burners or with automatic forced draught burners.

Both types of burner incorporate some form of automatic control and flame protection device.

## Steel Shell Boilers

These boilers are designed for higher pressures and greater heating capacities than cast iron sectional boilers. They normally operate at higher efficiencies. Shell boilers are generally used where heat outputs of above 586 kW are required. Models for high pressure hot water and steam are available with heat outputs up to 5,860 kW.

Although output ratings are high, the boilers are about 50% smaller in cross-sectional area than conventional sectional boilers of the same output rating.

Some types of shell boilers were described in Chapter 8. Another consists of two concentric shells enclosing an annular water space through which fire tubes are fitted. The combustion chamber is in the centre of the inner shell and the hot flue gases recirculate through the fire tubes, so giving a high thermal efficiency.

The common design of steam boiler is a three pass system, Fig. 9, where the flame travels down a central tube, perhaps a metre or more in diameter, into a reversal chamber. This reversal chamber is sometimes known as the combustion chamber. The hot gases then travel down a series of tubes back to the front of the boiler. Here they then reverse and go forward through a third set of tubes to the back of the boiler. The products of combustion may then go to the flue or sometimes pass through an economiser. An economiser is a heat exchanger which can be used to preheat the boiler feed water. Economisers enhance the efficiencies of boilers by about 5%.

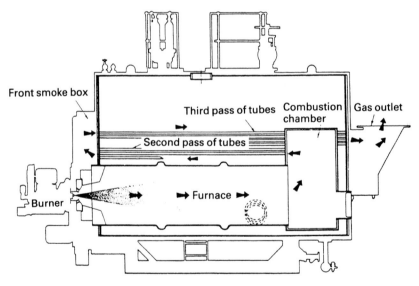

*Fig. 9    Three pass shell boiler*

*Modular Boilers*

There are various designs of modular boilers all of which generally possess similar characteristics as follows:

- low thermal capacity, quick response

- light in weight, suitable for roof top installations
- small size, to pass easily through doorways or roof traps and be transported in passenger lifts
- low noise level
- high thermal efficiency, low running costs
- easy to install and service.

Because a modular system generally consists of from two to about six boilers, there is relatively little loss of heat output when one boiler is shut down for servicing or when a breakdown occurs.

A number of modular boilers have natural draught burner systems and are fitted with down draught diverters. Comparatively recent designs have a premix burner and are connected to an open flue system but without a draught diverter. Schematic diagrams of a three module unit and an individual module are shown in Fig. 10.

The heat exchanger consists of finned copper tubes arranged in a circle around the perforated, cylindrical burner. Water enters at the front and passes through the lower tubes to the rear header, returning through the upper tubes to the flow.

The air/gas mixture is controlled by a zero governor which is impulsed to respond to fan pressure and to pressure in the flue chamber. So the governor will shut down if the fan fails or if the flue becomes blocked.

Because the thermal efficiency is remarkably high (85%), some condensation will occur in the heat exchangers when starting from cold and in the flue at that and other times. Provision must be made for condensate to flow to a removal point where it can be carried by a 22 mm pipe to a gulley. The flue should be insulated or twin-walled.

The modules will be shut down in sequence, normally starting at the top. Control may be by a step sequence controller or by boiler thermostats in each module. In this way, water passing through the upper modules is heated by flue gases from the lower.

The boiler must be fitted with a pumped circulation and the head loss is about 75 mbar at a temperature rise of 11°C.


## Condensing Boilers

Modern boilers are designed to operate at much higher efficiencies than in the past. If the flue gases are cooled down to their dewpoint of about 55°C much of the latent heat and the sensible heat can be recovered. The principle of a condensing boiler was explained in Vol. 2, Chapter 10.

Figure 11 shows a conventional boiler. For every 100 kWh of input, 20 kWh is lost through the flue (at about 200°C). Poor thermal

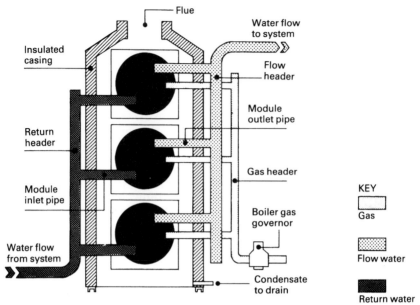

Schematic diagram of the three module Modecon 150.

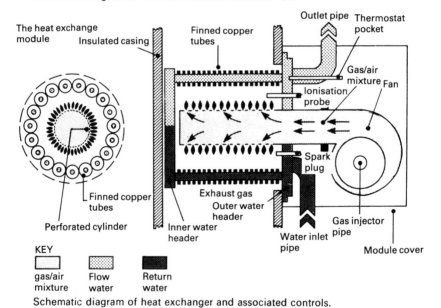

Schematic diagram of heat exchanger and associated controls.

*Fig. 10   Modular boiler*

insulation on these boilers, leads to casing losses of 3% or more. Efficiencies of these boilers are between 60 and 77%.

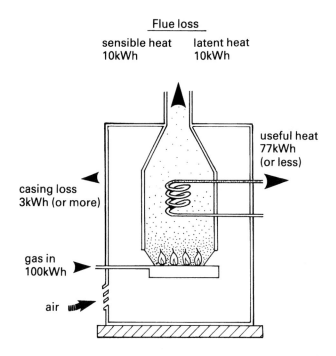

Flue loss

sensible heat        latent heat
10kWh                10kWh

useful heat
77kWh
(or less)

casing loss
3kWh (or more)

gas in
100kWh

air

*Fig. 11    Conventional boiler (The Building Services Research and Information Association (BSRIA))*

Figure 12 shows a boiler with a more efficient heat exchanger and casing insulation. Combustion air is preheated by being drawn in around the heat exchanger. Efficiency can be 84% without cooling the flue gases below dewpoint.

Figure 13 shows a boiler with an additional heat exchanger which cools the flue gases below dewpoint, releasing some of the latent and sensible heat and transferring it to the water in the heating system. The use of two heat exchangers with separate water connections provides flexibility to make the best use of variable temperature loads and gives an efficiency of 90% or more.

Figure 14 shows a boiler fitted with a further condensing heat exchanger which is suitable for circuits operating at temperatures below 30°C. This can be used to preheat a cold water supply to a domestic hot water cylinder. The additional heat recovered from the flue gases raises the efficiency to 94% or more.

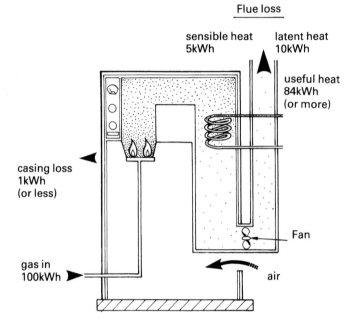

*Fig. 12    High-efficiency boiler (BSRIA)*

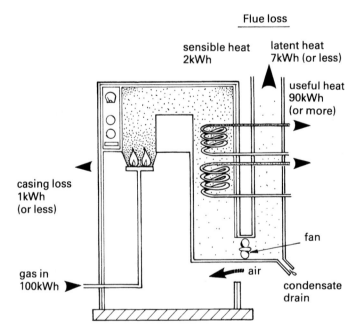

*Fig. 13    Single-condensing boiler (BSRIA)*

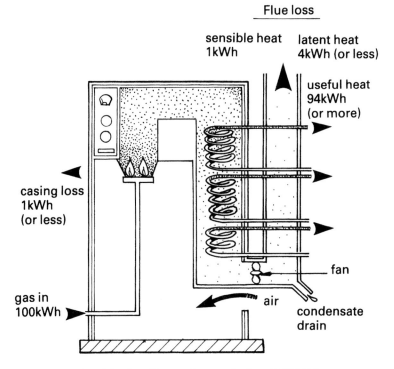

*Fig. 14    Double-condensing boiler (BSRIA)*

Figures 15 and 16 show double-condensing boilers with different heat exchanger configurations.

Modern high efficiency boilers can be used to good effect in virtually every commercial and industrial building. Single and double condensing boiler applications must be selected with more care to ensure any extra cost is fully justified. Condensing boilers should generally be part of a multi-boiler installation unless it is being used to generate both space heating and domestic hot water.

**Emitters**

Various types of emitter are available. Some provide heating principally by radiation, others principally by conduction. Most of these have been described before in previous chapters and it remains only to summarise the types and add a little further information.

*Radiant Heating*

This may be provided by:

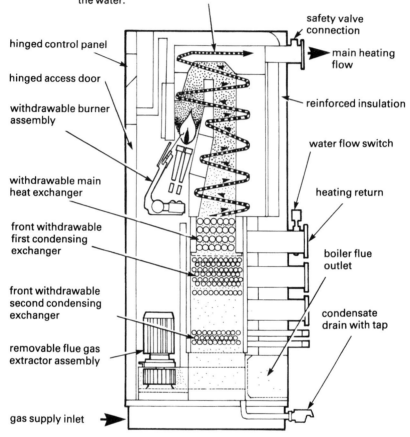

Dotted line indicates controlled water flow path from bottom to top of the boiler, giving a moving waterway around the heat exchanger area to draw the maximum heat from the gases into the water.

*Fig. 15   Double-condensing boiler (Atlantic Boilers)*

- heated suspended ceilings
- radiant panels
- radiant strips.

These were introduced in Chapter 9.

*Heated suspended ceilings.* A heating coil of comparatively small water content lies in contact cups on perforated ceiling panels of highly conductive metal. Pads of insulating material direct the heat downwards and also act as an acoustic insulator, Fig. 17. The system is slow to respond to changes in heating demand.

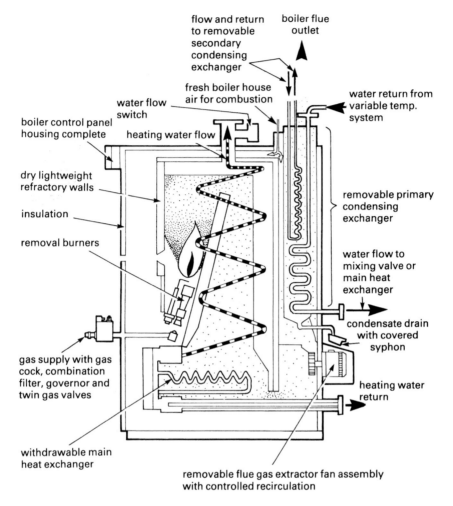

flow and return
to removable
secondary
condensing
exchanger

boiler flue
outlet

fresh boiler house
air for combustion

water flow
switch

water return from
variable temp.
system

boiler control panel
housing complete

heating water flow

dry lightweight
refractory walls

insulation

removal burners

removable primary
condensing
exchanger

water flow to
mixing valve or
main heat
exchanger

condensate drain
with covered
syphon

gas supply with gas
cock, combination
filter, governor and
twin gas valves

heating water
return

withdrawable main
heat exchanger

removable flue gas extractor fan assembly
with controlled recirculation

*Fig. 16    Double-condensing boiler (Atlantic Boilers)*

*Radiant panels.* These consist of a pipe coil fixed to a metal plate, usually of steel. There are many designs to suit different applications. They are usually suspended from roof trusses, as in Fig. 18, or fitted to walls at high level. Because they give localised heating they are suitable for entrance canopies, loading bays, workshops and exhibition halls.

*Radiant strip.* This is similar to the radiant panel but the pipe and the aluminium reflector form a long continuous strip suitable for heating corridors or large spaces where long uninterrupted runs are possible. The units may include lighting fittings and are usually

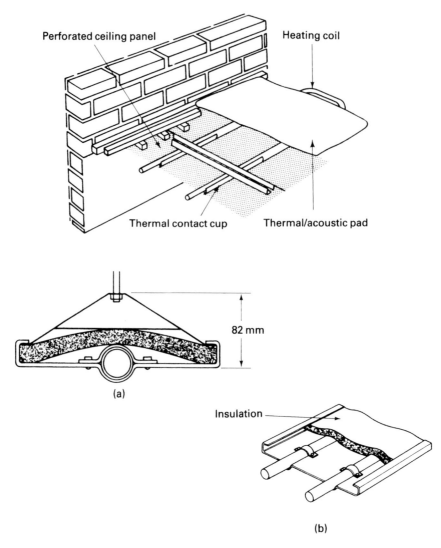

Fig. 17   Heated suspended ceiling

heated by medium or high pressure hot water. The higher surface temperatures produced reduce the length of strip required to about half that required with low pressure water. Fixing heights are given in Table 1.

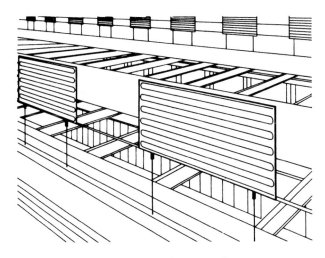

*Fig. 18    Radiant panels*

**TABLE 1  Fixing Heights for Radiant Strip Heating**

| *Hot water temperature* | *Minimum height of strip above floor level* | |
| | *One pipe* | *Two pipe* |
| *°C* | *m* | *m* |
|---|---|---|
| 70 | 2.4 | 2.7 |
| 95 | 2.7 | 3.0 |
| 115 | 3.0 | 3.3 |
| 140 | 3.0 | 3.6 |
| 160 | 3.3 | 4.0 |
| 180 | 3.6 | 4.2 |

## Convector Heaters

*Radiators.* The older types of cast iron column radiators have greater heating surfaces than the modern steel panels but are heavier and less attractive. They are still used where appearance is not the main consideration, principally in institutional premises. A cast iron hospital radiator is shown in Fig. 19. More recently it has been possible to get column radiators in cast aluminium, which are quite attractive but more expensive.

*Convectors.* Natural convectors consist of a water to air heat exchanger enclosed in a steel casing having louvres or grilles at the top and bottom. The output of warm air is usually controlled by a damper at the top of the unit, Fig. 20.

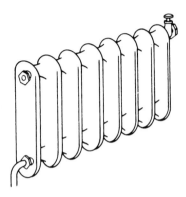

*Fig. 19   Hospital radiator*

*Wall strip heaters.* These are a continuous convector heater designed as a skirting heater or with additional heating tubes in heights up to about 1 m for under window heating in offices, Fig. 21.

*Fanned convectors.* Fan convectors are available with variable speed fans, usually low, normal and booster. Control is by thermostat which switches the fan on or off. Normally when sizing the heater, only output at low and normal settings is used. The use of such equipment needs a little thought in view of the noise which may be given off. The heaters are very compact and have a high heat output, Fig. 22(a). They may be fitted in recesses, Fig. 22(b) and are widely used in schools, offices, gymnasia and entrance halls. Sizes vary from 1 to 1.5 m in width and 1 to 2.5 m in height.

### Heater Batteries

These consist of heating coils of copper or steel finned tubes together with fans and filter. They may be housed in a framework incorporated into a duct system or form a complete heating unit as in Fig. 23.

Heater batteries may be used for:

- warm air heating
- air curtains
- make-up air heating
- air conditioning plants.

They may also be used in conjunction with radiators. In this case the warm air makes good the heat loss due to air change and the radiators compensate for the heat loss through the structure.

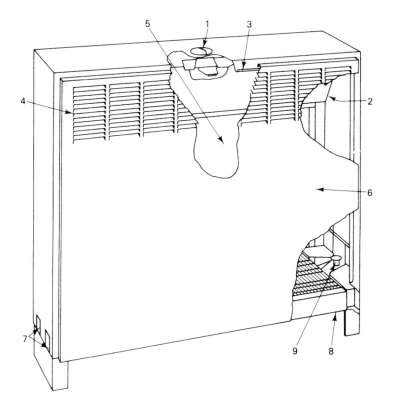

1 Damper control knob  6 Removable
2 Damper blade            front panel
3 Front panel           7 Knock-outs for
   support channel          connections (both ends)
4 Outlet grilles        8 Heating element
5 Insulated rear panel  9 Element support and
                           vertical adjustment

*Fig. 20    Convector heater*

## Unit Heaters

Unit heaters are small water to air heat exchangers with fanned convection. They are mounted overhead or at high level on walls. Most frequently used in factories or warehouses they are obtainable with horizontal or down flow discharge, Fig. 24.

## Control Systems

### Temperature Compensation

Whilst most domestic central heating systems require only simple control systems, the energy saving possible from a large commercial

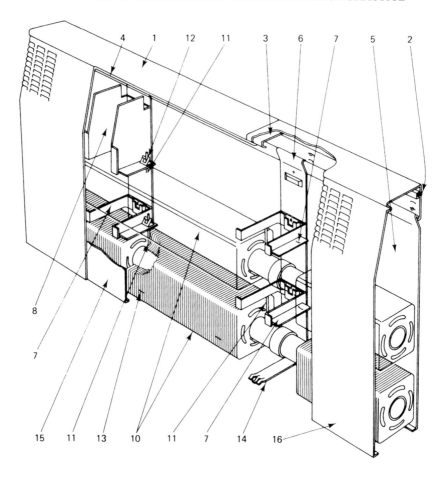

| | | |
|---|---|---|
| 1 | Top rail | |
| 2 | Plastic foam strip | |
| 3 | Inner joint rail | |
| 4 | Top rail stiffener | |
| 5 | Backplate | |
| 6 | Hanging strip | |
| 7 | Element support bracket | |
| 8 | Make-up plate/element support bracket | |

| | |
|---|---|
| 9 | Frontplate centre support (not shown) |
| 10 | Element |
| 11 | Element support hook |
| 12 | Spire clip |
| 13 | Noise suppression clip |
| 14 | Retaining clip |
| 15 | Make-up plate |
| 16 | Front plate |

*Fig. 21   Wall strip heater*

installation justifies the use of sophisticated equipment. Weather conditions vary considerably and plant is usually designed to give inside temperatures of 18 to 22°C with an outside temperature of about −1°C. However, the mean winter temperature in the south of England is about 12 to 13°C. This gives a load factor on the system of

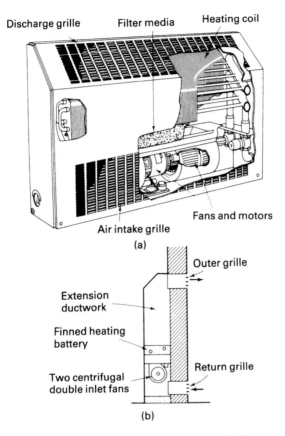

Discharge grille     Filter media     Heating coil

Fans and motors

Air intake grille

(a)

Outer grille

Extension
ductwork

Finned heating
battery

Return grille

Two centrifugal
double inlet fans

(b)

*Fig. 22    Fanned convector: (a) wall-mounted; (b) recessed*

less than 60%. If excess heating is to be avoided, and energy conserved, the boiler output must be related to the prevailing weather.

The second problem is that of regulating the heat output to meet day to day demands. This is normally done by splitting a large installation into a number of parts or zones, each under separate control. The pipework must be designed or modified to make this possible. When existing properties are being modernised it is often impossible to split the pipework into zones. For this reason the use of an outside temperature compensator for overall control became popular. By mixing flow and return water in a three port valve the heat input could be regulated in sympathy with the prevailing weather conditions, usually measured on the north wall of the building. Some compensators respond to solar gain and wind velocity as well as temperature. A programmed time switch may be incorporated to give day and night control and early morning boost.

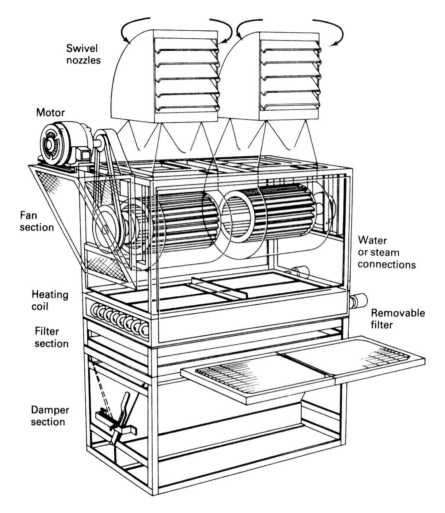

*Fig. 23   Heater batteries*

The outside temperature compensator controls the heating circuit independently of the hot water demand when both are fed from the one boiler. The hot water cylinder may be controlled by:

- cylinder thermostat controlling two port valve
- cylinder thermostat controlling separate pump on hot water circulation.

An alternative control system to the outside temperature compensator worth mentioning is known as an optimiser. The principle difference is that an optimiser also decides the time that the heating

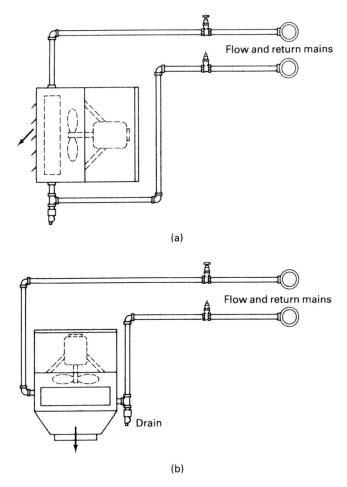

Flow and return mains

(a)

Flow and return mains

Drain

(b)

*Fig. 24    Unit heaters: (a) horizontal flow; (b) down flow*

installation should be brought on, on a day to day basis, that is if it takes into account the weather prevailing and the type of building in use.

It is now common for systems to incorporate the compensator and optimiser into a single device. The device can monitor and control all the building services including air conditioning, heating, energy consumption, lighting, access and fire alarms. These facilities are monitored and controlled by a computer system and are often known as 'Building Energy Management Systems' (BEMS). It is also possible in such systems to ensure that preset targets for electrical consumption are not exceeded by shedding non-essential loads to avoid maximum demand charges.

*Zone Control*

When pipework has been suitably arranged to provide separate zones, their control may be effected by:

- restricting the flow of water to the heating surfaces by two port valves
- modulating the zone heating surface temperature by means of mixed water using three port valves.

When restricting the water flow by room thermostats and two port valves, on/off or two position valves are usually suitable.

When heating is by blown warm air it is usually essential to use a modulating valve to avoid a blow hot/blow cold effect on the occupants.

In order to stabilise the pressure drop presented to the circulating pumps, three port diverter valves are sometimes used instead of two port valves. The heated flow water goes either to the heat exchanger or back to the common return but this increases system heat losses at times of low demand.

If the temperature of the heating surface is to be modulated, each mixing loop must either have its own pump or two port valve control.

Figure 25 is a schematic diagram of a building with a long east/west axis. The whole building is weather compensated from the north aspect. The southern faces are made into separate zones. In the example shown, the southeast zone is controlled by a three port valve which allows some of the flow water to bypass the heat exchangers. The southwest zone has a two port on/off valve which controls the flow through the zone's heat exchangers.

Restricting the flow to a number of heat exchangers provides a variable volume at constant temperature. When the flow is restricted the temperature drop across the system is increased. This can result in the emitters nearest to the source of heat always being hot, even under low load conditions. Those progressively nearer to the system return will be increasingly cooler. The emitters must be grouped in such a way that this effect is minimised, i.e. in parallel not in series.

A mixed water scheme gives constant volume at variable flow temperature and is preferable for large series/parallel groups of emitters, particularly when controlled by a single valve.

Figure 26 shows methods of applying three port valves.

(a) shows the valve as a mixer in the flow with two inlets and one outlet. This gives constant volume at variable temperature

(b) shows the valve as a diverter in the return with one inlet and two outlets. This gives constant volume at variable temperature

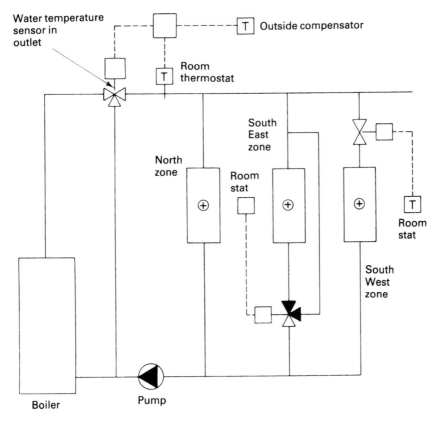

*Fig. 25    Zone control*

(c)    shows the valve as a diverter in the flow with one inlet and two
       outlets. This gives constant temperature with variable volume
(d)    shows the valve as a mixer in the return with two inlets and
       one outlet. This gives constant temperature with variable
       volume.

The effect of the valve on the system depends on whether it is used
as a mixer or a diverter and also on its position in the circuit.

Care must be taken when selecting a three port valve for use as a
diverter. Figure 27 shows typical valves. The 'lift and lay' types are
generally recommended for use as in Fig. 26(d) so that there will not
be any sudden reversal of pressure to cause the valve to bang on to its
seating. The rotary valve, Fig. 27(c), is not affected in this way.
Manufacturer's recommendations on the location and connection of
valves should always be followed.

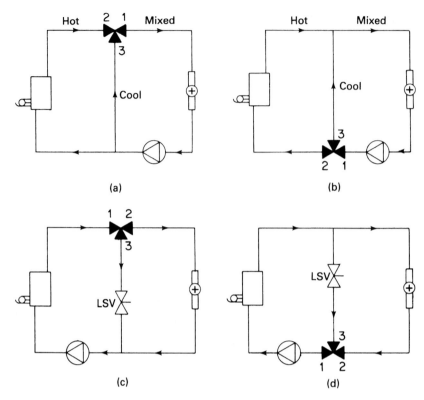

*Fig. 26*   *Use of three port valve: (a) mixer in flow; (b) diverter in return;*
*(c) diverter in flow; (d) mixer in return*

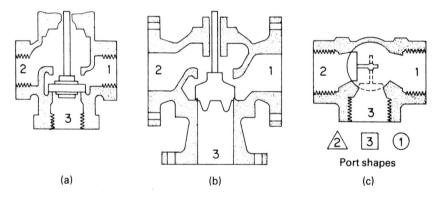

*Fig. 27*   *Types of three port valves: (a) lift and lay valve; (b) lift and lay*
*valve; (c) rotary valve*

## Rooftop Boiler Installations

The siting of boiler houses on the roofs of multi-storey buildings has become popular and offers architects and heating engineers a number of advantages over ground or basement floor locations.

### New Buildings

The principal advantages include:

- no flues running the full height of the building, so giving increased floor space and simplifying internal planning, Fig. 28
- no need for flue insulation to prevent condensation and the 'overheating' of adjacent rooms is eliminated
- the rooftop boiler room can be designed as a unit in conjunction with the water tank and lift housings
- the space not required in the basement or ground floor can be used for car parking or accommodation
- rooftop boilers may easily be drained down since they stand above the heating circuits and require only a relatively small quantity of water to be removed when the boiler needs attention
- rooftop boilers may be lighter and cheaper than basement or ground floor boilers which have to withstand high static pressures
- rooftop boiler rooms are more easily ventilated and generally do not require mechanical extraction.

### Existing Buildings

The foregoing advantages apply equally to existing buildings. In addition, a rooftop installation may offer the only solution to the problem of locating a boiler when one has not previously been fitted.

Modular boilers are particularly suitable since they may easily be transported by existing passenger lifts and manhandled into position.

### Types of Boiler

Most conventional gas-fired boilers are suitable for rooftop installation. However, because the static pressure on the boiler is very low, lightweight cast iron, sectional boilers which are unsuitable for basement locations may be used. Shell boilers, which are designed for high pressures, can be used in either roof or basement locations.

Modular boilers are particularly suitable for rooftop installation. They generally consist of a number of identical sectional boiler units, each with its own controls, coupled together to provide the desired output rating. Each individual unit is factory assembled and delivered

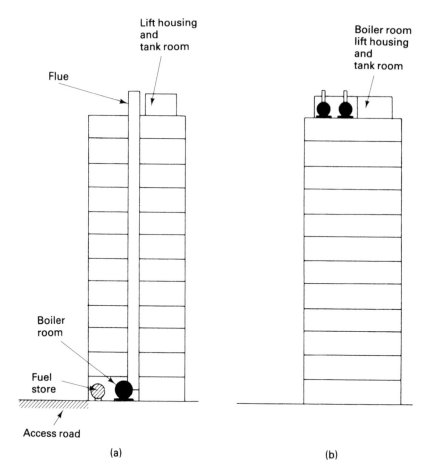

*Fig. 28   Comparison of installations: (a) conventional; (b) roof-top*

to site. Several units are then connected together, sometimes by means of standard headers but more normally the headers are manufactured on-site.

Control is usually by means of a step modulating system which maintains sufficient units in operation to meet the load. This eliminates cycling during periods when the full output of the installation is not required. A schematic diagram of the system is shown in Fig. 29.

When modular boilers are not used the rooftop installation usually consists of more than one boiler. Multi-boiler installations may consist of boilers of similar or widely varying outputs used in any combination to provide the total heat output required.

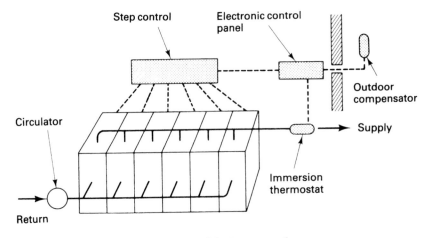

*Fig. 29    Step modulating control system*

## Flues

Figure 30 shows a typical arrangement of each boiler having its own individual flue. BS 5440: Part 2 2000 Installation of flues and ventilation for gas appliances of rated input not exceeding 60 kW.; should be referred to for installation of appliances with natural draught or forced draught, rated at below 60 kW, above 60 kW, BS 6644: 1991 Installation of gas-fired hot water boilers of rated inputs between 60 kW and 2 MW.; should be used.

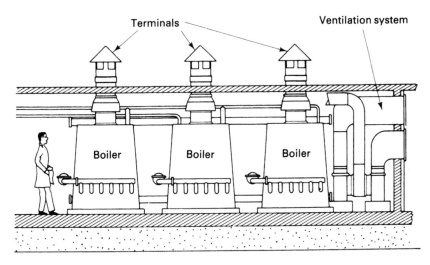

*Fig. 30    Rooftop boilers with separate flues*

*General Considerations*

Facilities must be provided for hoisting the boilers and associated plant into position. The hoists or cranes used in the construction of a new building may be retained to lift the central heating and water heating plant on to the roof. It must also be possible to replace the equipment subsequently without undue disturbance to the building or its occupants.

Because the boiler is at the top of the heating circuit, gravity circulation is not possible. All water flow through the system must be pumped.

The underside of the boiler bases must be thermally insulated to prevent the transmission of heat to the top floor of the building.

Anti-vibration measures are necessary to prevent noise and vibration from the boiler plant being transmitted to the structure.

To ensure a good standard of ventilation, rooftop boiler rooms should have at least two external walls.

## Water Heating

The sizing of boilers to meet commercial water heating demands has for many years been by rule of thumb. This has resulted in many boilers being oversized, sometimes by as much as 100%.

There are two sources of information for the accurate sizing of hot water plant for commercial buildings. These are, CIBSE Installation and Equipment data (Vol. B) and the British Gas guide to hot water plant sizing for commercial buildings.

The British Gas guide covers the two main demands for hot water: for catering and for service – washing, bathing and cleaning. The guide is based on twelve graphs Figs. 31 (a to l) which will meet most commercial requirements. The designer chooses the hot water recovery period most suitable for the system, these are shown on the graph curve in hours (h). He can then read off from the guide curve, the boiler power and storage capacity per person, or per meal, that is needed. The figures obtained represent the plant size required to produce the net hot water used, this excludes losses from the system. The boiler size has to be increased to compensate for heat losses from pipework and storage.

The storage volume must also be increased, by about 25%, to compensate for the mixing of incoming cold water and hot stored water. If hot water is required at a lower temperature than 65°C, then the boiler output must be adjusted by the ratio of the cold feed and hot water temperature differences. For example, if the hot water is required at 50°C and the cold feed temperature is 10°C

$$\text{New boiler output} = \text{calculated boiler} \times \frac{(50 - 10)}{(65 - 10)}$$

The use of the guide may be more easily understood by referring to the following examples:

*Example 1*

A designer requires the size of a hot water plant in a school which has 500 pupils and staff, serving 400 meals per day. The kitchens are in a separate building and require their own plant. For the purposes of this example the designer has chosen recovery period of two hours for service and one hour for catering.

From Fig. 31(a), Servicing requirement (2 hour recovery):

Boiler power          $0.035 \times 500 = 17.5 \text{ kW}$
Storage capacity     $1.1 \quad \times 500 = 550 \text{ litres}$

From Fig. 31(b), Catering requirements (1 hour recovery):

Boiler power          $0.1 \times 400 = 40 \text{ kW}$
Storage capacity     $1.6 \times 400 = 640 \text{ litres}$

Note: Allowances must be added to the storage capacity for mixing of incoming cold water with stored water and to the boiler power for system heat losses.

If a central hot water plant is required for this example, the service and catering requirements should be added together. This will give a recovery period of between one and two hours.

Total requirements:

Storage capacity     $550 + 640 = 1190 \text{ litres}$
Boiler power          $17.5 + 40 = 57.5 \text{ kW}$

$$\text{Recovery period} = \frac{\text{litres stored} \times 0.064^*}{\text{boiler power in kW}}$$

$$= \frac{1190 \times 0.064^*}{57.5} = 1.3 \text{ hours}$$

* The conversion factor 0.064 takes into account the specific heat of water (4.187 kJ/kg °C), a temperature rise of 55°C and the conversion from seconds to hours.

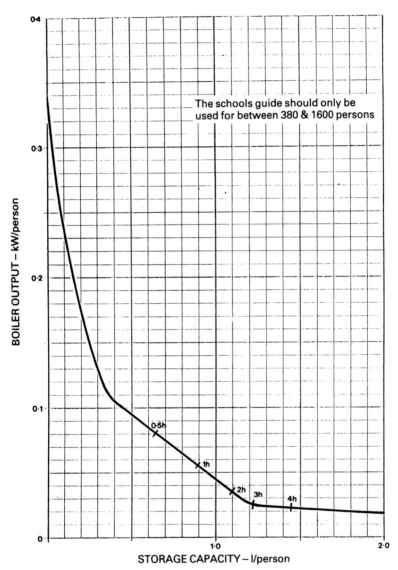

PLANT  SIZING  GUIDE  FOR  SCHOOLS – SERVICE / PERSON
(Adjusted  to  65°C)

*Fig. 31(a)*

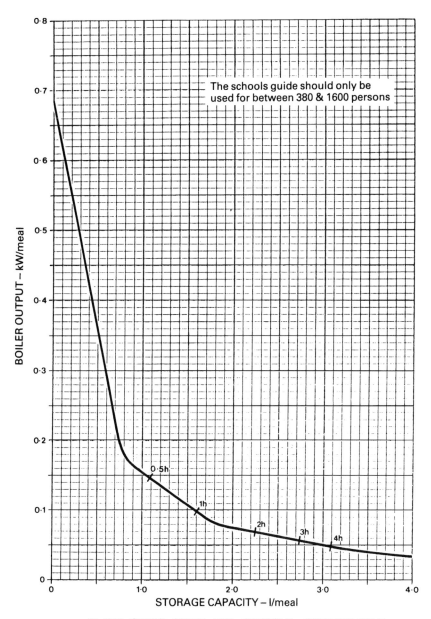

**PLANT  SIZING  GUIDE  FOR  SCHOOLS – CATERING / MEAL**
**(Adjusted to  65°C)**

*Fig. 31(b)*

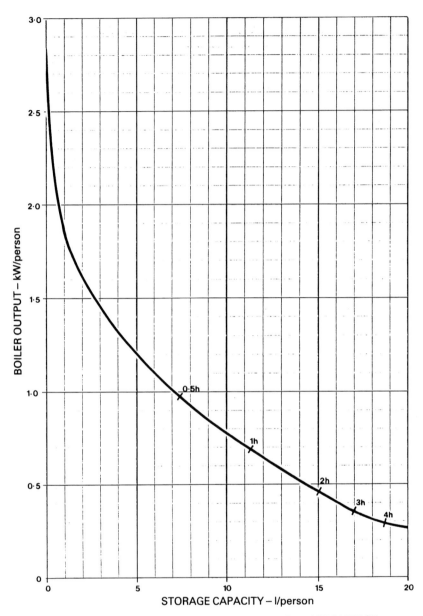

**PLANT SIZING GUIDE FOR HOTELS – SERVICE / PERSON**
**(Adjusted to 65°C)**

*Fig. 31(c)*

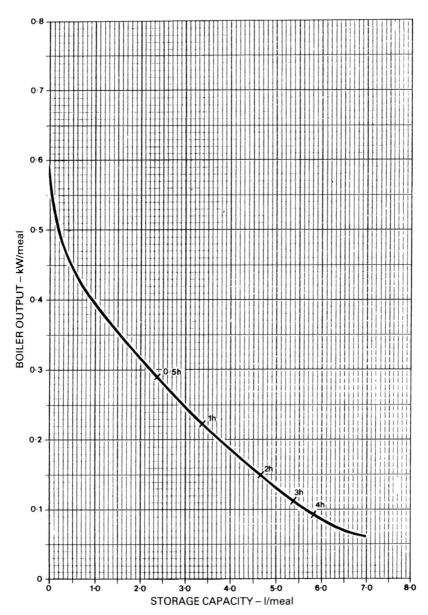

**PLANT SIZING GUIDE FOR HOTELS – CATERING / MEAL**
**(Adjusted to 65°C)**

*Fig. 31(d)*

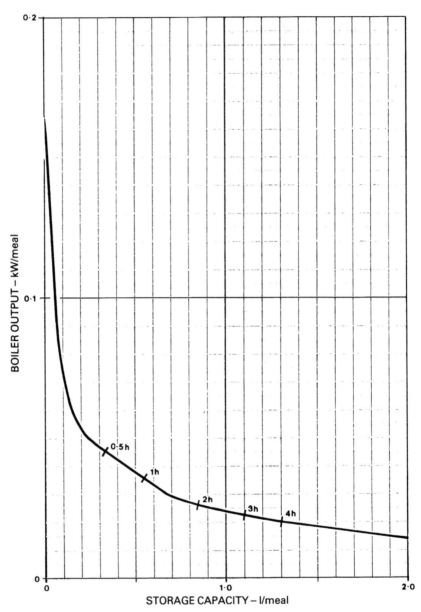

**PLANT SIZING GUIDE FOR RESTAURANTS—CATERING/MEAL**
**(Adjusted to 65°C)**

*Fig. 31(g)*

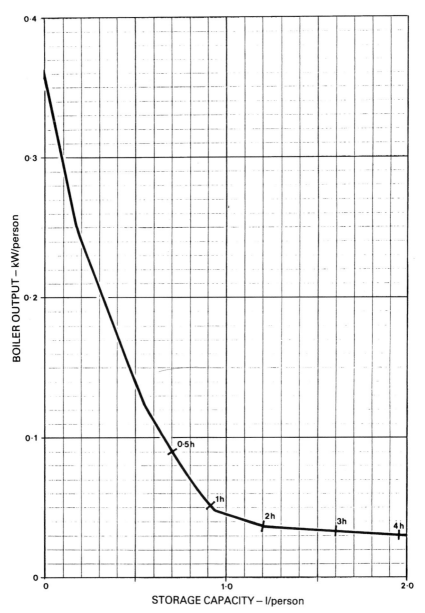

**PLANT SIZING GUIDE FOR OFFICES SERVICE / PERSON**
**(Adjusted to 65°C)**

*Fig. 31(h)*

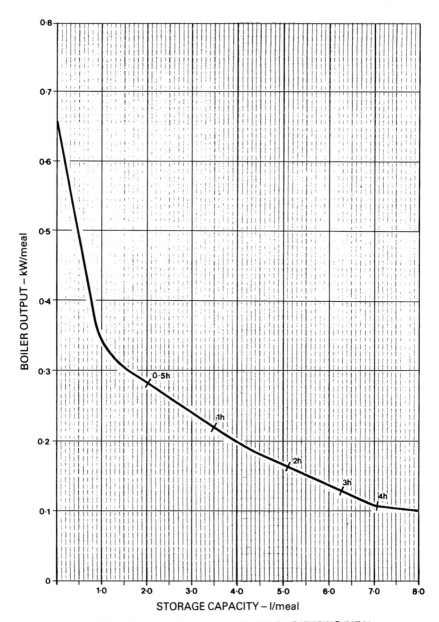

PLANT SIZING GUIDE FOR OFFICES–CATERING/MEAL
(Adjusted to 65°C)

*Fig. 31(i)*

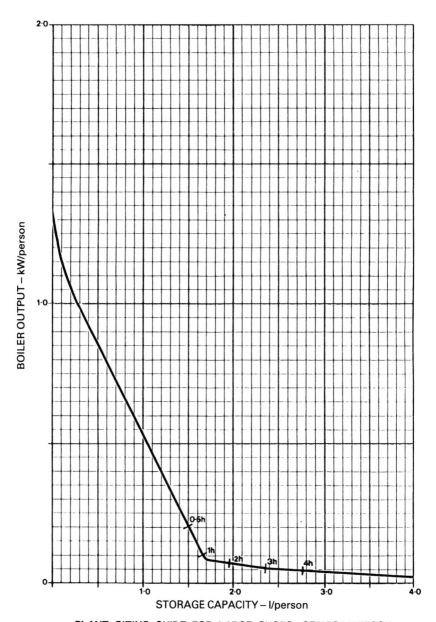

**PLANT SIZING GUIDE FOR LARGE SHOPS - SERVICE/PERSON**
(Adjusted to 65°C)

*Fig. 31(j)*

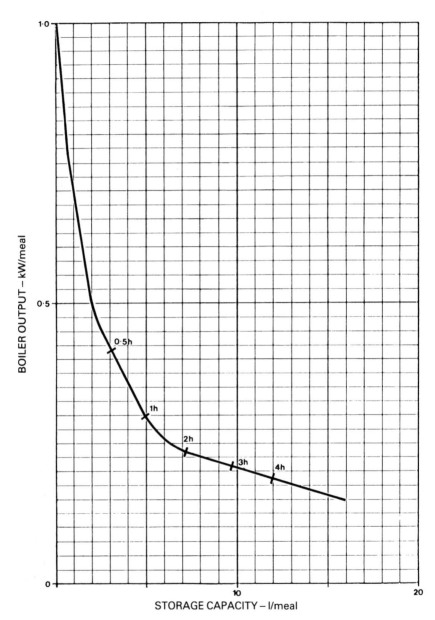

**PLANT SIZING GUIDE FOR STUDENT HOSTELS – SERVICE / PERSON**
**(Adjusted to 65°C)**

*Fig. 31(k)*

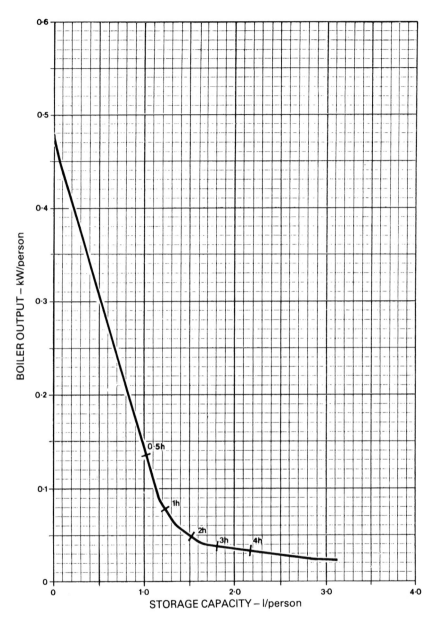

**PLANT SIZING GUIDE FOR LARGE SHOPS – CATERING/MEAL**
**(Adjusted to 65°C)**

*Fig. 31(1)*

*Example 2*

A student hostel is to be built to house 120 students and 10 members of staff. This is assumed to be one of many halls of residence on a campus where all meals are served at a separate restaurant. The building therefore only requires plant for service hot water. Assume the designer chooses 2.5 hours as a recovery period.

From Fig. 31(k) (Service requirement 2.5 hours recovery).

Boiler power        $0.23 \times 130 = 29.9$ kW
Storage capacity    $8.5 \times 130 = 1105$ litres

Note: Allowances must be added to these figures as described earlier.

*Example 3*

An existing restaurant serving 650 meals per day requires a new hot water system for its kitchens. The public toilets have basins already supplied by a multi-point water heater and the catering hot water is therefore the only requirement. Because of a restriction on space and a need for a fast heat-up time, the designer has chosen a storage water heater having a recovery period of 30 minutes.

From the restaurant graph in Fig. 31(g) (Catering requirements 30 minutes recovery):

Boiler power        $0.045 \times 650 = 29.25$ kW
Storage capacity    $0.37 \times 650 = 240.5$ litres

Note: Allowances must be added to these figures as described earlier.

*Multipoint Storage Water Heaters*

The domestic versions of these heaters were described in Vol. 2, Chapter 9. The commercial heaters have inputs ranging from 37 to 136 kW, Fig. 32.

They can be linked to a storage vessel to give increased storage capacity and are fitted with a loose draught diverter for connection to an individual open flue system, Fig. 33.

A multi-heater installation fitted to a common flue system is shown in Fig. 34.

These heaters can be centrally controlled as part of a computerised Building Energy Management System (BEMS).

*Instantaneous Water Heaters*

These heaters have an output range of 63 kW to 1,066 kW, they have a low water content and are fitted with a circulation pump. The heat exchangers are constructed of copper and the combustion chamber is

*Fig. 32    Multipoint water heater ( A•S Andrews)*

insulated to reduce radiation losses. They have full sequence automatic control with overheat cut-off, differential pressure operated
water flow switch, modulating control on gas and combustion air and
fault indicators. Figure 35 shows a heater fitted with a storage tank.

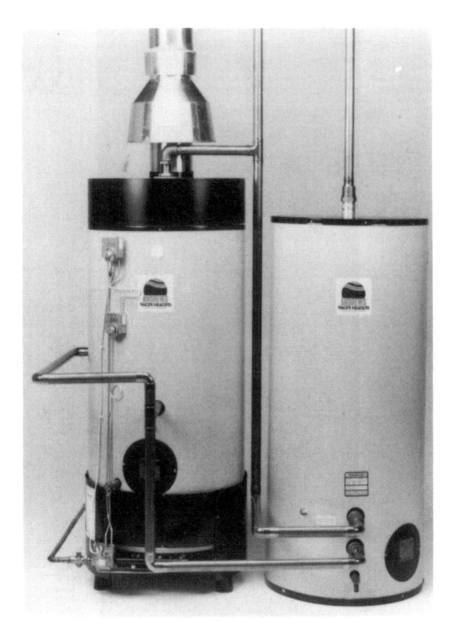

*Fig. 33   Multipoint storage water heater linked to an additional storage vessel (A•S Andrews)*

*Fig. 34    A multi-heater storage water heater installation (A·S Andrews)*

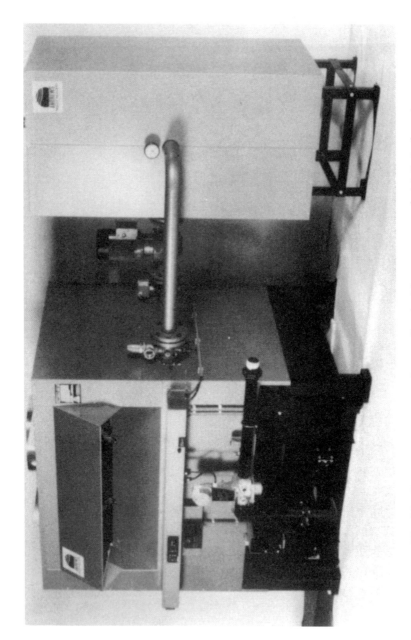

*Fig. 35   Instantaneous water heater fitted with a storage tank ( A•S Andrews )*

*Hot Water Boilers*

For large installations various types of boiler are used, often heating the water by a calorifier and also supplying central heating. A shell boiler with integral calorifier is shown in Fig. 36.

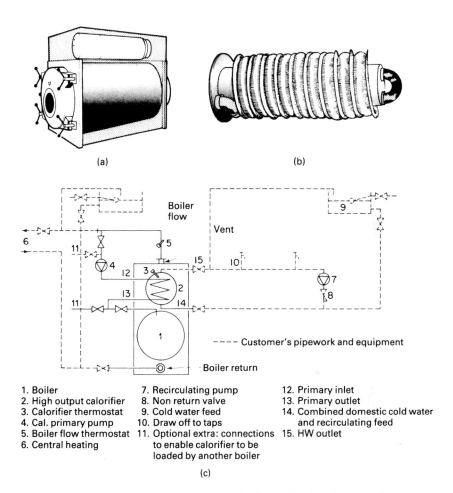

(a)          (b)

Boiler flow

Vent

6

--- Customer's pipework and equipment

Boiler return

1. Boiler
2. High output calorifier
3. Calorifier thermostat
4. Cal. primary pump
5. Boiler flow thermostat
6. Central heating

7. Recirculating pump
8. Non return valve
9. Cold water feed
10. Draw off to taps
11. Optional extra: connections to enable calorifier to be loaded by another boiler

12. Primary inlet
13. Primary outlet
14. Combined domestic cold water and recirculating feed
15. HW outlet

(c)

*Fig. 36    Shell boiler with calorifier: (a) boiler and calorifier complete; (b) calorifier; (c) pipework layout*

With most large systems the draw-offs are at a considerable distance from the source of hot water and a pumped secondary return is necessary. In multi-storey hotels a similar return would be necessary on each floor.

Where heating and hot water services are combined the boilers are generally controlled to give a variable volume of hot water at constant temperature. Whilst this is satisfactory for conventional boilers it gives problems in low water content modular boilers. At periods of little or no heating demand the mixing valve closes and all the heating water is recirculated. If, at the same time, the hot water demand is also minimal the calorifier will be bypassed and the boilers will be required to heat a very small volume of water. This is difficult to control and it is recommended that the heating and hot water services should be completely separate.

Independent systems, each with its own modular boilers, can be more effectively controlled and result in fuel savings.

## Changeover to Gas Firing

A large part of the expansion in commercial gas sales comes from the changeover of existing boilers to gas firing from other fuels.

The subject of boiler changeover is dealt with in detail in the British Gas publication, "Technical notes on changeover to gas of central heating and hot water boilers for non-domestic applications".

The notes apply to boilers with outputs above 60 kW on hot water or steam systems providing central heating or hot water for non-domestic premises. The changeover of boilers in domestic premises is not at present recommended.

Early changeovers were predominantly from solid fuel. Up to 1973 this was about 93% of the total. Since then changeover has been principally from oil firing, up to about 80% of the total.

### Advantage of Gas Firing

Although the decision to change to gas is basically a financial problem, the following factors may have some influence:

- there are no delivery or storage problems
- valuable space used for storage can be made available for other purposes
- no stock control necessary
- no advance payment for fuel
- no provision necessary for delivery vehicle access
- no ash or sludge removal
- lack of atmospheric pollution may allow shorter chimneys to be used
- reliability of fuel supplies
- gas boilers do not require an attendant except on very large installations

- easier to keep boilerhouse clean
- high thermal efficiency and controllability results in energy saving.

*Boiler or Burner Replacement*

The simplest method of changing to gas firing is to fit a gas burner with its ancillary controls and to carry out appropriate modifications to the existing boiler. This is usually less expensive than replacing the boiler itself.

However, it depends on the age and condition of the boiler and also on its size and type. A boiler might better be changed if it is:

- too small or too big for the system
- in poor mechanical condition
- nearing the end of its life (about 10 years)
- of a type for which a suitable burner does not exist
- better to use a condensing boiler or direct contact water heater because of efficiency.

Where there is any doubt it is usual for quotations for both courses of action to be requested by the customer. The difference in cost is least for the smaller boiler. Statistics show that boilers are more often replaced when more than 10 years old and less than 147 kW output. Overall, boilers were replaced in one third of the cases.

*Changeover*

The first step is to examine the boiler and the installation. The points to note are as follows:

Boiler      – make, type, output rating
         – condition, age, faults
         – compliance with relevant standards and codes
         – compliance with AOTC (Associated Offices Technical Committee) requirements for insurance

Installation – compliance with codes of practice
         – ventilation adequate
         – flue satisfactory
         – condition and faults
         – boiler size matched to heat demand
         – water treatment.

The second step is to select a suitable gas burner. In the past many solid fuel cast iron sectional boilers were fitted with natural draught packaged burners. However, difficulties have arisen in setting up and operating some natural draught burners and it is recommended that they should only be used where quietness is an overriding consideration. Forced draught burners are considered by the author to be preferred.

The forced draught burner must have the correct output, fan pressure and flame shape to suit the boiler. Burner manufacturers should be consulted and most can recommend a suitable model.

Where an oil boiler has an equivalent gas version the same burner can usually be used.

### Modifications to Boilers

Solid fuel boilers require a number of modifications when changed over to gas firing. Loose firebars in good condition may be covered with refractory tiles. Alternatively, they may be replaced by preformed linings or a steel plate to support the tiles. If the boiler has water cooled firebars, the base should be filled up with a refractory material to just below the bars. The floor temperature beneath the boiler should not exceed 65°C.

The efficiency can usually be improved by restricting some of the passages between the combustion chamber and the flue. Some passages may be blocked to make the combustion products take a longer route, always ensuring that combustion remains satisfactory. The use of such methods will however also increase the fan horse power requirements.

The inspection and cleaning doors must be replaced to make the boiler pressure-tight. They may be covered with steel plates or be replaced by composite panels which will fracture harmlessly if the pressure rises excessively.

The sighting flap should be replaced by a sighting port and the firing doors replaced by a steel plate on which the burners are mounted. The extent of the modifications will vary with the particular boiler. Figure 37 shows a sectional boiler, modified for an atmospheric burner. The refractory arch is used to cushion the boiler against thermal shock.

Oil boilers generally require less modification. The cleaning doors must always be replaced but if the refractory lining is in good condition and the gas burner fits on to the existing front plate, little else may be required.

The existing wiring must be inspected to ensure its suitability and its condition checked.

On changeover from oil the customer may wish to remove the oil tank so that the space may be put to other use. Removing an oil tank is a difficult job and it is best left to specialist firms who undertake this work.

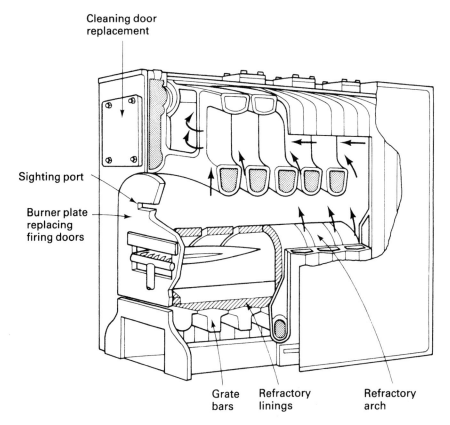

Cleaning door
replacement

Sighting port

Burner plate
replacing
firing doors

Grate    Refractory        Refractory
bars     linings           arch

*Fig. 37    Sectional boiler modified for change-over to gas firing*

Details of the procedure to be followed and the precautions necessary are detailed in various Health and Safety Executive publications.

*Flueing and Ventilation*

Ventilation requirements for gas-fired boilers are given in BS 6644: 1991 Installation of gas-fired hot water boilers of rated inputs between 60 kW and 2 MW (2nd and 3rd family gases).

It is not usual to line chimneys in commercial properties when changing to gas, however the flue should be swept clean.

Changeover may give rise to dissimilar boilers firing into the same flue. Gas boilers should not share a flue with solid fuel boilers but may share with oil boilers. Natural and forced draught burners should not normally share a flue.

Because the resistance in a boiler fitted with a natural draught burner is about 0.5 to 1.0 mbar, a down draught diverter or sometimes a draught stabiliser is usually fitted to ensure adequate flue pull. A double swing stabiliser can relieve both excessive up-draught and down-draught and Fig. 38 shows possible locations for its installation. Flapping draught stabilizers can however lead to pulsating and noisy combustion.

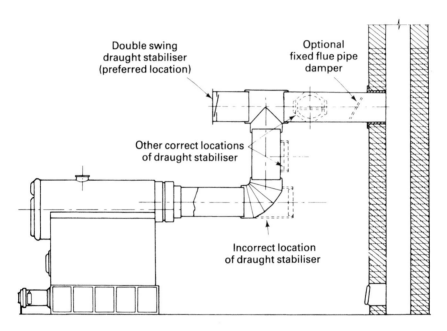

*Fig. 38    Location of flue draught stabiliser*

## FLUES FOR LARGE BOILERS

### Flues and Ventilation for Large Boilers

The design of flues generally is described in Vol. 2, Chapter 5.

The calculation of flue size is a complex procedure and only brief details are included here. The notes contain tables and nomograms which enable a satisfactory flue size to be determined. This is usually smaller than was obtained by previous methods but it has proved completely adequate.

### *Flue Systems*

The principle types of boiler flue systems are shown in Fig. 39 (a to f).

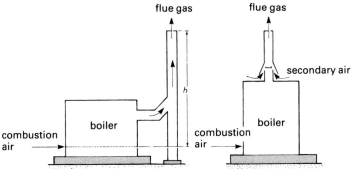

a. Natural draught (flue pull determined by 'h')   b. Natural draught (draught diverter)

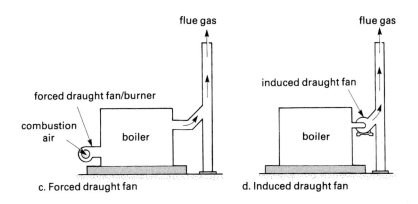

c. Forced draught fan                          d. Induced draught fan

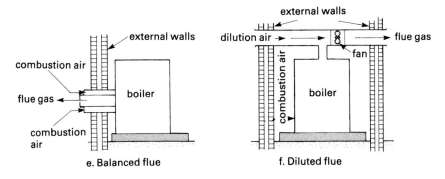

e. Balanced flue                               f. Diluted flue

*Fig. 39   Boiler flue types*

A balanced compartment is another method of installing an appliance, or appliances, in a closed compartment and arranging the flueing and ventilation so that some of the advantages of a balanced flue are obtained.

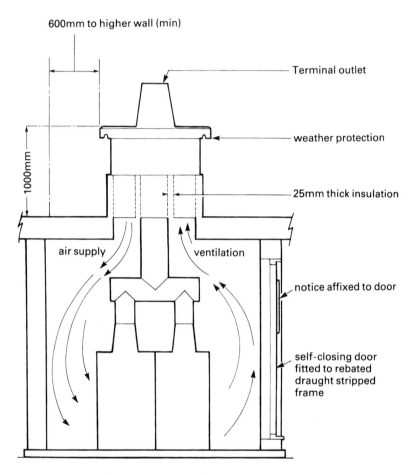

*Fig. 40   Balanced flue compartment*

The principal systems to be considered are:

- natural draught flues
- fanned draught flues
- fan-diluted flues
- modular boiler flues.

*Natural Draught Flues*

This system is usually fitted with a draught diverter which admits about 100% dilution and gives a $CO_2$ concentration in the secondary flue of about 4%.

*Induced Draught Flue Systems*

Many boilers use a combustion fan to provide forced or induced air to the burner. The fan is matched to the resistance of the appliance and the pressure at the flue spigot is usually zero. In some cases it is necessary for some flue pull to be provided. The flue is directly connected to the appliance so there is no dilution and the $CO_2$ concentration is about 8%. The flue could be natural draught or may be fitted with a fan.

With an induced draught flue system a fan is fitted into the flue to remove the combustion products. This system is normally used when:

- a natural draught system would not be satisfactory
  - the flue has insufficient height to give the required pull
  - the only possible flue run creates an exceptionally high resistance
- the natural draught system is unable to provide a satisfactory exit velocity for the flue products.

The major considerations, apart from size, when selecting and locating a fan are as follows:

- the fan may be required to withstand high temperatures, combustion products may be at more than 150°C
- with any fanned system a safe start draught proving switch or air flow switch must be fitted to shut off gas to the burner in the event of the fan failing or the flue becoming blocked
- if the fan is fitted on the outlet of the boiler, the entire flue is under pressure and combustion products may leak into the premises
- if the fan is located just below the flue terminal, the entire flue is under suction and any leaks on the joints will only allow air to be drawn in.

*Flue Design*

The British Gas publication gives two slightly different methods of sizing flues for natural draught or fanned draught systems. In brief, these methods are as follows:

Natural draught procedure:

- decide the route of the flue
- determine the vertical height to the terminal in metres
- determine the total length of the flue in metres
- determine the heat input rate of the appliance in kW
- decide on the $CO_2$ content of the flue gases.

The minimum flue diameter may then be read off from tables for 4% or 8% $CO_2$ respectively. An extract is shown in Table 2 as an example. Corrections must then be applied for flue losses if less than 20% of the gross heat input of the appliance is lost in the flue and for flue length if this is significantly greater than the vertical height. The corrections may be obtained from other tables and nomograms in the notes.

Induced draught procedure:

- decide the route of the flue
- determine the total length of the flue in metres
- determine the heat input rate of the appliance in kW
- decide on the $CO_2$ content of the flue gases.

TABLE 2  Flue Diameter (mm) for 4% $CO_2$ Concentration

| Appliance heat input, Q (kW) | Chimney height, $H_E(m)$ | | | | | | | | | | | | | |
| | 2 | 3 | 4 | 5 | 6 | 7 | 8 | 9 | 10 | 12 | 14 | 16 | 18 | 20 |
|---|---|---|---|---|---|---|---|---|---|---|---|---|---|---|
| 50 | 162 | 149 | 144 | 139 | 136 | 134 | 132 | 131 | 130 | 129 | 128 | 127 | 127 | 126 |
| 75 | 196 | 180 | 172 | 166 | 162 | 159 | 157 | 156 | 155 | 153 | 152 | 151 | 150 | 149 |
| 100 | 225 | 205 | 197 | 189 | 184 | 181 | 179 | 177 | 175 | 173 | 171 | 170 | 169 | 168 |
| 150 | 273 | 248 | 237 | 227 | 221 | 217 | 213 | 211 | 209 | 205 | 203 | 201 | 200 | 199 |
| 200 | 314 | 284 | 270 | 259 | 252 | 247 | 243 | 239 | 237 | 232 | 229 | 227 | 225 | 224 |
| 250 | 350 | 315 | 300 | 288 | 280 | 273 | 268 | 264 | 261 | 256 | 252 | 250 | 247 | 246 |
| 300 | 383 | 344 | 327 | 313 | 304 | 297 | 291 | 287 | 283 | 277 | 273 | 270 | 267 | 265 |
| 350 | 412 | 370 | 351 | 337 | 327 | 319 | 313 | 308 | 303 | 297 | 292 | 289 | 286 | 283 |
| 400 | 439 | 395 | 374 | 359 | 347 | 339 | 332 | 327 | 322 | 315 | 310 | 306 | 303 | 300 |
| 450 | 465 | 418 | 395 | 379 | 367 | 358 | 351 | 345 | 340 | 332 | 326 | 322 | 318 | 316 |
| 500 | 490 | 439 | 416 | 398 | 386 | 376 | 368 | 362 | 356 | 348 | 342 | 337 | 333 | 330 |
| 550 | 513 | 460 | 435 | 417 | 403 | 393 | 385 | 375 | 372 | 363 | 357 | 352 | 348 | 344 |
| 600 | 535 | 480 | 454 | 434 | 420 | 409 | 400 | 393 | 387 | 378 | 371 | 365 | 361 | 357 |
| 650 | 557 | 499 | 471 | 451 | 436 | 425 | 416 | 408 | 402 | 392 | 384 | 378 | 374 | 370 |
| 700 | 578 | 517 | 488 | 467 | 452 | 440 | 430 | 422 | 416 | 405 | 397 | 391 | 386 | 382 |
| 750 | 598 | 534 | 505 | 483 | 467 | 454 | 444 | 436 | 429 | 418 | 410 | 403 | 398 | 394 |
| 800 | 616 | 551 | 521 | 498 | 481 | 468 | 458 | 449 | 442 | 430 | 422 | 415 | 410 | 405 |
| 900 | 653 | 583 | 550 | 527 | 509 | 495 | 483 | 474 | 466 | 454 | 445 | 437 | 432 | 427 |

The volume of combustion products is then calculated or read from a nomogram and a flue diameter is selected to give a flue gas velocity of between 5 and 15 m/s.

The pressure to be provided by the fan can be calculated from the resistance of the flue, the fittings and the appliance. A fan which will deliver the required volume against the static pressure drop is then selected from manufacturers' data sheets. If no such fan is available a new flue diameter must be chosen and the calculations repeated.

## Fan Diluted Flues

These were described in detail in Vol. 2, Chapter 5. They are generally used for small or medium sized commercial installations where it is not practicable to use a natural draught flue. A typical example is in ground floor shop premises with offices or flats above. Large boiler installations normally have natural draught flues.

The main considerations for fan diluted flues are as follows:

- the induced air should reduce the $CO_2$ content to below 1%
- the duct dimensions should give a flue gas velocity of between 6 and 8 m/s
- the inlet and outlet of the duct should preferably be on the same wall
- the outlet of the duct should be remote from the ventilation inlet
- the outlet should normally be at least 3 m above ground level although this is not always possible
- the fan should be fitted to the duct by flexible connections and have rubber mountings to minimise noise.

Figure 41 shows a fan diluted flue system.

## Modular Boiler Flues

The types of flue used may be:

- individual flues
- common flues with natural draught
- common flues with induced draught.

Individual flues should be used, if possible. However, because of their improved appearance and economy, common flues are more frequently fitted.

With natural draught common flues, not more than six appliances should be connected to the same horizontal header and not more than eight to the same vertical flue. If more than eight boilers are installed the flue should be fan assisted.

The vertical flue or chimney should preferably be straight and taken from the centre of the header or group of appliances. The minimum chimney height should be 2 m. The header should be the

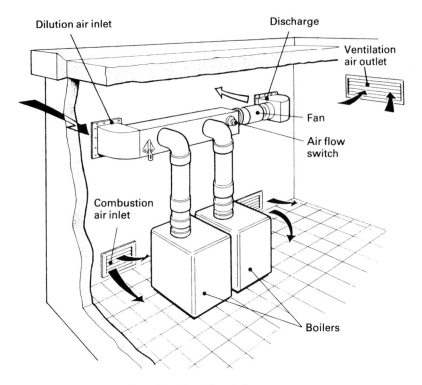

*Fig. 41   Fan diluted flue system*

same size as the chimney for all its length and be horizontal. Headers should be fitted as high as the boiler house headroom will allow.

The connector flues link each appliance to the header. They should be at least 0.5 m high from the base of the draught diverter. They are normally connected into the header via a bend and in some cases may be fitted with flow restrictors. Figure 42 shows a typical flue layout.

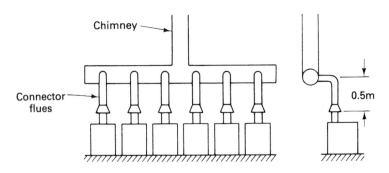

*Fig. 42   Modular boiler flues*

Natural draught flues are used where small numbers of boilers are installed. This is because, when some boilers are shut down, large quantities of cold air are drawn into the flue through their draught diverters. This limits the number of boilers which can operate satisfactorily since it reduces the pull of the flue.

With induced draught systems this problem does not occur. The volume of draught is dependent on the fan and is barely affected by the number of appliances in use. Fan-diluted systems may also be used where the installation is on the ground floor.

The sizing of natural draught modular systems is given in the British Gas publication. Briefly the procedure is as follows:

- decide the route of the common flue
- determine the vertical height from the top of the highest draught diverter to the top of the chimney
- determine the number of boilers and the total heat input rate in kW.

The basic diameter may now be read off the graph in Fig. 43. This is based on three appliances to each limb of the header, that is three appliances with an end chimney and six with a central chimney. A correction must be applied if four or more appliances are to be installed. The graph also assumes that the connecting flue has a 90° bend. If it enters the header vertically from below a further correction is required. These corrections may be obtained from the publication.

*Materials for Large Flues*

The main considerations when selecting the type of flue material are as follows:

- the inner wall of the flue must be capable of withstanding the highest temperature attained by the combustion products
- the flue wall must be resistant to condensation which is likely to occur when the appliance is started up from cold
- the flue should possess sufficient insulating properties or 'thermal resistance' to prevent the continuous formation of condensation during normal running.

The main consideration is the flue temperature and Table 3 gives the limiting temperatures of various materials.

Brick or concrete were the traditional materials used for tall flues. They were lined with insulating brick or clayware; flexible metal tubes are now commonly used.

The liner must be adequately sealed if the flue is under fan pressure. The space around the liner may be filled with loose insulating material, e.g., vermiculite.

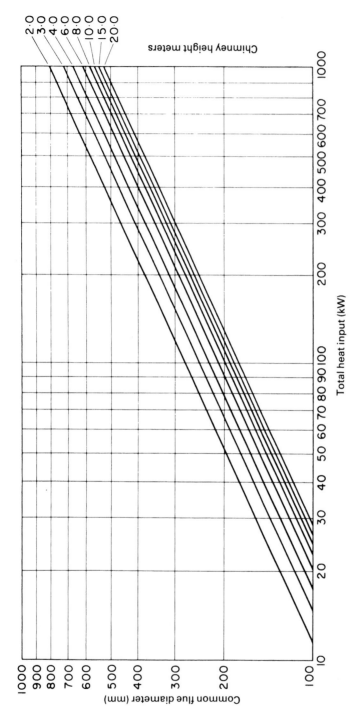

*Fig. 43   Variation of flue diameters with total heat input*

**TABLE 3 Temperature Resistance of Flue Materials**

| Material | Limiting temperature |
|---|---|
| *Non-metals* | |
| Brick lined with: | |
| Acid resistant brick | up to 200°C |
| Insulating brick | above 200°C |
| Glazed clayware | up to 200°C |
| Concrete – in situ or pre-cast | about 300°C |
| Asbestos cement – heavy quality | up to 260°C |
| Glass reinforced plastic | 150–200°C |
| *Metals* | |
| Mild steel | up to 450°C |
| Austenitic stainless steel | up to 750°C |
| Stainless steel | up to 500°C |
| Diffusion aluminised steel | up to 600°C |
| Aluminised steel | up to 600°C |
| Cast iron | up to 500°C |
| Galvanised steel | up to 360°C |
| Aluminium – high purity | up to 300°C |

Asbestos cement flues were popular and were obtainable in twin-wall form for additional insulation but they are no longer installed.

Metal flues are now more commonly used. Most metals are satisfactory if there is no condensation. Where condensation does occur, aluminised or stainless steels are the most resistant. Metal flues are also available in twin-wall form and, in some cases, insulating material is introduced into the space between the two walls.

Some metal flues, produced by specialist manufacturers, consist of a welded tube inside a self-supporting or guyed windshield. Several items of plant may be flued together within a common windshield, the tubes being insulated with loose fill material.

With all flues, facilities must be provided for their inspection and renewal when necessary.

## Heat Loss from Flues

To maintain the maximum flue pull and to avoid condensation the heat loss from the flue gases must be kept to a minimum. So flues are run inside the premises, where possible, or insulated when run externally. The object is to maintain the inner wall temperature above the dewpoint temperature of the combustion products. This is about 60°C.

Where this is not possible and condensation is inevitable the flue should be designed so that the water can flow freely to a point from which it can be drained to a gulley.

## Ventilation Requirements

Notes giving guidance and ventilation air requirements for large boilers are contained in BS 6644: 1991. In a boiler room/house, combustion and ventilation air is obtained by one of the four following methods:

(i)   By one or more low level openings and discharged by one or more high level openings.

(ii)  In a balanced flue compartment, air shall be supplied and discharged through a purposely designed flueing and ventilation system at high level, Fig. 40.

(iii) Air supplied by a fan at low level and discharged naturally at high level.

(iv)  Air supplied by a fan at low level and discharged by a second fan at high level.

The air supplied to the boiler house should keep the boiler house temperature at or below the following:

- at floor level (or 100 mm above floor level)          25°C
- at mid-level (1.5 m above floor level)                32°C
- at ceiling level (or 100 mm below ceiling level)      40°C

Where plant may be used to near its maximum capacity in summer, additional ventilation may be required.

### Boilers with Natural Air Supply

The main considerations are:

- ventilation grilles should open directly to outside air
- they should be located so that they cannot be easily blocked or flooded
- high level grilles should be as near as possible to the ceiling
- where only high level openings are possible it may be necessary to carry combustion air to floor level by means of a duct
- grilles should be designed to avoid setting up high velocity air streams in the boiler house
- where a boiler house is exposed, grilles should be sited on at least two sides and preferably on all four.

The sizes of ventilation grilles are given in Table 4.

**TABLE 4 Minimum Ventilation Free Area for Natural Air Supply**

| Low level (inlet) | 540 cm$^2$ plus 4.5 cm$^2$ per kW in excess of 60 kW total rated input |
| --- | --- |
| High level (outlet) | 270 cm$^2$ plus 2.25 cm$^2$ per kW in excess of 60 kW total rated input |

*Boilers with Fanned Air Supply*

The following points should be noted:

- two fans may be used, one to supply air, the other to extract. The extraction rate should not exceed the supply rate and they should not cause negative pressure, in relation to the outside pressure, to develop
- if only one fan is used it should supply air and extraction should be by natural ventilation

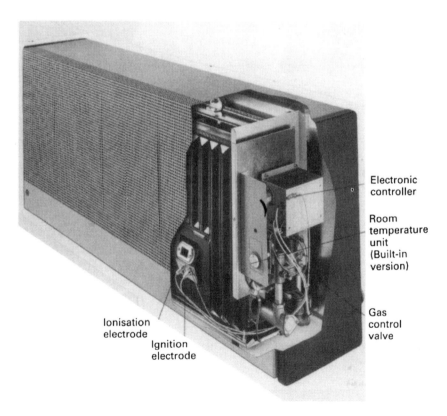

Electronic controller

Room temperature unit (Built-in version)

Gas control valve

Ionisation electrode

Ignition electrode

*Fig. 44   Commercial balanced flue convector heater (Drugasar)*

- supply air should enter at low level
- natural air vents should be at high level and sized as in Table 5
- an air flow switch must be fitted to shut off the gas to the burner should the fan fail.

The volume of air required for fanned air supply is given in Table 5.

## Boilers Changed Over from Oil to Gas

Where boilers have been changed over from oil to gas firing and where the air vents have proved adequate for the oil fired boiler, there is generally no need for them to be altered, even though they may be smaller than those recommended by Tables 4 and 5.

TABLE 5 Fanned Air Supply Ventilation Flow Rates

| Type of Boiler | Flow rate per 1000 kW total rated heat input | |
| --- | --- | --- |
| | Inlet air (combustion ventilation) | Extract air (ventilation) |
| | $m^3/s$ | $m^3/s$ |
| Natural draught boilers | 1.10 | 0.45 |
| Forced/induced draught boilers | 0.90 | 0.60 |

## Commercial Balanced Flue Convector Heaters

These heaters are fitted in churches, schools, offices, workshops, canteens, restaurants and other commercial premises. They are the same in principle as the domestic balanced flue convector but with much higher inputs and more sophisticated controls. Figure 44 shows a commercial balanced flue convector heater, available with a range of inputs from 6 to 14.2 kW. It is fitted with an electronic control which eliminates the need for a permanent pilot and has a thermostat which modulates the heat output. It has one control (built-in or remote) which sets the required comfort level and incorporates a night set-back device that automatically reverts day temperature to a lower night-time setting.

# Natural Gas and the Environment

Chapter 13 is written by R. Proffitt

## Introduction

Concern has been expressed about the environment for many years. The poor air quality resulting from the domestic open coal fires and the large-scale industrial burning of coal led to the 1956 Clean Air Act. This greatly reduced the 'smogs' of that time which were causing increasing health problems for the inhabitants of the industrial areas of the country. In December 1952 the great London smog lasted five days and caused some 4,000 deaths.

In the last ten years concern has increased to such an extent that most governments of the world are aware of the problems and many are taking action to reduce the effects of pollution.

The Intergovernmental Panel on Climate Change (IPCC) was formed in 1988 under the auspices of the United Nations Environment Programme and the World Meteorological Organisation. One of their tasks was to report to the World Climate Conference in 1990 on the scientific assessment of climate change.

The Vienna Convention on the Protection of the Ozone Layer was held in 1985 to deal with global climate changes, internationally through a framework convention. The Montreal Protocol laid down limits on the production and use of chlorofluorocarbons (CFCs) to a 50% reduction of those relative to 1986 levels.

The Environmental Protection Act 1990 Part 1 has been introduced and covers local air pollution control administered by local authorities. Integrated pollution control systems are enforced by the Environment Agency. There are in excess of 80 process guidance notes written at the present time covering a very wide range of industrial and commercial processes that cause pollution and these range from metal processes, crematoria, bitumen and tar processes, waste oil burners, etc. These prescribed processes under the Act can only be operated with the authorisation of the local authority and each process has limitations on the quantities of pollutants that can be emitted.

The techniques that are used to prevent and minimise the emissions are known as BATNEEC (Best Available Techniques Not Entailing Excessive Cost).

In 1992 the United Nations Conference on Environment and Development took place in Brazil involving over 100 countries. More recently the Kyoto Conference and Protocol was held in 1998, which proposed a quantified emission limitation or reduction for six green-house gases:

carbon dioxide ($CO_2$)          hydrofluorocarbons (HFCs)
methane ($CH_4$)                 perfluorocarbons (PFCs)
nitrous oxide ($N_2O$)           sulphur hexafluoride ($SF_6$)

The European Union agreed an 8% reduction based on the 1990 levels of these gases in the Commitment period 2008–2012.

The use of natural gas and LPG mainly affects combustion emissions to the atmosphere via flues and therefore affects air quality. However the combustion of alternative fuels such as heavy fuel oil, diesel, petrol, coal (and electricity generated by oil and coal at low efficiencies) cause much more pollution due to their chemistry and techniques for burning them. Thus the wider use of natural gas together with the modern utilisation techniques will contribute significantly to a better environment.


### The Greenhouse Effect

The sun is the only external source of heat for the earth which is heated differently on various parts of its surface, Fig. 1.

The sun's heat is greater at the equator than at the polar regions due to the shallower angle of incidence. Some of the visible sunlight is absorbed by the atmosphere and reflected from clouds and land – especially from deserts and snow. The remainder is absorbed by the surface, which is heated and in turn warms the atmosphere. The warm earth also radiates energy back into space, but being much cooler than the sun does so by giving off invisible infra-red radiation, Fig. 2.

The mean temperature of the earth is determined by the balance between the energy from the sun (sunlight) and invisible infra-red radiation leaving the earth. The atmosphere is relatively transparent to solar radiation but many atmospheric trace gases absorb some of the infra-red radiation emitted from the surface. As a result the atmosphere acts like a blanket, preventing much of the infra-red radiation from leaving the earth and its atmosphere; this makes the earth warmer.

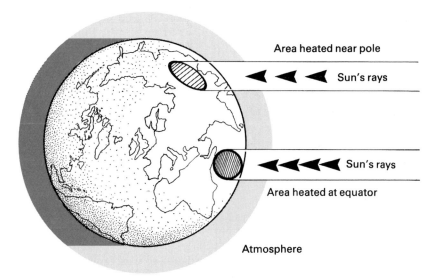

*Fig. 1   Distribution of the sun's heat on earth*

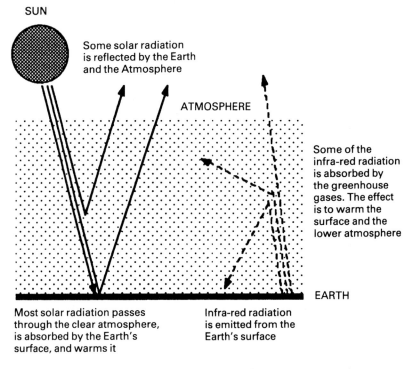

*Fig. 2   Effect of solar and infra-red radiation on earth*

This is called the greenhouse effect because in a greenhouse, glass allows sunlight in but keeps some infra-red radiation from escaping. The gases in our atmosphere with a similar effect are often called greenhouse gases and are trace gases such as carbon dioxide and water vapour.

The gases that contribute to global warming through the greenhouse effect and their relative contributions are:

| | |
|---|---|
| Carbon dioxide | 62% |
| Methane | 20% |
| Chlorofluorocarbons | 9% |
| Ozone | 5% |
| Nitrous oxide | 4% |

Another way of indicating the relative contribution to the greenhouse effect is shown in Table 1.

TABLE 1 Relative greenhouse effect of various gases relative to carbon dioxide

| Greenhouse Gas | Relative effect per molecule | Current Atmospheric concentration ppm |
|---|---|---|
| Carbon dioxide | 1 | 350 |
| Methane | 30 | 1.7 |
| Nitrous oxide | 160 | 0.31 |
| Ozone | 2 000 | 0.06 |
| CFC 11 | 21 000 | 0.00026 |
| CFC 12 | 25 000 | 0.00044 |

Thus from Table 1 it can be seen that whilst unburnt methane molecules are 30 times more damaging than carbon dioxide molecules, by far the worst are the CFC group, being over 20,000 times worse than carbon dioxide.

However the amount of carbon dioxide in the atmosphere is currently about 350 ppm and has been constantly increasing as shown in Fig. 3. It is expected to rise to 500 ppm by the year 2050.

Carbon dioxide in the atmosphere has increased from about 315 ppm in 1960 to 350 ppm in 1985. This has been determined by analysing air bubbles trapped in ice glaciers. Cores from ice have been taken from Mauna Loa, Hawaii for these tests.

The effect of these greenhouse gases is to increase global temperatures by 0.5–0.7°C. By the year 2050 the temperature is estimated to increase by between 1.5–4.5°C depending upon efforts made to reduce levels.

Whilst these increases in temperature may not appear much to the layman, it must be pointed out that scientific research indicates that rises in temperature of this magnitude have not happened in the past few thousand years!!

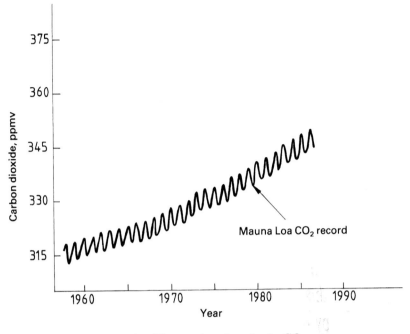

*Fig. 3    The accelerating rise in $CO_2$*

These temperature rises are potentially catastrophic since they may cause:

- major climatic changes causing more frequent extremes of floods, storms and heat waves
- a rise in sea level of 200–300 mm, which would cause flooding of coastal areas in the Thames Estuary and East Anglia; the Netherlands and Bangladesh would also be threatened
- changes in plant growth and consequently food supplies.

An increase in global temperature will mean a rise in mean sea level caused by the thermal expansion of the oceans and the melting of glaciers. It is difficult to estimate sea level rises because of change in snowfall and the uncertainty of predicting temperatures.

Increase in sea level lags behind global average temperature which in turn lags behind greenhouse gas concentrations. So whilst the actual rise may be 200 mm in say 2030 there would eventually be a much greater rise.

### Greenhouse Gases

The main greenhouse gases are carbon dioxide, methane, the CFCs, ozone and nitrous oxides.

*Carbon Dioxide*

Figure 4 shows the contributions to $CO_2$ emissions from human activities showing that the three main contributions are from electricity generation, transport and deforestation. Biological processes on land contribute 110,000 million tonnes of carbon as $CO_2$ to the atmosphere each year. This is largely balanced by an annual uptake of carbon during photosynthesis. The oceans are calculated to send out and absorb similar amounts. About 5,000 million tonnes are added to the atmosphere through burning fossil fuels and 1,000 million tonnes through land use changes, mainly through loss of tropical forests.

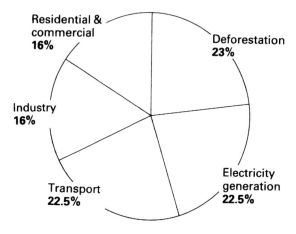

*Fig. 4 Percentage contribution to $CO_2$ emissions (1980–85) from human activities*

Human actions mean that allowing for losses in the oceans, an excess of 3,000 million tonnes of carbon in the form of $CO_2$ remains in the atmosphere.

*Methane*

Methane is the major constituent of natural gas but in the atmosphere it can come from many sources. There is some uncertainty about emission rates from each source but it is generally recognised that the major sources of methane in the atmosphere are paddy fields, swamps, marshes, ruminant animals and the burning of biomass. This is shown in Fig. 5 for the global situation.

Leakage of natural gas does occur from cracked pipes and leaking joints but this figure is difficult to obtain. Historically, in the statistics produced by British Gas plc, a figure for 'unaccounted for losses' was

### Global methane emissions (annual release – million tonnes)

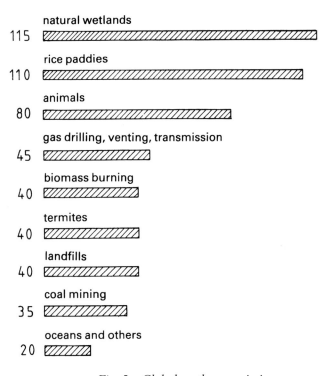

natural wetlands
115

rice paddies
110

animals
80

gas drilling, venting, transmission
45

biomass burning
40

termites
40

landfills
40

coal mining
35

oceans and others
20

*Fig. 5    Global methane emissions*

documented. This was the difference between gas metered into the system and that billed at the customers' premises. It was made up of several components such as metering errors, variations caused by temperature and pressure differences, differences between supplied calorific value and contracted calorific value, venting, leakage, theft and gas used for the industries own use. So only one component of this 'unaccounted for gas' is due to leakage. The leakage estimated from the transmission and distribution system is 10% of the national figure, which is about 345,000 tonnes per annum, Fig. 6.

Whilst the figure of 345,000 tonnes of methane may appear to be high, it takes considerable care to minimise loss of revenue that would arise from leakage. The natural gas distribution is through a system of 222,400 km of medium and low pressure pipeline in addition to its 17,600 km of high pressure pipeline. Cast iron pipe which had a tendency to be brittle and have suspect lead yarn jointing has been replaced and the extensive use of polyethylene pipe of various

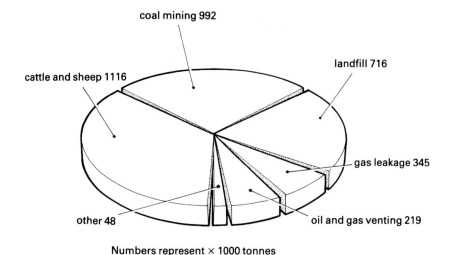

coal mining 992

landfill 716

cattle and sheep 1116

gas leakage 345

other 48

oil and gas venting 219

Numbers represent × 1000 tonnes

*Fig. 6    UK emissions of methane annually*

densities has greatly increased the integrity of the systems with the benefit of reduced natural gas losses.

## Nitrous Oxide

Natural vegetation gives off large quantities of nitrous oxide ($N_2O$) but the increase in the concentration of this gas is thought to come from combustion in power stations and cars and from agriculture. At global level oxides of nitrogen are reported as nitrous oxide. When used in context of air quality or combustion, oxides of nitrogen are referred to as NOx which is a mixture of nitrogen oxide (NO) and nitrogen dioxide ($NO_2$).

NOx modifies and reacts with other gases in the atmosphere to increase greenhouse gases.

## Chlorofluorocarbons (CFCs)

CFCs are man-made compounds containing chlorine, fluorine and carbon. They are non-toxic and inert making them very useful as aerosol propellants, refrigerants, insulators for blowing rigid and flexible foams and for cleaning electrical components. CFCs react to deplete the ozone in the atmosphere. The ozone helps protect the earth from harmful ultra-violet radiation. An international agreement – The Montreal Protocol on Substances which Deplete the Ozone Layer – was agreed in 1987 and many nations have agreed to halve their consumption by the end of the century.

CFCs are powerful greenhouse gases because they absorb infra-red radiation in a region of the spectrum where there is little absorption by other gases. CFCs have a lifetime of about 100 years in the atmosphere. Emission rates exceed the slow atmospheric decay rate so CFC levels are still increasing by 4–5% every year.

## The Ozone Layer

Ozone, $O_3$ is found mainly high in the atmosphere, in the stratosphere ozone layer. There is also a notable amount in the lower atmosphere where it acts as a greenhouse gas. The ozone layer in the stratosphere has the effect of shielding the earth from harmful rays from the sun. Of the solar energy coming into the earth's atmosphere, 31% is reflected by clouds and 23% is absorbed by water vapour, carbon dioxide, methane and ozone. Ozone absorbs solar radiation and has particularly strong absorption in the range 8–12 μm. Ozone in the stratosphere can be depleted by interaction with the chlorine atom from CFCs. Lack of ozone is thought to give rise to higher incidence of skin cancer and glaucomas due to an increase in ultra-violet light at ground level.

## Acid Rain

When fossil fuels are burnt, sulphur contained in them is, by and large oxidised into sulphur dioxide. The combustion process also forms oxides of nitrogen. In addition copius amounts of carbon dioxide are also produced. When these oxides are reacted with water they all produce dilute acid:

$$SO_2 \rightarrow \quad H_2SO_3 \quad \text{sulphurous acid}$$
$$H_2SO_4 \quad \text{sulphuric acid}$$
$$NO_2 \rightarrow \quad HNO_3 \quad \text{nitric acid}$$
$$CO_2 \rightarrow \quad H_2CO_3 \quad \text{carbonic acid}$$

Figure 7 depicts a model for acid rain, a tall stack emitting products of combustion high into the air and into the clouds. Precipitation of water in the clouds produces rain that is acidic in nature. The effect of this acid rain is to attack forests, plants, grasslands and upset nature's balance in lakes, streams and affected fresh water sites. The early examples of acid rain were in Scandinavia where acid rain from the power stations in England was causing problems with deforestation.

The production of the above pollutant oxides is discussed later in this chapter but it can be seen that fuels producing high levels of carbon dioxide and sulphur dioxide will produce the most acid rain.

*Fig. 7   A model for acid rain production*

## Air Quality

The standards for air pollution are concentrations of pollutants over a given time for purposes of measurement. The standards adopted by the UK are the National Air Quality Standards, which the Government put forward in 1997 as part of the National Quality Strategy. A summary is given below in Table 2.

**TABLE 2 Air Quality Standards**

| Pollutant | Standard Concentration | Measurement | Objective to be achieved by 2005 |
|---|---|---|---|
| Carbon monoxide | 10 ppb | running annual mean | 5 ppb |
| Nitrogen dioxide | 150 ppb | 1 hour mean | 150 ppb hourly mean |
| | 21 ppb | annual mean | 21 ppb annual mean |
| Sulphur dioxide | 100 ppb | 15 minute mean | 100 ppb measured as 99.9th percentile |
| Fine particles (PM 10) | 50 $\mu g/m^3$ | running 24 hour mean | 50 $\mu g/m^3$ measured as 99.9th percentile |
| Ozone | 50 ppb | running 8 hour mean | 50 ppb measured as 97th percentile |
| Lead | 0.5 $\mu g/m^3$ | annual mean | 0.5 $\mu g/m^3$ |
| Benzene | 5 ppb | running annual mean | 5 ppb |
| 1,3 Butadiene | 1 ppb | running annual mean | 1 ppb |

**Key**
ppm = parts per million
ppb = parts per billion
$\mu g/m^3$ = microgrammes per cubic metre

The National Air Quality Strategy Report March 1997 published information on the industrial emissions in the UK and this is shown as Table 3.

**TABLE 3 Industrial Emissions in the UK (National Air Quality Strategy, March 1997)**

| Pollutant | Total UK Emissions in 1995 (kilotonnes) | Industrial Emissions (kilotonnes) | Industry as % of total |
|---|---|---|---|
| Carbon monoxide | 5478 | 667 | 12 |
| Nitrogen oxides | 2293 | 852 | 37 |
| Sulphur dioxide | 2365 | 2112 | 89 |
| Particles | 232 | 135 | 59 |
| Lead | 1492 (tonnes) | 276 (tonnes) | 18 |
| Benzene | 35 | 6.9 | 20 |
| 1,3 Butadiene | 9.6 | 1.2 | 13 |

The Department of Health has issued guidance, which advises the public on the poor and very poor quality of air and the appropriate action to take. The quality is shown in Table 4.

**TABLE 4 Department of Environment Air Quality Categories (ppb. 1 hour mean)**

| | Nitrogen Dioxide | Sulphur Dioxide | Ozone |
|---|---|---|---|
| Very good | < 50 | <60 | < 50 |
| Good | 51 – 100 | 61 – 125 | 51 – 100 |
| Poor | 101 – 300 | 126 – 500 | 101 – 200 |
| Very poor | > 300 | > 500 | > 200 |

As we have seen earlier in this chapter, the other regular constituent is carbon dioxide at a level of about 350 ppm.

## Combustion Quality

### Carbon Dioxide

All fossil fuels contain carbon and hydrogen and burn to produce carbon dioxide and water vapour. The amount of carbon dioxide produced is different for each fuel. The higher the carbon to hydrogen ratio, the higher will be the carbon dioxide produced for a given heat release, Fig. 8.

Table 5 shows the actual figures for carbon dioxide for each megajoule of heat released.

**TABLE 5 CO$_2$ produced from energy and fuel**

|  | CO$_2$ (Kg/MJ) |
| --- | --- |
| Natural gas | 0.05136 |
| Gas oil | 0.06882 |
| Light fuel oil | 0.07200 |
| Medium fuel oil | 0.07264 |
| Heavy fuel oil | 0.07282 |
| Coal | 0.08984 |
| Electricity (from coal) | 0.2642 |

The carbon dioxide from combustion of coal will vary depending upon the classification of the coal. Anthracite is a high carbon coal

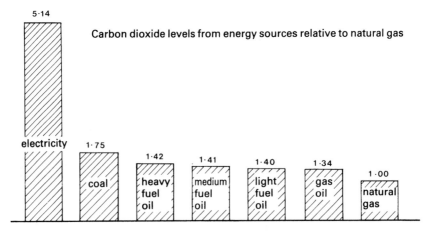

*Fig. 8    Carbon dioxide production from energy and fuel*

containing about 94% carbon whilst some general purpose coals contain 81%. From Fig. 8 it can be seen that coal produces 75% more carbon dioxide than natural gas. Chaper 10, Fig. 1 shows that the efficiency of producing electricity from coal via the grid is 34%. This means that the $CO_2$ equivalent for electricity produced from coal is 500% more than for natural gas.

If the thermal efficiency of other fuels is less than natural gas on any plant, then even more carbon dioxide will be produced because more heat will have to be generated to satisfy any efficiency gap.

Figure 9 shows how the production of $CO_2$ varies with the air/gas ratio during combustion. The maximum amount of $CO_2$ is produced at stoichiometric combustion but it must be remembered that at this point maximum heat is being generated and hence the fuel is being used most efficiently.

## Oxides of Nitrogen

The term NOx is used to describe the various oxides of nitrogen. The major oxide of nitrogen produced in flames is nitric oxide (NO) but significant amounts of nitrogen dioxide ($NO_2$) can also be formed under certain circumstances, for example in the low temperature region of the flame.

Three routes to nitrogen oxide formation have been identified in flames. 'Thermal NOx' is due to thermal fixation of atmospheric nitrogen with oxygen supplied to the burner. Chemically bound nitrogen in the fuel is known to react very readily with the oxygen in the air to produce 'fuel NOx'. So called 'prompt NOx' may also be significant in flames with low levels of thermal and fuel NOx.

In natural gas flames, NOx is produced by the thermal and prompt routes, the former predominating when using preheated air. 'Fuel NOx' is often more significant in oil and coal flames. For this and other reasons associated with their physical state, as liquid drops or solid particles drop within flames, a much higher volume of NOx is emitted than by an equivalent natural gas flame.

The formation of thermal NOx is strongly dependent upon temperature. Also significant is the residence time of potential NOx forming chemicals within these high temperature zones. Figure 10 shows how the formation of NOx is influenced by the air/gas ratio.

At higher air preheat temperatures, Fig. 11, the NOx increases from about 500 ppm to levels in excess of 4,000 ppm, both conditions being about 25% excess air.

The levels of excess air, the air preheat temperature and the way that the air is mixed with the gas are all important factors in the production of NOx in industrial burners.

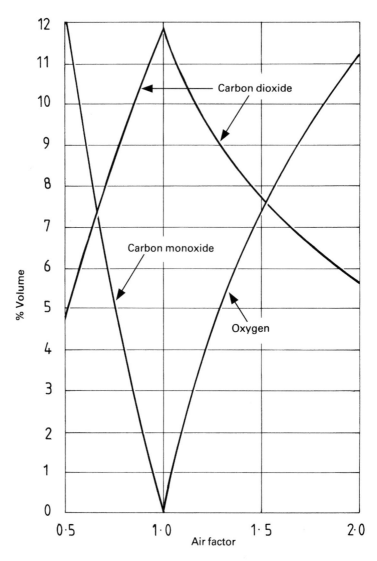

*Fig. 9    Carbon dioxide variations with air/gas ratio*

It has also been established that NOx levels can be reduced by recirculation of cooled flue gas with the combustion air supply to the burner. Up to 50% NOx reduction can be achieved on boilers and furnaces in this way, Fig. 12. This inert diluent limits the peak flame temperature and helps control NOx.

The oxides of nitrogen produced by various fuels over a range of fuel uses is shown in Table 6.

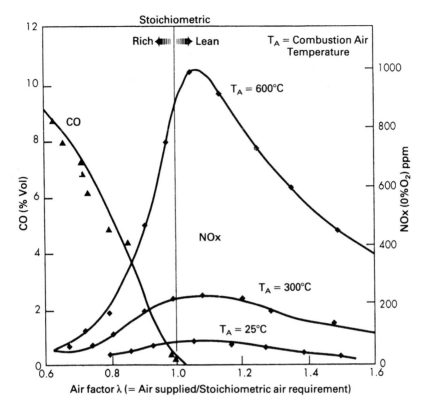

*Fig. 10    Oxide of nitrogen formation with air/gas ratio*

**TABLE 6 NOx emissions (mg/MJ) for various fuels**

|                                        | *Natural gas* | *Oil*    | *Solid fuel* |
|----------------------------------------|:-------------:|:--------:|:------------:|
| Domestic space heating/water heating   | 40            | 65       | 135          |
| Industrial processes                   | 40 – 125      | 70 – 160 | 245          |
| Power stations                         | 115           | 155      | 245          |

It can therefore be seen that natural gas is beneficial in producing low NOx figures and is the less polluting fuel. The actual values produced for any given process are difficult to predict and should be established by measurement.

See also 'Low NOx Burners' on page 177.

## Sulphur Oxides

These are produced when fossil fuels containing sulphur are burnt in air. The main oxide of sulphur produced is sulphur dioxide although some sulphur trioxide may be produced also.

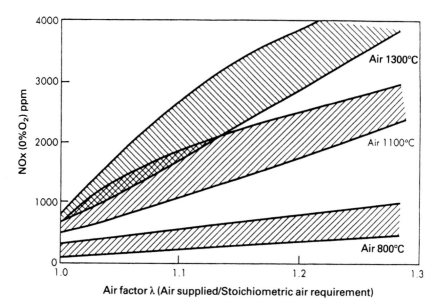

*Fig. 11    Oxides of nitrogen formation with preheated air*

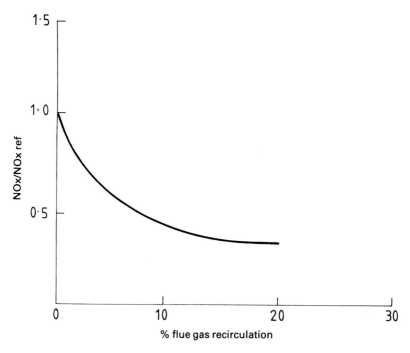

*Fig. 12    Reduction of NOx with flue gas recirculation*

Sulphur oxides are the most polluting by-product of all fossil fuels and are main contributors to acid rain. Figure 13 shows the amount of sulphur dioxide emitted per unit of energy for the various fossil fuels.

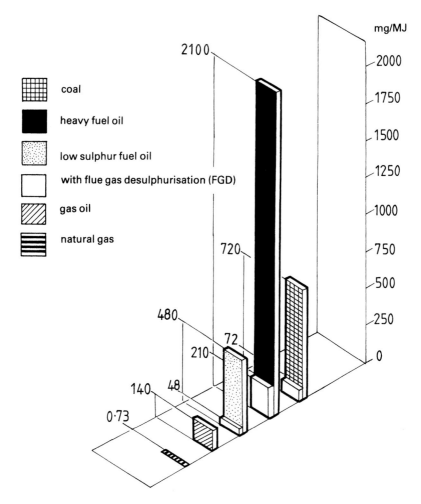

*Fig. 13    Sulphur dioxide emission factors*

An indication of the $SO_2$ produced from power stations, industrial boilers and domestic sources is shown in Fig. 14.

Natural gas from most North Sea gas fields contains little or no sulphur. Where hydrogen sulphide is found as part of the gas produced, it is removed before the gas enters the natural gas transmission and distribution system. Minute quantities of organic sulphur com-

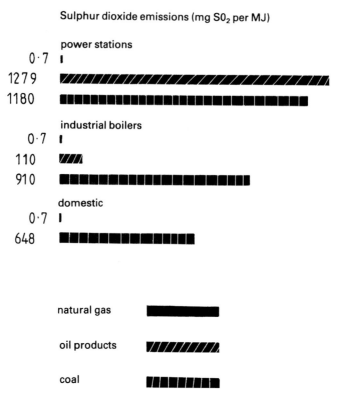

*Fig. 14   Sulphur dioxide production sources*

pounds with strong odours are added to natural gas to give it a distinctive smell for safety reasons. This still leaves the amount of sulphur in natural gas at an extremely low level compared with other fuels. To produce the same amount of energy, burning coal conventionally produces over a thousand times more sulphur dioxide than burning natural gas. Even with the addition of flue gas desulphurization, which can be fitted to large coal burning plant such as power stations, coal still produces 100 times as much sulphur dioxide as natural gas.

## Carbon Monoxide

Carbon monoxide is an unwelcome product of combustion. It occurs when there is insufficient air to completely combust the gas. Figure 9 shows how the carbon monoxide increases as the air for combustion decreases below the stoichiometric line air factor 1. At 50% combustion air, carbon monoxide has risen to an alarming 11–12%. Carbon monoxide is poisonous and the time weighted average for exposure to CO for eight hours for a normal person is 50 ppm. This means that a healthy person must not work in an atmosphere of more than 50 ppm for eight hours. Exposure for one hour of 0.15% would cause death and 0.5–1.0% exposure causes death in a few minutes.

The effect of carbon monoxide in flue products is to decrease dramatically the thermal efficiency of combustion and is therefore to be minimised. It is often difficult to eliminate small amounts of carbon monoxide in industrial burners. Carbon monoxide production is the function of poor mixing of the air and gas. Figure 15 shows the curve for a good burner and a poor burner. It is therefore imperative that burners are designed, commissioned and operated correctly to minimise pollution and excessive use of fuel.

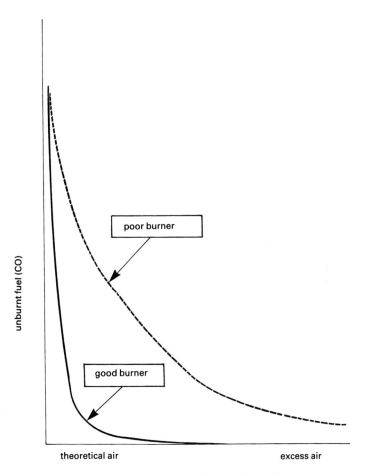

*Fig. 15    Carbon monoxide from industrial burners*

## Opportunities to Reduce Emissions

The previous pages have shown that the use of natural gas compared to other fossil fuels will reduce carbon dioxide, sulphur dioxide and oxides of nitrogen. Another method of reducing emissions is to use more efficient and effective ways of using energy resources.

### Steam Replacement

Most large factories have large centralised steam plant consisting of several water tube or shell steam boilers. These boilers produce steam continuously for heating and for process plant, using large amounts of energy. Whilst the combustion efficiency is quite high, often 75–80%, examination of other losses show that the overall efficiency can be quite low. Figure 16 shows the net heat from the process to be only 48% of the original fuel input and indicates where the energy losses have occurred. The matching supply/demand losses are made of zero load losses – that is, when boilers are warming up and part load losses when the boiler is operating at levels lower than the nominal.

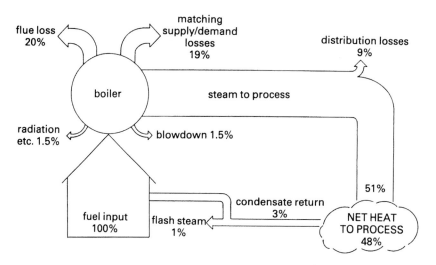

*Fig. 16   Typical energy loss from a steam system*

It has been proved many times practically, that energy savings of 40–50% can be made by replacing central site steam generating boilers. Point of use heat application has been employed for the various stages of process and local heating methods – overhead radiant heaters, convectors and local boilers have all been employed for heating. This has the advantage that (a) each application of direct local heating has its own high efficiency and (b) each application of gas can be tightly programmed and turned off when not in use. This cannot be done with a centralised steam boiler plant. Process steam heating can be replaced with either direct firing or a small local boiler if it is essential to have steam.

## Combined Heat and Power

Chapter 10 on this subject has shown how efficient this process can be in producing on-site electricity and useful waste heat. The reduction of $CO_2$ is about 70% compared with conventional steam raising and the use of power from the national grid. The use of gas for CHP as an environmentally friendly fuel has also been shown earlier. The range in size of prime movers to generate almost any quantity of electricity has also been shown.

## Process Integration

The energy saving and environmental benefits which arise from the application of CHP can be further enhanced if the customers' process heat and power requirements are optimised prior to the specification of CHP. The development of process integration by 'Pinch Technology' provides this systematic approach. This technique involves the application of thermodynamic principles whilst taking account of the practical engineering and operation constraints. In the brewery process, detailed studies have shown that energy saving of 45% in the heating requirement and 20% in the power requirement can be made.

## Recuperative and Regenerative Burners

The high thermal efficiencies achieved by modern gas firing techniques can be enhanced still further particularly when applied to high temperature industrial processes. These high efficiencies have often been achieved by preheating the combustion air from the waste heat in the flue products. Recuperative and regenerative burners, of which thousands have been sold, can produce savings of between 30 and 70% on high temperature plant with a further reduction in carbon dioxide emissions to the atmosphere in direct proportions to these savings.

Whilst oxides of nitrogen increase with the use of preheated air, it is possible to apply techniques to reduce this level. These techniques include staged combustion, flue gas recirculation and in some cases the injection of water vapour. Table 7 shows the comparison of emissions that can be achieved.

TABLE 7 Comparison of flue gas emissions resulting from a gas-fired, an oil-fired and an electric furnace

| Energy Source | Gas-fired Furnace | | | Oil-fired Furnace | Electric Furnace |
|---|---|---|---|---|---|
| | cold air burner | Recup. Burner (no NOx control) | Recup. Burner (with NOx control) | Cold air burner | Coal-fired Power Station |
| Fuel consumption (kW gross) | 300 | 200 | 220 | 288 | 300 |
| Efficiency % | 20 | 30 | 27 | 21 | 20 |
| NOx vpm (3% $O_2$) | 150 | 400 | 200 | 175 | 316 (6% $O_2$) |
| NOx emitted g/h | 83 | 148 | 81 | 100 | 233 |
| CO emitted kg/h | 54 | 36 | 40 | 72 | 98 |

The table shows by applying typical NOx emission data and allowing for changes in the efficiency of primary fuel use, that the emissions of the pollutants are lowest from a recuperated natural gas furnace.

## Transport

Transport contributes one of the highest levels of $CO_2$ from human activities. It also is a major source of nitrogen oxides, hydrocarbons, carbon monoxide, lead, particulate matter and sulphur dioxide. These are derived from petrol and diesel vehicles. Worldwide there are about 1.5 million natural gas powered vehicles offering the lowest levels of pollution for transportation. For small vehicles they are bi-fuel, operating on petrol or natural gas at the touch of a switch and for panel vehicles, lorries, refuse wagons, buses, etc., custom-made gas engines are used. The Government is assisting the introduction with a range of grants and the taxation of gas, based on the propensity to pollute, makes these vehicles very attractive. In the UK there are probably about 700 such vehicles (as of year 2000). The European Union have set standards for pollution levels which will improve the environment from this source.

## The Future

The future of the environment is in the hands of the human race. Results to date of progress have been to cause widespread devastation and pollution of land, air and sea. Fortunately there has been enough reaction to make governments aware of the problems and they are taking action to try to control the deterioration of the environment.

Figure 17 shows the observed global temperatures, expressed as a departure from the mean and predicted for the next 60 years.

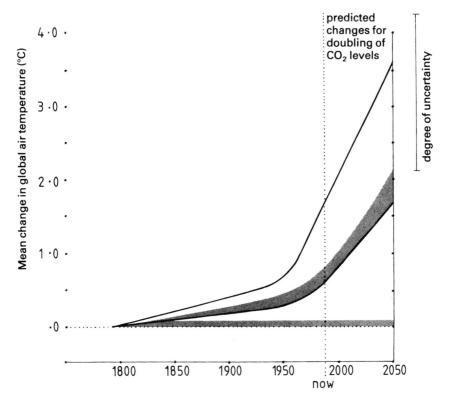

*Fig. 17   Predicted global temperatures for the next 60 years*

One result of these changes is shown in Fig. 18 which indicates the predicted changes in global sea level relative to the present and shows the upper and lower estimates.

The potential impact of sea level rise was demonstrated dramatically by the 1953 North Sea Flood Surge. Three hundred people lost their lives when large areas of East Anglia and the Thames Estuary were flooded and a further 2,000 people died in the Netherlands.

Changes in vegetation and agriculture could affect crops, including a change in yield and quality and suitability for human or animal consumption.

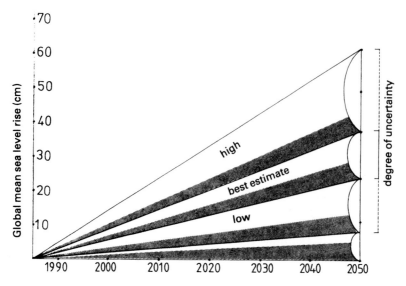

*Fig. 18    Predicted changes in global sea level showing upper and lower
estimates*

Changes to ocean currents could alter fish populations by altering temperature, salination and available nutrients.

There will undoubtedly be some impact on the urban environment and infrastructure with repercussions for economic and social well-being. It is hoped that natural gas will be applied wherever appropriate and will be burned efficiently and safely to achieve minimum pollution to create a better environment.

# Index